CLIMATE RISK AND THE WEATHER MARKET

CLIMATE RISK AND THE WEATHER MARKET

Financial Risk Management
with Weather Hedges

Edited by
Robert S. Dischel

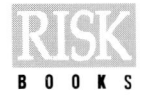

RISK
BOOKS

Published by Risk Books, a division of the Risk Waters Group.

Haymarket House
28–29 Haymarket
London SW1Y 4RX
Tel: +44 (0)20 7484 9700
Fax: +44 (0)20 7484 9758
E-mail: books@riskwaters.com
Websites: www.riskbooks.com
www.riskwaters.com

Every effort has been made to secure the permission of individual copyright holders for inclusion.

© Risk Waters Group Ltd 2002

ISBN 1 899 332 52 9

British Library Cataloguing in Publication Data
A catalogue record for this book is available from the British Library

Head of Risk Books: Conrad Gardner
Risk Books Editor: Kathryn Roberts

Typeset by Mark Heslington, Scarborough, North Yorkshire

Printed and bound in Great Britain by Bookcraft (Bath) Ltd, Somerset

CONDITIONS OF SALE

All rights reserved. No part of this publication may be reproduced in any material form whether by photocopying or storing in any medium by electronic means whether or not transiently or incidentally to some other use for this publication without the prior written consent of the copyright owner except in accordance with the provisions of the Copyright, Designs and Patents Act 1988 or under the terms of a licence issued by the Copyright Licensing Agency Limited of 90, Tottenham Court Road, London W1P 0LP.

Warning: the doing of any unauthorised act in relation to this work may result in both civil and criminal liability.

Every effort has been made to ensure the accuracy of the text at the time of publication. However, no responsibility for loss occasioned to any person acting or refraining from acting as a result of the material contained in this publication will be accepted by Risk Waters Group Ltd.

Many of the product names contained in this publication are registered trade marks, and Risk Books has made every effort to print them with the capitalisation and punctuation used by the trademark owner. For reasons of textual clarity, it is not our house style to use symbols such as TM, ®, etc. However, the absence of such symbols should not be taken to indicate absence of trademark protection; anyone wishing to use product names in the public domain should first clear such use with the product owner.

CONTENTS

	List of panels	vii
	List of contributors	ix
	Dedication	xv
	Preface	xvii

ELEMENTS OF THE WEATHER RISK MARKET

1	**Introduction To The Weather Market: Dawn to Mid-Morning** Robert S. Dischel of Weather Market Observer, LLC	3
2	**Financial Weather Contracts and Their Application in Risk Management** Robert S. Dischel of Weather Market Observer, LLC and Pauline Barrieu of Université de Paris VI and Doctorate HEC, France	25
3	**Hedging Precipitation Risk** Thomas Ruck of Entergy-Koch Trading, LP	43
4	**Weather and Climate – Measurements and Variability** Steve Smith of ACE Tempest Reinsurance Ltd	55
5	**Weather Data: Cleaning and Enhancement** Auguste C. Boissonnade of Risk Management Solutions and Lawrence J. Heitkemper and David Whitehead of Earth Satellite Corporation	73

CLIMATE FORECASTS, MANAGING VARIATIONS, AND DERIVATIVE PRICES

6	**The Nature of Climate Uncertainty and Considerations for Weather Risk Managers** Mark Gibbas of Applied Insurance Research, Inc.	97
7	**Weather and Seasonal Forecasting** Mark S. Roulston of the University of Oxford and Leonard A. Smith of the University of Oxford and London School of Economics	115
8	**Weather Derivative Modelling and Valuation: A Statistical Perspective** Anders Brix, Stephen Jewson and Christine Ziehmann of Risk Management Solutions	127
9	**The Accuracy and Value of Operational Seasonal Weather Forecasts in The Weather Risk Market** Jeffrey A. Shorter, Todd M. Crawford and Robert J. Boucher of WSI Energycast Trader	151
10	**Use of Meteorological Forecasts in Weather Derivative Pricing** Stephen Jewson, Christine Ziehmann and Anders Brix of Risk Management Solutions	169

11	**The Weather in Weather Risk** John A. Dutton of the Pennsylvania State University and Weather Ventures Ltd	**185**

INVESTOR ISSUES

12	**Weather Risk Management in the Alternative Risk Transfer Market** Julian Roberts of Aon Capital Markets Ltd	**215**
13	**Weather Note Securitisation** Frank Caifa of Swiss Re	**231**
14	**Managing a Portfolio of Weather Derivatives** Lixin Zeng of Willis Re Inc. and Kevin D. Perry of the University of Utah	**241**

EXPERIENCE IN APPLICATION

15	**A Case Study of Heating Oil Partners' Weather Hedging Experience** Paul J. Forrest of Heating Oil Partners, LP	**265**
16	**Weather Indexes for Developing Countries** Panos Varangis of World Bank, Jerry R. Skees of the University of Kentucky and Barry J. Barnett of the University of Georgia	**279**
17	**Weather Risk Management for Agriculture and Agri-Business in Developing Countries** Ulrich Hess of IFC, Kaspar Richter of World Bank and Andrea Stoppa of Procom Agr, Rome	**295**
	End Piece **Speculations on the Future of the Weather Market** Ulrich Hess, Kaspar Richter and Andrea Stoppa Panos Varangis, Jerry R. Skees and Barry J. Barnett J. Scott Mathews Stephen Jewson Julian Roberts Robert S. Dischel	**311**
	Index	**319**

LIST OF PANELS

CLIMATE RISK: KEY THEMES AND EXAMPLES

Throughout this book special panels introduce key themes and offer illustrative examples:

Ambient temperature and the demand for power 4
Global warming and the US weather market 15
German winters 18
Centuries of temperature history 20
End-user strategies 31
Alternative derivative strategies 36
Measuring precipitation – an imprecise science 49
Global climate change 67
Synoptic vs climate data 76
Air density, pressure and winds 98
General circulation of the atmosphere 107
Computer models: grid-point vs spectral 117
Observing the atmosphere 118
Evaluation of forecasts 123
Deriving the payoff distribution 130
Ensemble forecasts 154
Statistical tools 157
The convolution process 161
Beating the 10-day barrier in Europe 170
Atmospheric general circulation models 171
Forecasts, trends and anomalies 172
Why do forecasts go wrong? 177
Meteorological indexes 179
Weather and climate risk categories and strategies 189
A forecast experiment 199
The availability of weather and climate information 203
Pluvius insurance 217
Current securitisation activity 234
Investing in weather-linked instruments 258
Terminology 266
Rounding techniques 270
The weather derivative 272
Fondos de aseguramiento 285
Possible government strategies in weather and
catastrophe risk sharing 289
A market for new WRM 298

LIST OF CONTRIBUTORS

Dr Barry J. Barnett is associate professor of agricultural and applied economics at the University of Georgia. He teaches undergraduate and graduate courses in agricultural public policy as well as an undergraduate course in microeconomic theory. Dr Barnett has conducted research on behalf of the US Department of Agriculture's Risk Management Agency and the US Department of State's Agency for International Development. He has published research findings on farm-level crop insurance, area-yield insurance and weather derivatives. Dr Barnett earned a BS degree in agricultural economics, a BBA degree in finance and PhD in agricultural economics from the University of Kentucky. Prior to accepting his current appointment at the University of Georgia, Dr Barnett held assistant and associate professorships at Mississippi State University.

Pauline Barrieu is currently finishing her double education: a PhD in applied mathematics at the University of Paris VI and a PhD in finance at Doctorat HEC (both in France). She has already gained a certificate from ESSEC Graduate Business School and a post-graduate diploma in probability from the University of Paris VI. Pauline will start a lecturer position at London School of Economics in September 2002. She is particularly interested in illiquid and incomplete markets, and derivatives with a non-quoted underlying asset. She has also special interests in real options and environmental economics.

Auguste C. Boissonnade, PhD, is a principal scientist at Risk Management Solutions (RMS) and is vice president of model development for the weather risk team. He was the original architect of RMS' industry-leading hurricane catastrophe model and has extensive experience developing risk models for a broad range of applications. Prior to joining RMS, Auguste was a project leader at Lawrence Livermore National Laboratory with responsibility for developing probabilistic seismic, wind and flood hazard guidelines for the US Nuclear Regulatory Commission and for the US Department of Energy. Auguste holds PhD and MS degrees from Stanford University. He is a member of the American Meteorological Society and of the American Society of Civil Engineers and a reviewer for the National Science Foundation. Auguste has authored more than 50 publications, including one book.

Robert J. Boucher is an atmospheric scientist in the Forecast Research Center at Weather Services International (WSI). He received his BS degree in meteorology from Plymouth State College and his MS degree in atmospheric science from Texas Tech University. His research interests are in the area of seasonal forecasting and the utilisation of seasonal forecasts in the area of risk management. Mr Boucher splits his time between the production of twice-monthly operational seasonal forecasts, researching methodologies in enhancing WSI's forecasting accuracy, and managing the weather information infrastructure that allows clients quick access to real-time weather information.

Anders Brix is a lead modeller based in the London office of Risk Management Solutions (RMS) London office, with responsibility for researching and implementing advanced modelling techniques for weather risk. Prior to joining RMS, he developed pricing models and conducted dynamic financial modelling in the Instrat actuarial services unit of reinsurance broker Guy Carpenter. Anders received a PhD in mathematical statistics from the Royal Veterinary and Agricultural University in Denmark, and has conducted research in a wide range of areas within statistics and probability theory. He has worked at several academic European institutions, and he holds a Candid. Scient. degree in mathematical statistics from the University of Copenhagen.

Frank Caifa is an associate director of Swiss Re in New York as a member of the Global Weather Desk. He is responsible for the development, marketing and underwriting of simple or complex weather products, which combine traditional risk transfer, financing and capital market techniques, primarily for large US corporations and insurance companies. His main focus and operational authority is for the US and Canada, although he does work closely with the European team located in Zurich, Switzerland. Swiss Re has been a major player in the weather business since September 1998 and Mr Caifa was at the genesis of its operations. Mr Caifa has spent six years working in or with the insurance industry, and has developed programmes for corporate and insurance company clients while serving in positions at AIG Risk Finance prior to joining Swiss Re. He received his BBA from Adelphi University. He is a certified public accountant, licensed in New York and also a licensed securities representative having passed both the Series 7 and Series 63 exams.

Dr Todd M. Crawford, PhD, is a meteorologist in the Forecast Research and Development Center at Weather Services International (WSI). He received his BS degree in Meteorology from Wisconsin, his MS and PhD degrees from the University of Oklahoma, and performed his post-doctorate work at the National Severe Storms Laboratory. In 1998–9, Dr Crawford finished second nationally in the National Collegiate Weather Forecasting Contest. His research interests have spanned many facets of meteorology from convective initiation and energy budget modelling to improving temperature forecasts by inserting real-time vegetation and soil moisture data into a mesoscale model. Currently Dr Crawford focuses his research in the area of seasonal forecasting and the utilisation of seasonal forecasts in the area of risk management. He splits his time between the production of WSI's twice-monthly operational seasonal forecasts and leading the research effort aimed at continuing to improve these forecasts. Dr Crawford has written over ten papers in refereed archival journals.

Dr Robert S. Dischel, is currently the president of the Weather Market Observer, a weather risk consulting venture in New York City. Bob gained his financial quantitative skills working in the capital markets for almost two decades, resulting in a position as a managing director in the portfolio department of a multi-billion dollar US insurer. Before his work in the capital markets, Bob was a university professor. Bob is a frequent speaker at weather risk, energy and financial conferences, and has published several articles on weather risk practices. He is certified by the American Meteorological Society as a Consulting Meteorologist, and is a member of the National Council of Industrial Meteorologists in the US. Bob earned a PhD in oceanography from New York University in 1975.

Dr John A. Dutton is professor of meteorology and dean of the College of Earth and Mineral Sciences at Pennsylvania State University. He has served as chair of the board of Atmospheric Sciences and Climate and as chair or member of other committees of the National Research Council (NRC) related to atmospheric science and aviation. He is a fellow of the American Meteorological Society and the American Association for the Advancement of Science. Dr Dutton holds three degrees in meteorology from the University of Wisconsin and served as an officer in the Air Weather Service of the US Air Force. His expertise includes dynamic meteorology, spectral modelling, climate theory and global change. Dr Dutton is a principal in Weather Ventures Ltd, a firm providing strategic advice and information for the management of weather and climate risk.

Paul J. Forrest is the chief financial officer of Heating Oil Partners, LP. He has also served as the CFO of Cap Gemini America, Inc., treasurer of Josephson International, Inc, controller of Morgan Stanley & Co. Inc. and manager of long-range planning for TWA. A former US Navy aviator, Mr Forrest holds an undergraduate degree from Brown University and a MBA from the Wharton Graduate Division of the University of Pennsylvania. Mr Forrest has been a frequent presenter at risk management conferences in the US and Bermuda.

Mark Gibbas is a Senior Research Scientist and Meteorologist at Applied Insrance Reasearch Inc. (AIR). Mr Gibbas has worked extensively in many areas in meteorology including remote sensing, numerical weather and climate prediction, algorithm development, forecast system design and time-series analysis. Mr Gibbas is responsible for AIR's long-range weather/climate forecasting and associated products for insurance, reinsurance and weather derivative interests. Prior to joining AIR, Mr Gibbas led the team at Litton-TASC responsible for developing long-range forecasting systems to service energy interests. Additionally, Mr Gibbas has conducted research for the World Meteorological Organization as part of the Ibero-American Climate Project, where he assessed the meteorological and climatological capabilities of numerous Latin American countries. Mr Gibbas earned his BS degree at Plymouth State College, graduating *summa cum laude* with a major in meteorology and minors in mathematics and computer science.

Lawrence J. Heitkemper is director of Earth Satellite Corporation's weather and crop division. He manages the risk weather, energy weather, agricultural weather, aviation weather and surface transportation weather activities for EarthSat. He was a primary developer for EarthSat's crop monitoring system, long-range forecast model, and weather risk management data programme. Mr Heitkemper joined EarthSat in 1975. He was previously employed at St Louis University and Environmental Quality Research as a researcher. Mr Heitkemper has a BS degree in meteorology from St Louis University, and an MS degree in resource economics from the University of Maryland. He also has completed MBA training and completed the Commodity Trading Futures Commission (CFTC) Commodity Trading Advisory Course Series. Mr Heitkemper is an active member in the American Meteorological Society, the American Agricultural Economics Association and the National Mathematics Honor Society.

Ulrich Hess works for the agri-business department at the International Finance Corporation, (IFC) World Bank Group. He initiated IFC's work on weather risk management and currently manages a weather index insurance project in Morocco and Mexico, and a global weather risk facility as well as traditional agri-business projects such as coffee projects. He joined the World Bank's Ghana country unit in 1998 and then worked for the Commodity Price Risk Management International Task Force and IFCs Middle East North Africa department. Prior to joining the World Bank Group as a young professional, Mr Hess was working in management and development consulting in Italy, Germany, France and Denmark on enterprise development, fisheries and environmental services. His main research interest is in agricultural risk management. Mr Hess holds a masters degree in economics from Bocconi University, Milan and studied at Institut d'Etudes Politiques de Paris, Freie Universität Berlin and Yale Law School.

Dr Stephen Jewson is director of business development at Risk Management Solutions (RMS) in London. He is responsible for marketing and sales of RMS's weather derivatives products in Europe. Prior to joining RMS, Stephen was an academic researching climate variability. He worked at the universities of Oxford, Reading, Monash and Bologna and published a number of academic papers. He has a PhD in climate modelling, degrees in mathematics and is a regular speaker and writer on weather risk quantification.

J. Scott Mathews began his risk management career in 1978 with the commodity brokerage unit of the Continental Grain Company. Later joining Oppenheimer & Co., he was assigned to the

Middle East in 1985, where he managed an investment fund for the ruling family of Kuwait. From 1991 to 1997, Mathews held two positions at Citibank, initially as the associate director of energy futures for Citicorp Commodities Corporation and later as the director of energy sourcing for Citibank's global commercial real estate portfolio. In 1998 he founded an independent consultancy advising banks, brokers, energy companies and re-insurers in the practice of weather risk transfer. In January 2002, Mr Mathews established a partnership for introducing investors to the exchange-based weather market. Mr Mathews is a graduate of Colgate University.

Dr Kevin D. Perry, PhD is currently an assistant professor in the meteorology department at the University of Utah and is an expert in the areas of meteorological instrumentation, air pollution and global climate change. He has participated in several international field projects involving airborne, shipborne and ground-based instrumentation and has published more than a dozen articles in the refereed literature. Dr Perry earned a BS degree in meteorology from Iowa State University of Science and Technology and a PhD in atmospheric science from the University of Washington. Prior to accepting his academic position, Dr Perry served as a research scientist at Crocker Nuclear Laboratory at the University of California at Davis.

Kaspar Richter works as World Bank economist on East Asia. Since early 2001, he has assisted planning ministries and statistical offices in East Timor, Lao PDR and Thailand in building capacity for effective poverty reduction policies. Prior World Bank assignments include helping to design a debt relief package for Benin, and promoting social sector reforms in Tajikistan. Other professional experience includes working for an EU economic policy think tank in Moscow, freelancing for the Economist Intelligence Unit, and consultancy work for the European Bank of Reconstruction and Development and the German economic ministry. After studying both economics and political science at the Freie Univeristaet Berlin, he pursued his PhD at the Economics department of the London School of Economics and Political Science.

Julian Roberts is a director of Aon Capital Markets and is a member of the London-based team actively operating in the so-called 'convergence' market between insurance and finance. Aon Capital Markets has been successful in specialising in insurance-linked securities, contingent capital and various derivative risk management products, including weather and credit. Julian's background is in the provision of risk consultancy, analytics and modelling leading to the structuring of non-traditional solutions in the capital and insurance markets. Julian joined Sedgwick as a Lloyd's broker before establishing the specialist risk management consultancy, ARM Ltd. When ARM was acquired by Aon, Julian transferred to establish and develop Aon Risk Consultants (actuarial and analytical resource for wholesale and reinsurance brokers) and subsequently joined Aon Capital Markets. Julian has an honours degree in natural sciences from the University of Oxford and a masters degree in agricultural economics from the University of London.

Dr Mark S. Roulston is a junior research fellow at Pembroke College, Oxford University and associated with the Oxford Centre for Industrial and Applied Mathematics. He is also a visiting research fellow in the Centre for the Analysis of Time Series, London School of Economics. Mark's research interests include techniques for generating and evaluating probabilistic weather and climate forecasts, and the integration of such forecasts into decision making and risk management. Dr Roulston gained his BA degree in physics from Cambridge University and a PhD in planetary science from the California Institute of Technology.

Thomas Ruck is currently senior vice president for weather structured products, at Houston-based Entergy-Koch Trading and has been in the weather market for two years. Tom has traded and structured energy derivative transactions since 1989. He also has experience in agricultural commodities, fertilizers, coal, petroleum coke and base metals. Tom joined Koch Industries in 1995 following several years at Wall Street investment banks

(UBS, JP Morgan and Morgan Stanley). He began his career in petroleum refining and international supply with Exxon C USA. Tom holds bachelors and masters degrees in chemical engineering from Manhattan College, New York.

Dr Jeffrey A. Shorter, PhD, is a vice president of marketing at Weather Services International (WSI). Dr Shorter earned his PhD in physical chemistry from the Massachusetts Institute of Technology and his BS degree with honours in chemistry from the University of Southern California. Dr Shorter is responsible for all aspects of WSI's Energycast Trading business, which includes PowerTrader, GasTrader, WeatherTrader, and MetTrader. Prior to his marketing position, he led the technical development of WSI's energy trader decision support tools, which included the development and improvement of WSI's seasonal forecasting capabilities. Dr Shorter is active in the meteorological research community, has over a dozen archival publications, and also serves as one of five members on the American Meteorological Society Admissions Committee. Dr Shorter has given numerous presentations at both technical and professional conferences.

Jerry R. Skees is H B Price Professor in the department of agricultural economics at the University of Kentucky. He is also president of GlobalAgRisk, Inc., a firm that performs policy and risk work for the World Bank and USAID. Mr Skees has made numerous contributions in research and education on the US crop insurance programme. He was the primary architect outside the Federal government in helping design and rate the USDA Group Risk Plan insurance. Mr Skees has worked in academia, government and for private-sector firms. In recent years, his consulting activities have involved work for reinsurers and weather market makers. His recent international experience has included projects in Morocco, Mexico, Argentina, Mongolia and Romania. Many of his recent papers can be found on the Internet: www.cifarrm.com

Leonard A. Smith is a reader in statistics for the London School of Economics (LSE) and a senior research fellow at the Pembroke College, Oxford. He is also director of the Centre for Analysis of Time Series at LSE. Leonard's research interests include the ultimate limits to predictability, the generation and evaluation of estimates of forecast uncertainty in physical and mathematical systems, non-linear modelling techniques and the nature of the relationship, if any, between physical and economic systems and our models of them, the definition of noise, and non-linear time-series analysis. Dr Smith attained his BS degree in physics, mathematics and computer science from the University of Florida and his PhD in physics from Columbia University. Dr Smith's previous appointments include post-doctoral or visiting scientist/professor positions at Cambridge University, Ecole Normale Superieure (Paris), Warwick University and University of Potsdam.

Dr Steve Smith is a quantitative analyst with ACE Tempest Reinsurance Ltd, where he directs research activities. Before joining ACE Tempest Re, Dr Smith held positions as a risk manager at Commerzbank Global Equities and as a derivatives analyst at ING Barings. Dr Smith's current research interests include climate diagnostics and hurricane and earthquake damage modelling. Dr Smith holds a first class honours degree in physics and a doctorate in atmospheric physics, both from the University of Oxford, where he also conducted post-doctoral research into the atmosphere of Jupiter. Dr Smith is a Fellow of the Royal Meteorological Society and a Chartered Physicist.

Andrea Stoppa is a consultant in agriculture economics. He is active mainly in the fields of risk management in agriculture and agricultural policy analysis. He holds a first class degree in agricultural sciences from the University of Viterbo, Italy and masters degrees in agriculture economics and economics from the Universities of Naples, Italy and Iowa State, USA. Formerly responsible for the risk and insurance research area at ISMEA (Italian Research Institute for Agricultural Markets) he now consults for various public and private institutions that include the Italian Ministry for Agricultural Policies, the World Bank, the International Finance Corporation and the Commission of the European

Union. He teaches classes on risk management in agriculture at the University of Rome 3. He is currently involved in an IFC project aimed at developing a rainfall insurance programme in Morocco.

Panos Varangis, a Greek national, is a senior economist at the World Bank's rural development department. His primary responsibility is to look into issues related to commodity and weather risk management. Mr Varangis joined the World Bank in 1987. He has also worked at the Development Research Group where he was involved in research areas related to commodity markets, and in particular risk management and finance, weather index insurance and commodity market liberalisation. He has also served at the commodities policy and analysis unit. Mr Varangis has worked extensively in the areas of agricultural policies, risk management and commodity marketing and trade finance systems. He has initiated a project to examine the application of weather risk management products to agriculture in developing countries. Mr Varangis holds a MA degree in economics from Georgetown University and a PhD degree in economics from Columbia University, New York.

David Whitehead is a research meteorologist in the CROPCAST/Weather group with Earth Satellite Corporation (EarthSat). Mr Whitehead is currently involved in the management of the Climetrix project, a joint venture between Risk Management Systems (RMS) and EarthSat. Mr Whitehead is educated and experienced in meteorology/technology, economics, risk management and communications in operational problem-solving environments. Mr Whitehead received a bachelors degree in science from Cornell University and a masters degree in meteorology from Florida State University.

Lixin Zeng is senior vice president at Willis Re, responsible for analytical product development. Previously, he was a vice president at the risk analysis group of Benfield Blanch and the principal researcher for applied underwriting research at Arkwright Mutual Insurance Company. Since 1997, Lixin has worked in a variety of roles related to weather risk management and weather derivative trading. He has published numerous weather-related articles in both meteorological and finance journals and has frequently spoken at conferences in this area. Lixin holds a PhD in atmospheric sciences from the University of Washington, and is a member of the American Risk and Insurance Association and the American Meteorological Society.

Christine Ziehmann is senior modeller at Risk Management Solutions (RMS) where she is responsible for the development of weather derivatives pricing algorithms and software. She received a masters degree in meteorology from Free University of Berlin, Germany. In her PhD thesis she investigated the predictability of non-linear systems in general and the predictability of the atmosphere in particular as one of the most complex examples in this class of systems. Christine has published papers in physical and meteorological journals. She worked with the German Weather Service and the University of Oxford in several projects on the validation of medium-range ensemble forecasts.

DEDICATION

Credit is due to the 28 authors of this book for their collaboration and contributions.

I express my genuine appreciation to Kathryn Roberts of Risk Books, for a partnership that embraced the complete process to the last edit on the closing page, and to my wife and life companion, Phyllis Dischel, for supporting me throughout this project.

For my development as a scientist, I owe thanks to Professor Jerome Spar, who taught me to describe atmospheric–oceanic physical events with mathematics, and Professor Willard J. Pierson, Jr., my mentor and colleague, who taught me some decades back to search for new answers to old questions.

PREFACE

I am the daughter of Earth and Water,
And the nursling of the sky;
I pass through the pores of the ocean and shores;
I change but I cannot die.
              Percy Bysshe Shelley, *The Cloud* (lines 73–6)

If you are a weather risk neophyte seeking a starting point for your journey to discovery, you will find it here. We describe the reasons why this new market exists, how it works, and where there is a benefit for your business in using weather risk products. As every chapter stands on its own, you can read this book in any order. You may, however, want to begin at the beginning. If you are a seasoned contender seeking an advantage, you will find that in each chapter the authors have written something that will give you a competitive edge.

This book, as does a good meal, serves up a sequence of courses beginning with the uncomplicated and progressing to the complex.

The first section on market fundamentals embarks on a five-chapter primer, moving from a survey of issues important in today's market (Chapter 1), to basics of weather derivatives and how to use them (Chapters 2 and 3), to an evaluation of meteorological and climatological issues of concern to the market (Chapter 4), and rounds out with a review of where to find climate data and what must be done with it (Chapter 5).

No one knows all there is to know about weather, climate or the weather risk market, so we gave the podium in the next section to a few meteorologists and modellers. The combined offering of the first five authors (Chapters 6–10) presents a diversity of views and a richness of topics not found elsewhere. For example, they all agree that climate is a complex process and that new understanding of the role of the Tropical Pacific Ocean in this process has advanced the state-of-the-art of seasonal forecasting, but each presents an individualistic view on how to use forecasts in pricing weather-indexed contracts that differs from the others. The discussions on whether or not we should accept global warming as the state of climate for the future and assume that next year will be warmer than last year provoke debate. Next we learn how to enrich the conclusions drawn from the short few decades in the record of the past, and make a better estimate of the probability distribution of weather (Chapter 11).

In the third section our authors present ideas developed in other markets and applied to this one. We are reminded that weather protection with financial structures is not limited to derivatives and has been part of insurance practices for more than a century. We are also shown where the capital and insurance markets have separate identities and where they converge in the art of alternative risk transfer (Chapter 12). From the first, there has been the suggestion that capital market investors could bring liquidity to the weather market if only the market offered them an attractive reason to do so. Securitising weather derivatives into weather notes would offer them a familiar asset structure not correlated to other assets in their portfolio (Chapter 13). Weather asset originators became weather portfolio managers even when that was not their intention, but many investors now believe they can gain return while managing risk, and here they might learn how to do this (Chapter 14).

The last set of chapters presents the views of a few experienced practitioners and their practical applications. There is a discussion of the core activity of the market – warm winter protection for an energy supplier (Chapter 15). Finally, there are two discussions of the weather market's promise for developing countries (Chapter 16 and 17).

The End Piece offers a few speculations on the weather market's next events.

BIBLIOGRAPHY

Shelley, P. B., 1820, 'The Cloud', http://www.library.utoronto.ca/utel/rp/poems/shelley6b.html.

Elements of the Weather Risk Market

1

Introduction To The Weather Market: Dawn to Mid-Morning

Robert S. Dischel

Weather Market Observer, LLC

Mark Twain, the American humorist, wrote of the immoderate climate in the US Southwest:

"Sometimes we have the seasons in their regular order, and then again we have winter all the summer and summer all the winter."

"The climate is good," he noted, "what there is of it."

Twain's style, of course, was to exaggerate real events to bring to his readers a sense of the sometimes upside down quality of life, whether in politics, in social exchanges or in weather. What would he say of the back-to-back record-breaking winters in the US, when the mildest winter of the century, 1999–2000, was followed by one of the coldest of the century, 2000–1?[1]

The extremely mild US winters of 1997, 1998 and 1999 lessened the need to heat homes. Reduced need for power over successive years produced bonus savings for consumers and a threat to power suppliers. This thrust the developing US weather risk market into action, and it soon spread across the ocean in an environment of expanding deregulation[2] and concern about extremes in European climate.[3]

The foundation of today's financial weather contracts is in the US power market, and the impulse for this activity to grow is the result of the confluence of several factors:

❑ The need for weather protection arises from the weather-induced variability of financial return that energy suppliers experience within a heating or cooling season. Generally, revenue from residential sales is strong in hot summers with high demand for air conditioning, and in cold winters with high demand for heating. Revenue is generally weak in mild seasons – both cool summers and warm winters.

❑ The expansion of deregulation of the power markets has deprived many utilities of the usual practice of passing the costs of adverse weather along to customers in the form of higher rates. This formerly allowed suppliers to "smooth out" cashflow variations; now, deregulation fosters a competitive environment among suppliers for customers who had been captive, and those who control costs better are at an advantage.

❑ The anxiety among US suppliers who experienced unusually warm winters and

INTRODUCTION TO THE WEATHER MARKET: DAWN TO MID-MORNING

PANEL 1

AMBIENT TEMPERATURE AND THE DEMAND FOR POWER

The weather derivative market developed in the US around the need to service energy suppliers with a product not found elsewhere. Beyond a base level, the residential use of power rises and falls with outdoor temperature: cold winters encourage consumers to use more energy for heating and hot summers require more cooling than mild seasons. A risk for power providers is that weather-sensitive revenues can fluctuate unacceptably with the rise and fall of weather-sensitive sales and costs. The American power marketers were the first to offer weather derivatives to their power customers as a hedge for that risk.

The differing demands for residential electricity for three US locations of dissimilar climate and population are shown in the figure below. The two horizontal axes are the monthly average outdoor temperature (in Celsius and Fahrenheit). The vertical axis is a normalised scale on which the demand for electricity in a month in a region is divided by the annual demand in that region.

The three locations are:

❑ Southeast Florida, a coastal area on a peninsula of the eastern coast of the US that is surrounded by warm ocean (the Gulf of Mexico is to the west and the Atlantic Gulf Stream Current is to the east);
❑ Southeast Wisconsin, a location interior to the North American continent and far from an ocean (although it is just south of Lake Superior and just west of Lake Michigan); and
❑ Southeast Washington State in the northwestern corner of the US, not far from the western Pacific Ocean with its cool southward flowing currents.

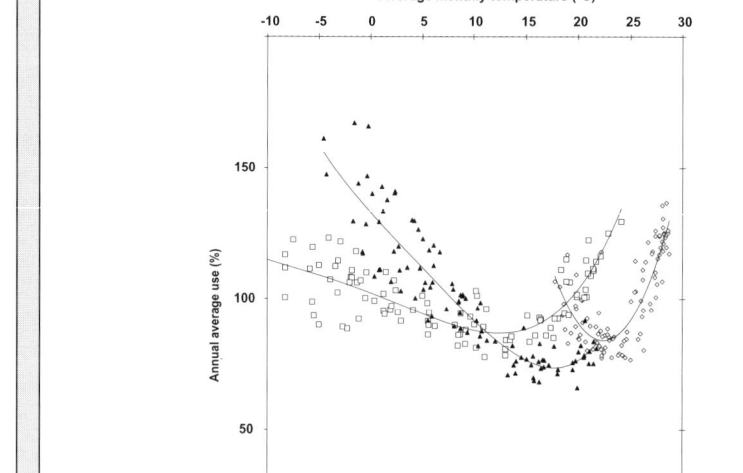

Residential use of electricity in response to outdoor temperature

○ Southeast Florida   □ Southeast Wisconsin   ▲ Southeast Washington

Source: US National Climatic Data Center (NCDC) (temperature) and US Department of Energy (DOE) (electricity).

The effect of bodies of water on local climate can be profound; specifically, the closer the ocean, the more pronounced the effect on long-term average temperatures, as is explained in all basic texts on meteorology, oceanography and climate. Of the three sites, Florida, the set of data on the right of the figure, has the narrowest range of temperatures reflecting the narrow range of nearby ocean temperatures. Washington State temperatures reflect the influence of the nearby Pacific Ocean that has a broader annual cycle and generally lower temperatures than the ocean adjacent to Florida. Wisconsin, in the interior of the continental shield, has a wider range of temperatures than the other two sites, reflecting the wider seasonal extremes that result when there is no moderating effect from the ocean.

A polynomial regression line fitted to the Southeast Florida data forms a smile pattern; the two upward corners of the smile indicate the different uses of electricity in different seasons – for cooling in warm months and heating in cool months. (In practice, separate linear regression lines would be fit to the data for each season for hedging purposes.)

Importantly, the transition from a heating to a cooling season – the time when there is neither a strong demand for heating or cooling – can be associated with a range of inter-season temperatures. It is at these inter-season temperatures that heating and cooling energy demand is at a minimum and indicates the demand for non-temperature-related activities such as lighting and cooking. The inter-season range for Southeastern Florida is between about 71°F and 77°F.

The widest range of temperatures of the three sites is for the land-locked Wisconsin site (whose low temperatures reach the left side of the figure). The normalised demand for electricity is lower in winter than in summer, and the lowest of the three sites. Fuel oil and gas are more common method for heating in this region than is electricity. (See the Hot Air Gas Company discussion at the end of this chapter.) The inter-season temperatures dividing the seasons in this location centred at about 54°F.

The seasonal variation in electricity demand for the site in Washington State differs from that of Wisconsin: electricity is used for heating in winter but is not needed as much for air conditioning this region with its cooler summers. The inter-season temperatures in this location are centred at about 64°F. (Of the three sites, this comes the closest to the 65°F inter-season reference temperature used in most definitions of degree-days.)

The difference in inter-season temperatures among different locations measures the particularity of the division between heating and cooling seasons. The location-specific transition temperatures indicate that the division into heating and cooling seasons depends on the local population and its collective behaviour, as well as on climate.

Note: If variations in temperature were the only factor affecting variations in electricity demand, then all the points for each site would fall neatly on a smooth curve for that site. There is, however, a scattering of points around the demand curve at each site. This is due to at least two factors:

1. No effort was made to distinguish between, say, a warm February and a cool April in any year: although these months might have the same average temperature, they will have a different demand for indoor light as their length of day differs.
2. Also not isolated in this analysis are the influences of wind and rain that cool structures, sunlight that warms structures, high humidity that causes physical discomfort and prompts higher energy use, snow cover on a roof that reduces the loss of heat from the structure, and more. Nevertheless, it is common practice in today's weather risk market, for the sake of convenience, to disregard these other variables and to see the demands for both cooling and for heating being approximately linear with outdoor temperature within a season.

INTRODUCTION TO THE WEATHER MARKET: DAWN TO MID-MORNING

cool summers in the late 1990s. The growing awareness of the influence of El Niño on seasonal weather in the US, and the worry that a prolonged El Niño would cause a prolonged mild season to continue, led some energy providers to take protective measures by swapping weather risk with power marketers.

The energy providers were always aware of the habits of their customers to demand less heating in warmer than normal winters, and more heating in colder than normal winters. These and the demand variations in summer correlate well with outdoor temperatures and allow providers to easily estimate their needs for energy volumes. They also found that they could use the same demand-vs-temperature relationships to estimate the cost of weather variations (see Panel 1). As the regulatory rules were changing in favour of competition over customers, suppliers became more concerned about variations in weather-related costs and more alert to managing costs.

The energy supplier's concerns initially focused on declining revenue in mild seasons, and the earliest weather derivatives addressed only this issue. Nevertheless, rising sales in extreme seasons can also be a problem if the collective demand for power from customers exceeds a supplier's capacity to deliver it. To supplement its capacity at these times, the supplier must purchase energy in a deregulated market or pay the increased expense of generating it with supplemental capacity. Competition over supply when other suppliers also seek more capacity makes market prices rise. If the sales price it can ask of its customers is capped but cost is not, the revenue from price differences – sales price minus purchase price – will decline. Weather derivatives make it possible to manage, at least in part, the concerns related to both mild and extreme seasons by paying the supplier when a season is adverse to revenue – this is the concept of weather hedging.

Servicing the energy market is still the primary focus of the weather market, but the potential to provide weather protection for other sectors of the economy is an encouraging prospect. Once an active market for hedging weather risk was established, everyone thought that other potential end-users who have different weather-related needs would also get involved and the market would grow significantly. This expansion of the weather market to non-energy sectors has been slow and the market is still struggling to meet these needs.[4] It turns out that few enterprises can so easily translate weather variations into costs as the energy market does: the cost of weather to farmers, nut growers, golf course and amusement park operators, and many others, is often more complex and brings more weather variables into the equation than does the energy market; where the energy market focuses mostly on the opposites of warm or cool, others focus also on wet or dry, sunny or cloudy, windy or calm, and more.

The wide appeal of the central concept of the weather market is to provide a new kind of protection for weather-exposed enterprises and this new risk protection concept seems so useful, that we are left wondering why the weather market is not universal. Why has this fledgling market failed to soar as it was expected to do? We begin to understand why when we look at the way the market stumbles over fundamental issues.

❏ Principally, there is an information handicap. Weather risk analysis is founded on comprehensive climate data. Good analysis requires a relatively complete and adequately long time-series of weather measurements. We also need the description of all changes in measurement methodologies and all changes in measurement locations. (The description of the history of changes in measurement methods and measurement location is known as "metadata.")

❏ Access to climate data and metadata in the US is practically unlimited: for more than 200 US measurement sites the data quality is excellent, but coverage of the vast geographical expanse of the US is mostly in major metropolitan areas. The market has advanced well in the US for those locations with good data, but in the

US and elsewhere, the market expansion has been inhibited by data difficulties. For example, access to data in Canada and Australia is good but the metadata is spotty. Elsewhere, we learn, there is little or no access, or often it is expensive (as in some European countries), and the data may not be complete enough as is the case in much of Africa and Asia. (See Chapter 8 for a more complete statement.)
- ❏ Climate and weather events by themselves are difficult to describe, even without trying to build financial products tied to them. But when meteorological arguments are wrapped in the fabric of financial derivatives, the explanations can be daunting and discouraging for the non-expert.
- ❏ The market does not yet offer diverse and economical derivative products. For example, end-users outside the power sector want new products such as temperature combined with rainfall, and non-standard products can be expensive if available at all.
- ❏ The market has yet to reach a level of robust activity to support the liquidity that attracts capital market investors. Institutional investors who anticipated a successful launch of a weather-linked bond are still waiting.[5]

The weather market is so new that we are still at the frontier of discovery. For example, there is no shared practice of how much climate data to use and how to use it. There is no widespread pricing technology, like the Black–Scholes approach used so widely in other markets. Without such a point of reference, personal perspectives dominate; there is no consensus on where to begin contract negotiations, and many trades are never consummated because of differing views of price and value.

Spectators and speculators of the weather market hope that these hurdles are temporary and will ultimately be overcome. Contract originators see an opportunity to speculate in weather for profit, end-users see an opportunity to protect weather-exposed profit, and some governments realise the importance to their nations of their role is in providing information to support the private sector and safeguard the weather-exposed economy; various offices within US the National Oceanic and Atmospheric Administration, for example, meet often for conferences with the weather industry.[6]

Is it a weather risk market or climate risk market?
"Climate is what we expect, weather is what we get."[7]

The weather risk market should probably be called the climate risk market, as climate was and is its principal focus.

At its inception, the market concentrated on measures of average temperature and the accumulation of degree-days in multi-month seasons.[8] Now there is rising interest in contracts such as cumulative rainfall. Averages and accumulations are issues of climate rather than weather. To a much lesser degree, the market asks about specific events and averages over the next few days. These almost immediate events are weather issues; the meteorological community describes the current condition of the atmosphere and its near-term evolution as "weather". It understands that the view it has of evolving weather is constrained by the amount of information it possesses, and believes that there are immeasurable events that will drive the atmosphere in ways that cannot at present be foreseen. This means that there is a limit to predictability. Meteorologists believe that this limit on weather predictability is somewhere between a few days to two weeks into the future, and depends on the current weather patterns and on what is being predicted.

Climate is generally taken to mean the average state of the atmosphere, however, this is a simplistic view. Climate is a complex and ever-changing event that is described by the average conditions, and departures from the average of conditions over the planet. Climate includes the conditions of the atmosphere, the ocean, ice cover on land and sea, land surfaces and the full range of living things (correspondingly, the atmosphere, hydrosphere, cryosphere, surface of the lithosphere and biosphere). The interplay among these physical and biological "spheres", driven principally by the sun's energy, is the focus of climate and its variations. This interplay (called interactions or exchanges) measured from a local to the global scale, over periods of months to geologic time, shapes the climate.

The weather market wants both short-term forecasts of the weather and long-term forecasts of the seasons. These both have a place in market practices. Some weather contracts cover the immediate future, a period for which weather forecasts are useful. Most contracts, however, cover a period of a few months beginning a few months in the future. Both weather forecasts and seasonal forecasts are used in some of the common weather derivative pricing methods.

The weather market watches the developments in the Pacific Ocean closely, as it is now known that the ocean cycle in the tropics influences the US climate in ways that are in some measure predictable, at least under some conditions. Not just the weather market but people around the Pacific watch the irregular evolution of the cycles of El Niño, La Niña, and related large-scale atmosphere–ocean events with keen interest.[9] Combined atmosphere–ocean events affect the weather and the seasons around the Pacific. The tropical atmosphere and its weather, shift in response to shifting patterns of the ocean that lies beneath.

Financial weather contracts
A financial weather contract is a weather contingent contract whose payoff will be determined by future weather events (see Chapters 2 and 3). The contract links payments to a weather index that is the collection of values of a weather variable measured at a stated location during an explicit period. The average temperature at Charles de Gaulle airport, Paris, during November, December and January is an example of a weather index; the accumulated heating degree-days (HDDs) for the same period at the same location is related but different index.[10]

Financial weather contracts are available to supplement or hedge earnings when adverse weather might otherwise reduce returns. The purpose of weather hedging is to own a contingent cashflow that counters the declines in weather contingent business returns. A weather-exposed business that hedges well reduces its weather sensitivity.

WHY THERE ARE WEATHER DERIVATIVES[11]
Weather derivatives are a new solution to an old problem. The old problem is that businesses can comfortably absorb only some of the increased costs and decreased revenues caused by adverse weather. The weather variations of most interest to the market are departures from the average weather in a coming season from the average of selected past seasons. Weather variations do not have to be extreme to interrupt normal business cashflows. In fact, while small departures of a season from the norm are generally absorbed into the costs of operation and large departures and catastrophes are generally managed with insurance, moderate departures are difficult to absorb or manage, especially if this happens year after year. The insurance industry has not always been as interested in events where payments are as likely as, say, one in five, as they often are with the moderate departures that are hedged with weather derivatives.

Most businesses develop their practices based on the average of many past seasons, yet difficult seasons for an enterprise can be those that differ only moderately from the average. Until now, with few exceptions (see Chapter 12 for a discussion of the Pluvius contract), they have been limited in their ability to manage moderate weather variations except with insurance that required a demonstration of loss, or with alterations in usual practices such as reducing shipments to locations where sales may be uncertain. Weather derivatives offer a wider alternative based solely on weather events.

IS THE CLIMATE RISK MARKET A TEMPERATURE RISK MARKET?
Trades in the weather market in its first couple of years were all temperature-based, almost exclusively indexed to cooling degree-days (CDDs) and heating degree-days (HDDs) (see Chapter 15).[12] Even the early contracts in Europe were indexed to HDDs (the first European trade was an HDD swap agreement between Enron and Scottish Hydro in October 1997). The US continues to trade mostly in degree-day contracts while the global weather market trades in degree-days and average temperature contracts.

The Weather Risk Management Association (WRMA) initiated two surveys to estimate the size of the weather risk market.[13] The WRMA reports that there were nearly 4,000 weather transactions for the year ending March 31, 2002, with a total notional value (the maximum potential loss on each contract) of over US$4.3 billion. These are significant increases over the prior year's activity.

Degree-day contracts continue to draw interest because degree-days have been a traditional measure of the demand for power, and their derivative format is one with which energy providers and speculators are comfortable. Unfortunately, the degree-day legacy may have inhibited the rate of growth of the weather market, as degree-days are not particularly useful for retailers, entertainment enterprises, building contractors, truck dispatchers and others who measure their weather exposure in terms of temperature, rainfall, snowfall, freezing, extreme heat or some other variable not well-tracked by degree-days.

This may be changing, however, as average temperature becomes an accepted alternative measure of the demand for power. Those who developed analytical skills with degree-days can easily work with average temperature, and the databases from which they calculate degree-days are temperature databases. There are other reasons to consider moving from degree-days to temperature, or at least having both. For example:

❑ Temperature-indexed contracts appeal to potential end-users not in the power sector because average temperature is a continuous spectrum with no cut-off – there is no reference temperature (65°F or 18°C) dividing the spectrum into positive values and zero.
❑ The concept of average temperature requires little explanation, whereas the concept of degree-days sometimes raises eyebrows.

The transition to temperature-indexed contracts has already begun: they are more available from originators, especially outside the US; in December 2001 the London International Financial Futures Exchange (LIFFE) began offering weather futures indexed to average temperature in Berlin, London and Paris.

Weather trades with variables other than temperature are on the rise, if growing slowly. Rainfall and snowfall contracts have been consummated for retailers, ski resort operators, restaurateurs and, of course, electricity suppliers. The market provides streamflow contracts as an alternative to, and in addition to, precipitation contracts for applications such as water supply: some think streamflow is the better measure of water to fill the lakes behind dams than is precipitation. (Streamflow is the flow of surface water within well-defined banks, from narrow streams you can step over to the great rivers.) There is enough preparatory activity to think that wind contracts will soon take off.

The weather exposure of many potential end-users is linked to more than one weather variable, but the market has yet to learn how to structure and price contracts indexed to two or more weather variables, and to do so in ways that are economical for the end-user. Structuring multi-variable contracts is a frontier activity for the market.

INTRODUCTION TO
THE WEATHER
MARKET: DAWN
TO MID-MORNING

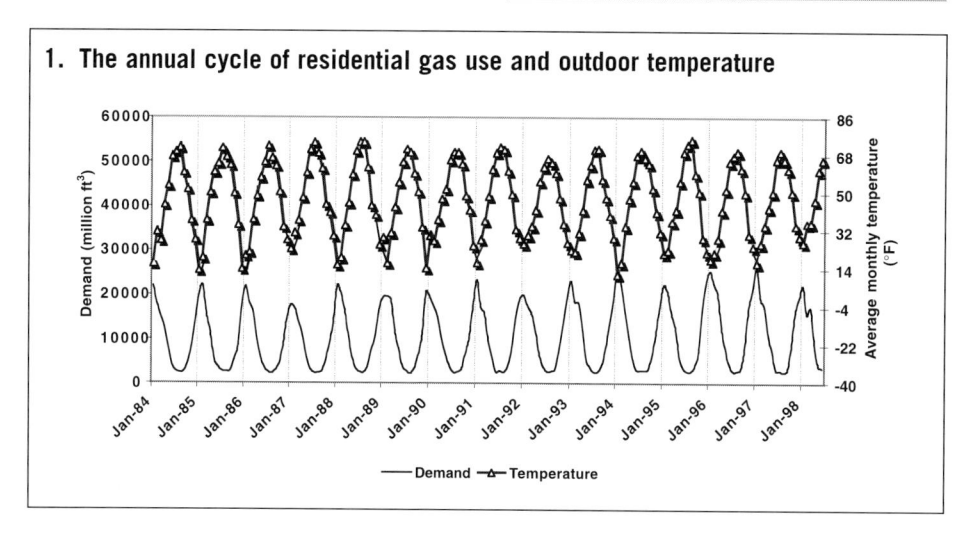

1. The annual cycle of residential gas use and outdoor temperature

If the market increases its expansion into non-temperature contracts it could better service the non-power sectors of the economy where it will find many times the potential end-users it has found in the energy sector. This represents an enormous growth opportunity for the market.

WEATHER VARIATIONS AND THEIR IMPACT ON REVENUE
Most businesses can operate comfortably with weather events that are very near the long-term average, but somewhere along the spectrum of departures from average conditions, weather events cause financial discomfort. To discriminate just which variations can be tolerated from those variations that cannot is a complex task. A thorough analysis could provide different answers at different times, because business conditions change. But the first step is always to understand and quantify the cost of past weather variations, and to cast this into present financial terms. This means correlating the demand for electricity, snowfall amounts, wind speeds, or whatever is the weather exposure, into revenue and cost.

Calculating the cost of weather
The Hot Air Gas Company (HAGC) is a proxy name for a real supplier of gas for home heating near the US and Canadian Border. Its effort to develop appropriate risk protection provides an insightful case study.

HAGC performed a weather cost analysis in three steps. First it reviewed the demand for gas that it experienced and the simultaneous monthly temperatures, for each month over the recent 30 years. A portion of these two time-series is presented in Figure 1. The demand for gas is read on the vertical scale at the left of the figure, and the outdoor temperature is read on the vertical scale at the right.

HAGC is located in an interior and cold region of the North American continental shield, where temperatures sometimes fall below 14°F (–10°C). The 30-year average winter temperature is 30.5°F (–0.8°C) and its standard deviation is 2.3°F (1.3°C).

Comparing the two time-series we see how gas demand and temperature are negatively correlated throughout the year. There is a maximum demand in winter and a minimum demand in summer. The minimum demand in summer is greater than zero because there is always a demand for some gas. This base level is more or less the same from year to year. Mild winter months result in mild demand; severe winter months result in strong demand. Sometimes the annual minimum monthly temperature is in January and sometimes it is in February. A cold year can be followed by either another cold year or a warm year. There is no apparent warming or cooling trend in this temperature record that would have to be removed for the analysis of the time-series to be useful in estimating weather costs and probabilities.

The relationship between these two parameters can be quantified when the full

30 years of each series is plotted against the other, as they are in Figure 2. In this figure, demand values are read on the vertical axes and temperature values are read on the horizontal axes.

In the HAGC supply region, the demand for gas falls to the base level when average temperatures in a month are above roughly 62°F (16.7°C). In warmer months (>62°F), customer usage is not related to temperature or the need to heat, but rather to other activities such as cooking or fuelling hot water heaters. The strong negative correlation (–0.94) of demand with temperature is in months with average temperatures below 62°F.

Many possible formulations could have been used to quantify this relationship; the simplest of course, is a linear regression. Straight lines do seem to fit the data well in Figure 2, capturing the rising demand with falling temperature in cooler months and the basic non-heating demand for gas in warmer months (the "hockey stick" pattern), although some scatter of points around the line remain (see Panel 1). The scatter of the points that do not fall on the regression lines is partly because some gas is used for other purposes, but, importantly, also because many conditions were not included in the analysis for this chart. Among these are the chilling effects of wind, the insulating effects of snow cover, and the amount of time people spend indoors as the sunlight waxes and wanes in an annual cycle. (The demand for electricity for air conditioning in warmer months is not currently a concern for the HAGC, and is not considered in this analysis.)

In some winter months in which strong revenue is hoped for, the demand is low when the weather is warm. The HAGC could look for revenue protection for these months. Also, stronger than expected demand in colder than normal months could exceed that planned for. This is difficult to manage with normal business practices in a deregulated power market and the company could look for protection here as well. Both under- and over-supply can have financial impacts and are thus candidates for hedging.

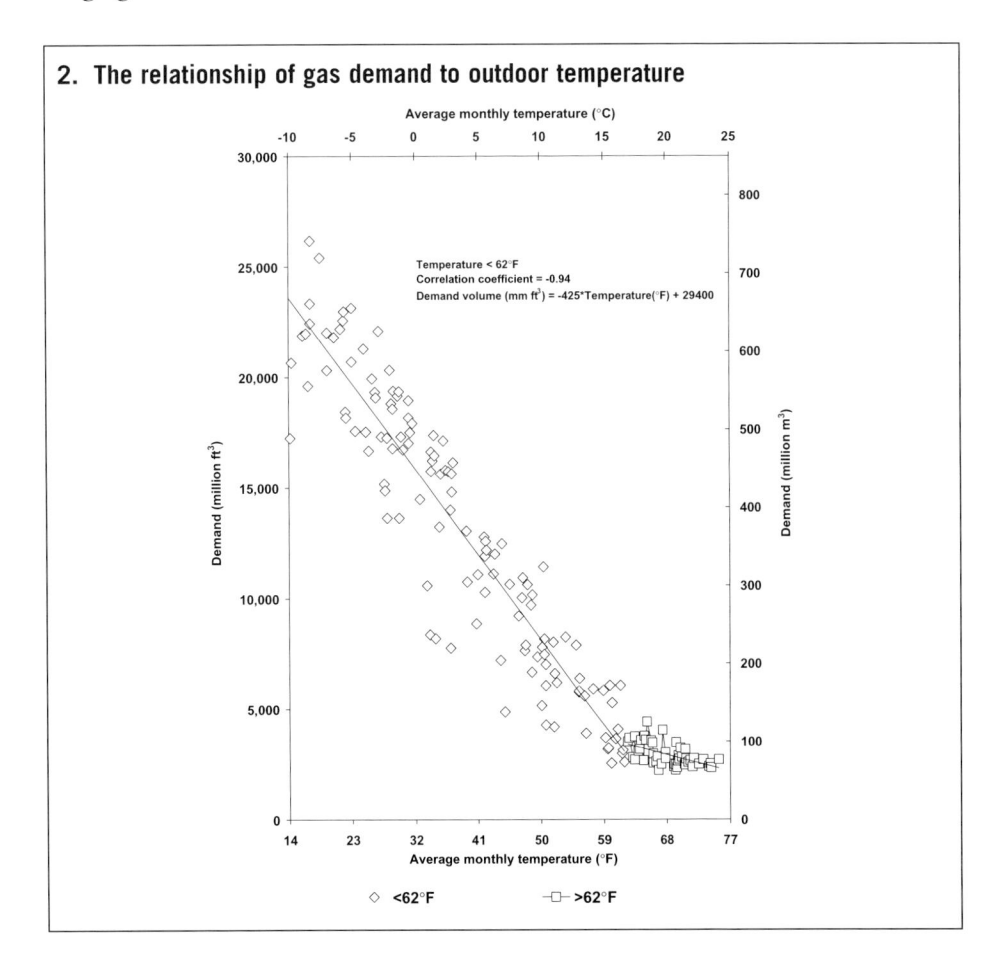

2. The relationship of gas demand to outdoor temperature

Temperature < 62°F
Correlation coefficient = -0.94
Demand volume (mm ft³) = -425*Temperature(°F) + 29400

INTRODUCTION TO THE WEATHER MARKET: DAWN TO MID-MORNING

Historical gas demand in the heating season (defined here as November–March), weather-related costs, past price variations and the expected impact of deregulation are a few of the factors that were considered when the demand data was converted into the "cost of weather."

Figure 3 shows two weather-dependent cashflows. The one marked "Unhedged revenue" is the revenue from gas sales less operating costs, including the cost for the expected purchase gas in a deregulated market. The other is the weather-contingent cashflow of a call option that might be used to offset the risk to revenue. (Call options are also discussed in Chapters 2 and 3.)

As can be seen in Figure 3, unhedged revenue increases as temperature declines, at least to the point where costly purchases erode the sales-less-cost margin. Based on history, HAGC expects that net revenue would be positive in any winter with an average temperature lower than 34.5°F (1.4°C) – the 30-year average plus almost two standard deviations but higher than 24.3°F (–4.3°C) – the 30-year average minus roughly three standard deviations. At temperatures higher than this, additional supply may have to be purchased. Net revenue would also be negative when sales decline in a mild winter with an average temperature higher than 34.5°F (1.4°C).

The company must discover its tolerance for weather risk to build an appropriate programme and to understand the value of weather derivative choices. It must decide, for example, if revenue must be greater than zero, or if some other level (presumably higher than zero) is better for sustained financial health. Figure 3 provides the HAGC with the ability to see the average temperatures about which it should be concerned. Knowing its risk tolerance and estimating weather probabilities has allowed HAGC to select among derivative alternatives and gain weather protection.

(The HAGC weather risk programme example is continued in Chapter 2.)

Weather probabilities and derivative prices

The price of a weather financial contract is based principally on the following three factors:

1. weather probabilities;
2. a subjective view of anticipated weather in the contract period; and
3. market supply and demand.

The first step in estimating future weather probabilities is to analyse the history of the weather or climatology. Climatology is only the starting point for the data analysis, as climatological data must be contiguous (no holes), data must be adjusted for changes

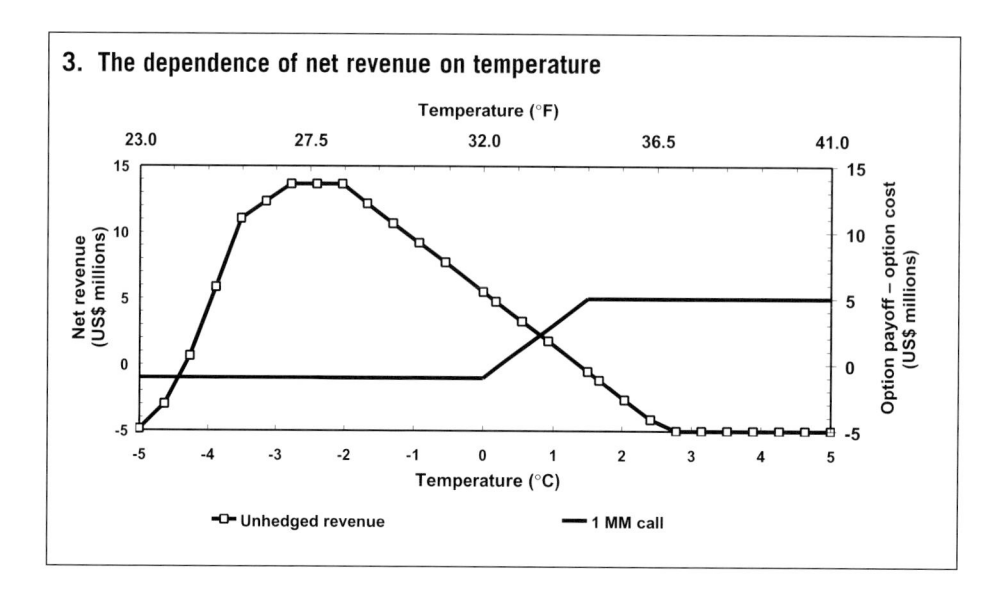

3. The dependence of net revenue on temperature

in measurement practices (described by metadata), and adjustments might be needed to remove trends from evolving environmental conditions of many scales (see Chapters 5, 6, 8, 9 and 10). Once the data is in a suitably enhanced state, statistical models are often applied to the modified climatology to augment the few decades of data into a more generalised climate distribution.

The analysts' subjective view of the anticipated season is often no more than a weather or climate forecast. Each participant may have his or her own view of the near-term weather or season and how it might differ from climatology. This view is used to modify the climatology, as the forecast is thought to sometimes carry recent and relevant information about the future.

Market supply and demand is partially driven by end-user needs, but in these early stages of the market, speculators hold the largest share of the market's value, giving them the stronger influence. It is the speculators, who are continuously trying to improve the risk positions of their portfolios that set the prices and drive most transactions.

Weather is erratic and climate evolves

The time-series of winter temperature for New York City's Central Park are presented in Figure 4. The data spans 113 years from the winter of 1869–70 to the most recent winter at the time of this printing, 2001–2. The figure is a favourite exhibit at weather market conferences because it illustrates so many of the issues and dilemmas in managing climatological data. For example, analysts debate the length of the data record to use in developing the representative statistics that will govern future climate probabilities. New York has warmed from the time of the start of this record in what may have been two stages: before and after 1930. Notable is an apparent trend of higher temperature in the last few decades – but with the frequent appearance of the lower temperatures of cold winters. The presence of these cold winters makes it difficult to assert, as some do each year, that next year will be warmer than average.

Some part of this temperature increase in New York City is likely due to the demands on the environment of New York's expanding population. More pavement over previously green areas, more air conditioners and vehicles, and generally higher energy use, are just a few of the reasons that make cities everywhere warmer than their surrounding areas (meteorologists call this the "urban heat island effect").

Perspectives differ on whether the apparent trend in temperature series such as the one for New York City are the effect of global warming, the impact of an

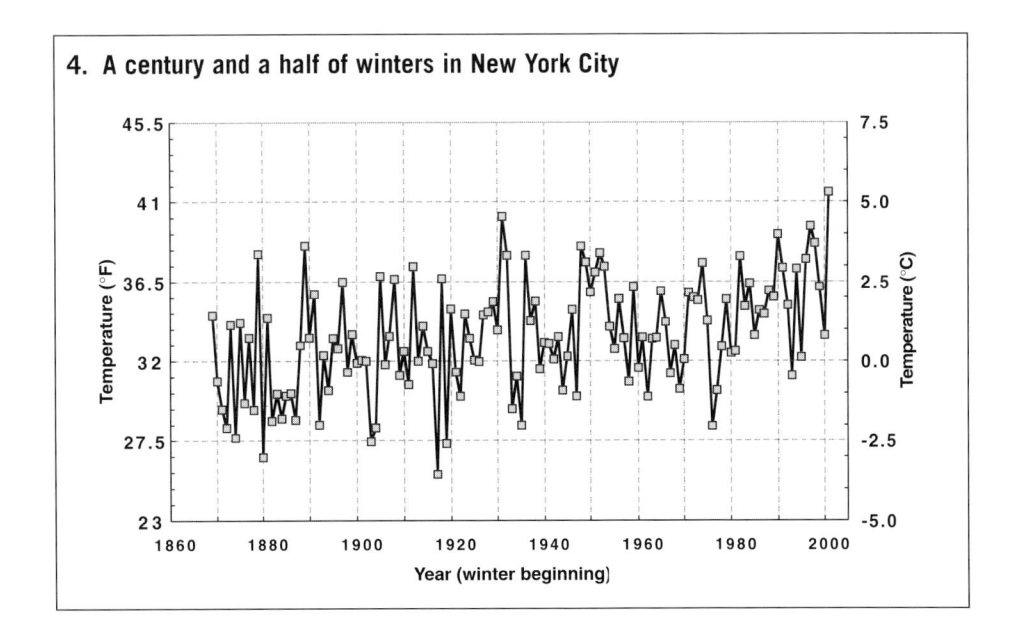

4. A century and a half of winters in New York City

INTRODUCTION TO THE WEATHER MARKET: DAWN TO MID-MORNING

expanding urban population, or the cyclic evolution temperature, or some combination of all of these (see Panels 2, 3 and 4).[14] It is difficult to translate the climatological measurements simply into market-useful weather probabilities, and there are many views on how it should be done (see Chapters 5, 6, 9 and 10).

We know that local conditions strongly influence weather probabilities. The analysis of temperatures for Freiburg, Germany, in Figure 5 illustrates how using data without understanding the local climate could potentially be misleading as to probabilities. To make this point, the 30-year record has been split into two periods with a histogram for each: one period is the recent 15 years; the other is the 15 years immediately prior. (In general, 15 years is a short period over which to draw conclusions about climate – nevertheless, casting prudence aside, it is common weather market practice to do so.)

The recent 15-year period (the darker bars) appears to have shifted to higher temperatures than the earlier period. This feature is also apparent at other sites in Germany, as well as at many sites in Europe (see Panel 3). Recent winters in mid-latitude Europe have been warmer than the long-term average winter (see endnote 3). Freiburg, in the Rhine River Valley, is in a region that is influenced by Foehn winds – warm dry winds that irregularly descend from the mountains into the Black Forest and raise temperatures. The Foehn is driven by larger-scale weather patterns and is more frequent in some years than in others. Each of the two 15-year periods in Figure 5 show the effect of the cooler temperatures sometimes being replaced by the warmer Foehn temperatures – there are two groups of temperatures in each period. Curiously, a simple average for each period falls between the Foehn mode and the cooler mode, at the place of lowest event occurrence: the average is a misleading descriptor of this data.

FORECASTERS SEE SKILL IN FORECASTS, BUT DO OTHERS?

The weather market needs at least two different kinds of forecasts: short-term weather forecasts that look out a few days, and long-term seasonal forecasts that look out a few months.[15]

The goal of a weather forecast is to describe the evolution of current weather, especially the weather patterns (or systems) over regions, as far into the future as there is skill to do so (meaning with accuracy and reliability). When the forecaster skilfully predicts the movement and evolution of a weather pattern, it is also possible to correctly estimate the time, location and intensity of specific events, as they can be interpolated from the larger pattern. Even if the weather pattern is forecast well, the

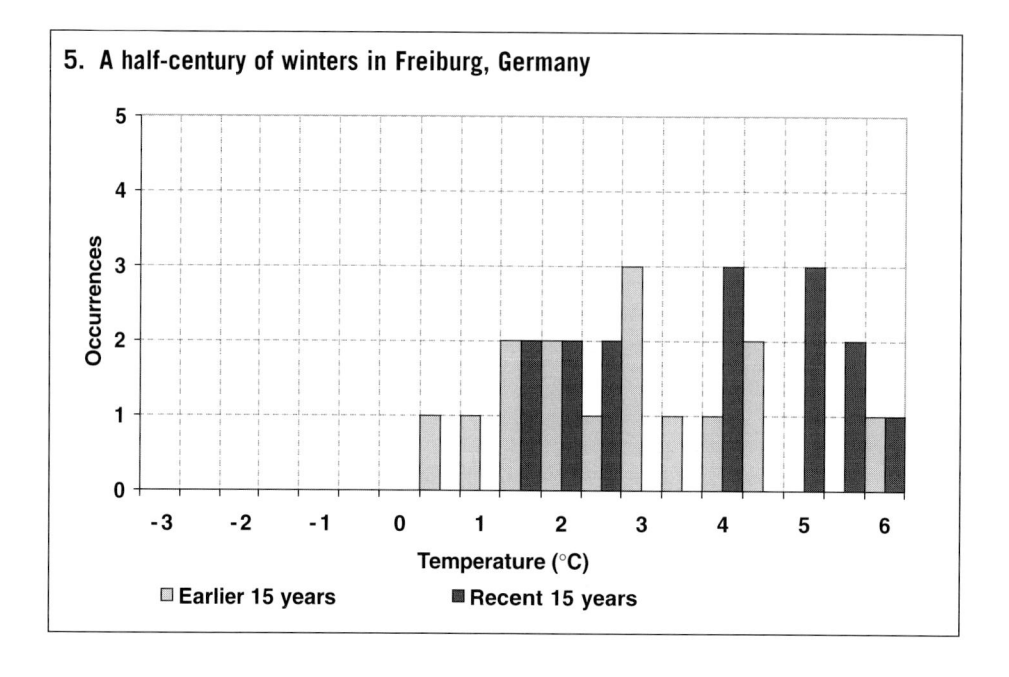

5. A half-century of winters in Freiburg, Germany

PANEL 2

GLOBAL WARMING AND THE US WEATHER MARKET

There is little doubt we are in the midst of a period of rapid global warming of the troposphere (although the stratosphere has cooled).[16] Globally averaged atmospheric temperature measured at the earth's surface (land and ocean) rose an estimated 0.6°C during the 20th century, and the 1990s were the warmest decade in this period; there is evidence that the upper layer of the ocean has also warmed.[17]

Regardless of the reasons for the warming, all scientists would agree that the climate varies in complex ways that are only partly understood and are only partially predictable (see Chapters 4, 6, 7, 9 and 10). Society must deal with the question of the potential for global warming, and so must the weather market.

It is typical for weather market participants to estimate average temperature at a location using a recent decade, or a few decades at most. They do this, they say, because the warming trend will continue, and recent years are a better example of the near future than the fuller record. This means that most weather derivatives trade based on average temperatures that are higher than for earlier decades. It is also possible that the current trend is part of a natural cycle that could reverse itself into a cooling trend. This has happened before (see Panel 4).

Longer records aid in developing a view of what the future may hold. The sequence of winter temperatures averaged over the US in the figure below is for the period 1895–2002, almost 11 decades. The year indicated on the horizontal axis is the year in which the winter began (that is, in this chart, 2000 means winter 2000–1). Average temperatures are indicated in both degrees Fahrenheit and Celsius on the vertical axes.

There are two segments of winter represented in the figure:

❏ early winter, the average for the months of November and December, indicated by the triangles; and
❏ an estimate of the heating season, the average for November–March, indicated by the connected circles.

Weather derivative trading in the US focuses on only a few locations around the country, not on a national average. Nonetheless, a few details relevant to weather trading and pricing in the US emerge when the winter data in the figure is reviewed:

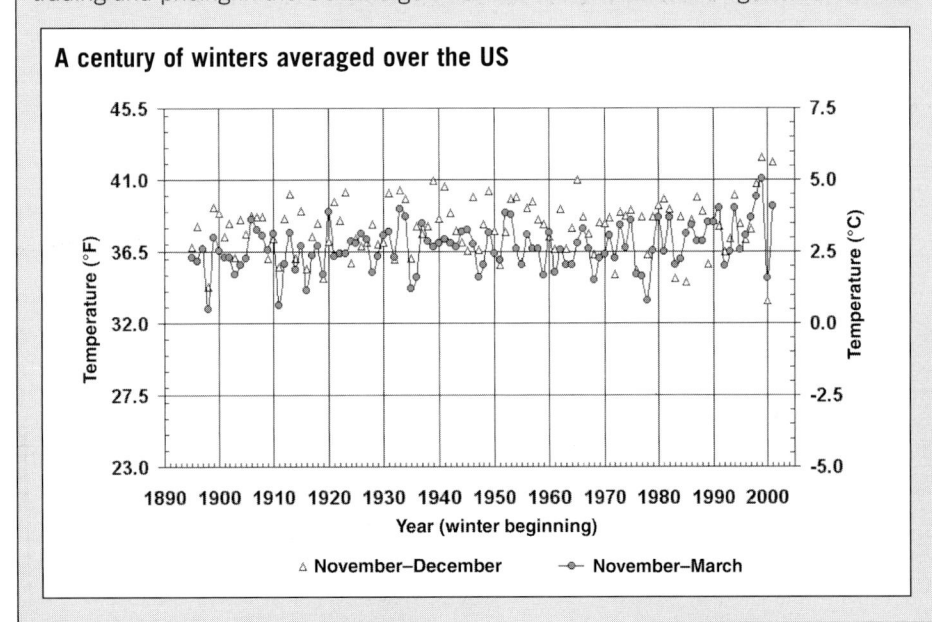

A century of winters averaged over the US

INTRODUCTION TO THE WEATHER MARKET: DAWN TO MID-MORNING

❑ A rise in November–March temperatures of about 0.8°C or about 1.4°F can be estimated from linear regression analysis of the full record. If this trend is valid, it represents an increase of about 120 HDDs for the heating season over the period of the record.
❑ The standard deviation of year to year differences, a measure of changes from one year to the next, is also 0.8°C, the same as the estimated temperature rise. As the standard deviation is an average of differences, this means that changes from one year to the next frequently exceeded the estimated magnitude of the warming trend.
❑ November–December temperatures were mostly higher than those of November–March until the 1980s. Since then, early winter has often been colder, or at least as cold, as the heating season.
❑ As of 2002, the warmest November–December in the record was in 1999, and it was followed by the coldest in the record in 2000. The warmest November–March was also in 1999, and November–March 2000 was among the coldest.
❑ Four winters in a row, 1996, 1997, 1998 and 1999, were each warmer than the one before.

This last point, one warm year after another, was part of what gave the weather market a push to activity. Most end-users wanted warm winter protection (HDD puts) and derivative originators used the averages of recent years to structure these weather derivatives. With each year being warmer than the prior, even though they used the warm average of recent years, the originators often still had to pay. Warm winter protection became increasingly expensive. Then came the winter of 2000–01! Most forecasters predicted that the winter 2000–01 would be somewhat colder than the prior few warm winters. Speculators were of two opposing views: warmer than normal and colder than recent years. No one was correct. The year began with a record warm winter but soon turned bitterly cold, and the nationally averaged December–February period was among the coldest (and driest) periods on record (much of the West and Midwest US had near-normal temperatures, but that too was colder than recent years).[18] It was a risky winter for some speculators.

End-users who had purchased warm winter protection in 2000 in spite of the forecast (knowing that forecasts are not always accurate) received no payoff, and those who sold them these contracts profited – quite a different outcome from the prior few years.

Weather market analysts argue among themselves about selecting the length of historical record to best represent future seasons. Should one use the most recent five, 10, 20, 30 or 50 years of temperatures for the average? Should the length be different for the standard deviation of temperature? Should the apparent trends in some data series (only some of which are warming trends) be mathematically removed (see Chapter 8)?

With recent years being warmer than long-term averages, a short record often yields a higher average temperature than does a longer record. Someone who believes we are in a period of warming would look to the shorter record. If, on the other hand, these are cyclic changes and that normal volatility applies, then the longer multi-decade record might be a better choice. These are difficult decisions. Different perspectives yield different answers, and the answers might be different for different locations.

With US winters breaking records for both high and low temperatures in successive years, everyone asks what is causing this volatility - and will it continue? No one has the answer. Forecasters cannot see far enough into the future to tell us - we will have to wait and see. What we do know is that volatility is a normal part of climate history. Thus, if end-user hedging is important this year, it is probably going to continue to be important.

specific smaller events may or may not be, as they are interpretations from the larger pattern.

The weather forecaster will say that short-term forecasts are better than ever, and most scientists would agree; the weather forecaster's skill is in predicting the evolution of a pattern, and this skill has measurably improved. However, the forecaster cannot accurately place the event (say snowfall and snow depth) precisely at a specific intersection of a named avenue and street at a specific time. The meteorological community has not explained this limitation well enough to the forecast-consuming public. As a result, many do not derive the full value of forecasts, and may have low estimates of their verity.

The goal of a seasonal forecast is to describe the average weather that will be experienced either at a single location or a region. The most useful seasonal forecast for the weather market is a forecast of probability distributions. Forecast distributions include all the possible values of the variable of interest (temperature or rainfall, for example) presented with a measure of their likelihood. (The derivative analyst uses the forecast distribution to modify the one calculated from climatology.) While the forecast probability distribution may be for a specific location or region, it is the average event over a period that is being forecast, not a specific event. The expected number of specific events might also be forecast if it is done without assigning specific times for the events.

Seasonal forecasts are drawn from a larger pattern, as are weather forecasts, but look further into the future than does a weather forecaster; the seasonal forecaster looks at a larger geographical scale: the current and evolving conditions on a global scale. Seasonal forecasts can also be interpolated to make site-specific seasonal forecasts. The climate forecaster would assert that forecasts have improved over recent decades. The skill in the forecast, however, depends on current global conditions such as the presence or absence of a strong oceanic El Niño or La Niña, and the season for which a forecast is made. Scientists think that climate forecasting might be at the edge of new and important discoveries. The skill of a forecast is measured in many ways, from the anecdotal experience of rain-soaked people who understood the forecast to be for a sunny and dry day, to the statistics that meteorologists and climatologists have developed to objectively measure forecast skill (see Chapter 11). This can explain the difference in opinions on weather forecast skill.

Forecasts for specific conditions beyond a few days, say two weeks at the maximum, are thought to be beyond the current skill of weather forecasters. The scientific community mostly believes that there are variations in the earth–atmosphere system that will grow large enough within two weeks or less so as to make any forecast beyond this period simply speculation. There are, of course, some forecasters who will assert the predictability of specific events far into the future. For each one of these *long*-long-term weather forecasters there are a thousand who say it cannot be done. Most of the meteorological community frowns on predictions of specific events beyond a couple of weeks because it has never seen a convincing demonstration that there is a scientific basis to these assertions. Long-term forecasts generally do not extend beyond a year, and are usually limited to a few months. An exception to this is the prediction that global warming will continue as long as society continues to burn fossil fuels. The evidence for anthropogenic global warming is inferred from recent events and computer models, and many find it convincing – some do not (see Panel 4).

FOR THE WEATHER-SENSITIVE END-USER, NOT TO HEDGE IS TO GAMBLE ON THE WEATHER
Enterprises with a natural weather exposure – one that inherently affects the enterprise – carry a natural weather bet: when the weather is favourable, they win, if the weather is unfavourable, they lose. It has been the practice of weather-sensitive

PANEL 3

GERMAN WINTERS

The year to year seasonal variations in the US and Europe do not follow the same course. The figure presents German winters for the period of 1953–2001, almost a half-century, and the temperatures are averaged over 10 locations over the months of December-February.[19, 20] The year indicated on the horizontal axis is the year in which the winter began (that is, in this chart, 2000 means winter 2000–01). Average temperatures are indicated in both degrees Fahrenheit and Celsius on the vertical axes.

The events in Germany are different from those in the US in a few ways. These are:

❑ No German records were broken in recent years.
❑ The cold German winter of 1995 occurred midpoint in an otherwise warm European decade.
❑ The time of the cold German winter of 1995 was a time of an average winter in the US.
❑ The time of the very cold US winter of 2000–01 was a time of a warm winter in Germany.

The differences in temperature time-series between Germany and the US indicate that climatic forces drive the seasons differently in these two regions (see the discussion on ENSO vs NAO in Chapters 4, 6, 7 and 9).

Forty-nine years is not a large sample from which to draw conclusions about climate. Nevertheless, a regression line fitted to this short time-series could be interpreted as indicating a rise in temperature of a couple of degrees Celsius in the half-century of the record, much of this is in the last five years. This possible warming pattern is consistent with the conclusions on global warming of the Inter-governmental Panel on Climate Change.[21]

The standard deviation of the year to year changes in German winters appears larger than for the US time-series, but this is likely a result of the small number of stations used in the area average: averaging over more stations would likely smooth this apparent volatility. Note that the short record and small number of available stations is characteristic of acquiring national data for Europe (see Chapter 8).

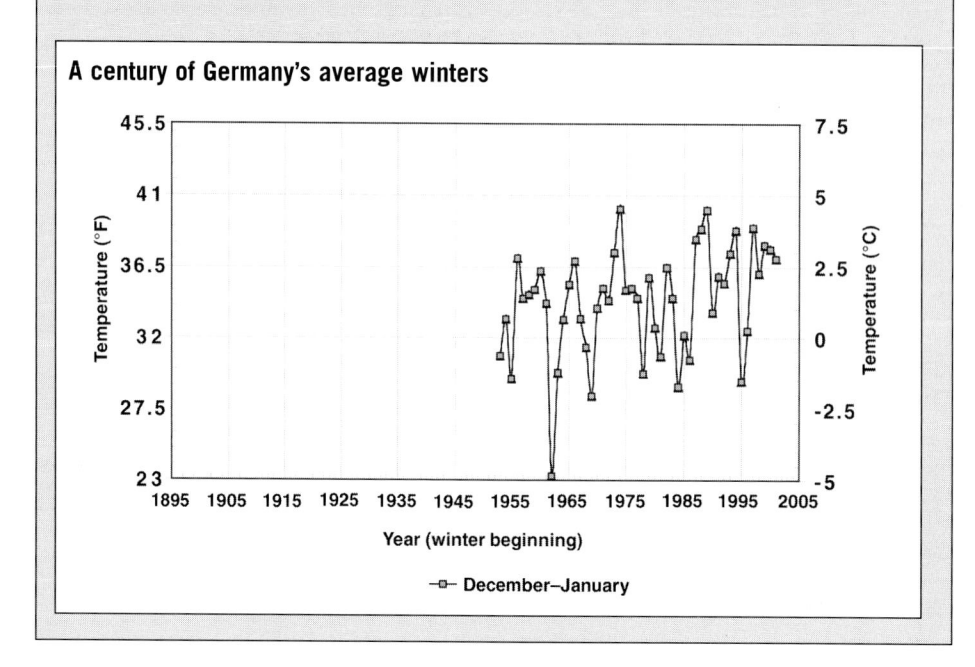

A century of Germany's average winters

— December–January

businesses to defend against weather variations by adjusting business activities, such as deciding how many air conditioners to ship to each city, or to limit expansion in uncertain conditions. These practices are based on a look back at long-term averages, and perhaps on forecasts. This does nothing to relieve them of the weather gamble, as there is little or no information on which to make these decisions. Buying commodity contracts in futures markets protects against price variations but not against a drop or a surge in demand. Another alternative is to set capital aside to manage the downturns in weather-related activities.

Most end-users continue to bet on the weather by being passive: they refrain from beginning a programme of hedging weather risk. This occurs even though there is now an expanded choice of financial weather hedges with which the weather bet can be neutralised. However, a hedged business is in a better place to improve service to existing customers, to enlarge a customer base, and even to expand the business through acquisition.

END-USERS ALWAYS WIN WHEN THEY HEDGE WELL
The principle fiducial responsibility of management is to assure the financial health of an enterprise. If this health is subject at least in part to changeable weather, then management must consider the benefits of weather hedging. Engagement in a weather risk programme can always reduce earnings' volatility, however, the challenge is to achieve this reduction with appropriate weather protection at an economical cost.

The goal of hedging is to be less concerned, or not concerned at all, about the impact of weather on cashflow or return. Management achieves freedom from the weather when it engages in a hedge that would pay supplemental cash (or delivery of a commodity) should weather be adverse to earnings. This freedom can avoid fluctuations that raise investor concerns. It also includes the ability to use capital that might otherwise be held in reserve against the possibility of adverse weather. Capital freed from reserve can be used for acquisition, development or investment, regardless of the coming seasonal weather events.

Of course, freedom from the weather comes at a price: hedging must be paid for with cash or with a promise that could reduce returns. Cash is paid in advance for a call or put option. Collars and swaps carry the promise to pay if a season is favourable, and the expectation to be paid if a season is adverse (see Chapter 2).

Some end-users argue that the cost of hedging, which is based on long-term averages, reduces returns in the long term through the continuing year to year cost of the hedge. They quite correctly observe that speculators offer hedges that are priced to provide them with a positive return. Clearly, paying for a hedge each year costs, but this view ignores the freedom a hedge offers the end-user to reinvest capital reserve and, very importantly, discounts the intangible benefit of confidently knowing financial disasters have been avoided.

Economically justifiable involvement in a well-structured weather risk programme reduces exposure of weather-related revenue and cost. A hedge smoothes cashflows and enhances freedoms so it can be said that "hedging is winning."

THE VALUE OF END-USER HEDGING IS REALISED OVER YEARS
All derivatives traders, actuaries and speculators "play the odds." They calculate a "probable payoff" by multiplying each possible payoff by the probability of that payoff. It is the same with weather derivatives: the payment for each possible weather event is multiplied by an estimate of the probability of that event (as described in the section "Weather probabilities" above and in Chapter 5).

The originator of a call option, for example, would use an estimate of probabilities, one for each possible outcome, to calculate what is called the "fair price" for the option. Then the originator would add a "risk premium" to the fair

> **PANEL 4**
>
> ## CENTURIES OF TEMPERATURE HISTORY
>
> One record of the temperatures for the past three and a half centuries, is the time-series of temperatures known as the Central England Temperature (CET) series. The CET temperature data are available from the UK Met Office Hadley Centre (http://www.metoffice.com/research/hadleycentre). The Hadley Centre states that "The monthly series began in 1659, and to date is the longest available instrumental record of temperature in the world." Other records compiled from less direct measurements of temperature span millennia.
>
> As can be seen in the figure, this record begins with a sharp cooling, seen as a drop of almost 2°C in annualised temperature. This is followed by an almost three quarter-century warming of about 3°C, from about 7.25°C to 10.5°C. In this warming period, the temperature rose from a point lower than those of the 20th century warming – to a point close to those of today. This earlier period in central England witnessed a larger warming in a shorter time than that of the last century. It was abruptly terminated with a dazzling drop of 3.5°C in only a few years. Then temperatures recovered almost to the previous high, again, in only a few years. In terms of volatility, the warming, then precipitous cooling, then warming of the 18th century was more dramatic than the changes of the current period.
>
> Since the last glacial epoch 11,000 years ago, there have been many warm and cool periods.
>
> **The Central England Temperature series**
>
> *Source*: UK Met Office Hadley Centre

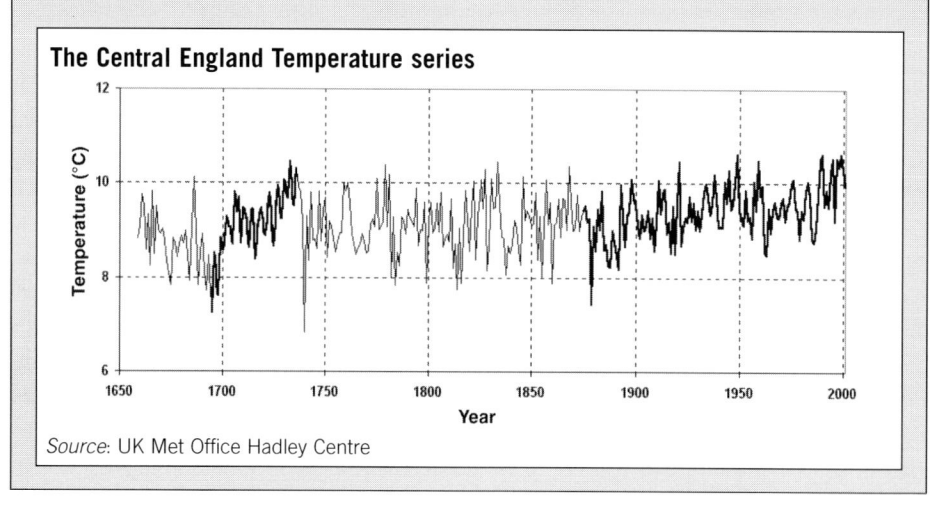

price to compensate for the risk it takes of wrongly estimating probabilities and the risk of having to pay. This can be compared to the odds and payoff of roulette. A casino includes the numbers one through 36, zero, and double-zero (00) as possible outcomes in the game. There are then 38 possibilities, each with equal probability of occurring, 1/38. Each has the same payoff, but it is only 36 times the bet, not 38. The "probable payoff" is then 36 multiplied by 1/38, which is less than one. Said another way, one round of roulette has a 1 in 38 chance of paying US$36 for each US$1 bet.

A one-season visit to the weather market is like one play at the roulette table! Hedging in any one year might also have a probable payoff of less than one if the originators estimate of weather probabilities is reliable. The difference between roulette and weather hedging is that the latter has a definite, valuable and measurable business advantage to an end-user, where roulette is simply a game of chance. End-users who recognise the value of hedging in one year generally see it as a continuing process, requiring year after year involvement. A weather risk management programme is analogous to business insurance and there are some liabilities a business would never allow to be uninsured. If there is a need to hedge in any one year, then it is likely that there is a need to hedge in all years.

Should an end-user hedge even when the seasonal forecast appears favourable to the business? Seasonal forecasts are not that reliable yet and give only marginal assurance over the expected outcomes based on climatology. Speculators do not bet on specific outcomes but rather tilt historical probabilities in the direction of the forecast probabilities. Forecasts indicate likelihood, strongly or weakly, but do not assure it.

Some good reasons for an individual enterprise to not hedge include the following:

❑ Weather risk is not large compared to other enterprise risks and limited internal resources should be focused where there are big problems; there just is not enough talent in the enterprise to justify the cost and effort to understand the weather market as well as do the speculators and consultants, and then to build an appropriate weather risk programme.
❑ After balanced and informed analysis, it is thought that the derivative cost is not a good economic choice; the financial structure is too expensive for its probable payoff compared to the probable revenue decline, that is, the potential cost of weather variability.
❑ It is not possible to buy a derivative structure that appropriately hedges the weather risk; for reasons of speculator preference, the required structure is just not available. There is nothing offered with the needed payoff or is not specific enough to the region where protection is needed. A derivative might not be valuable even if its price is low if it does not provide the required protection.

It has been said that if others in competing corporations do not hedge the weather, then there is no "punishment" for not hedging either. How long will unhedged weather risk escape the attention of business analysts and investors now that there are suitable means to protect from weather variations and reduce earnings volatility?

Capital market investors and market liquidity
The first phase of the weather market has been to provide hedging opportunities to end-users and enhanced return to successful speculators. More end-users and more originators come to the market each year, and trading is more active than it was. Nonetheless, the products offered appear not to be what is needed to attract capital market investors even though the weather market offers them a diversifying asset.

One reason is that evaluating the potential return and risk of a weather derivative demands a significant commitment of resources to interpret weather information and understand new kinds of pricing methods. This is especially true for a portfolio of weather assets with a variety of disparate structures each indexed to weather at different locations. Another reason is that internal policies in some enterprises sometimes limit the placement of derivatives in portfolios.

Nevertheless, there continues to be interest from large investors to enhance their mix of assets by holding weather-indexed securities. Weather derivative returns are not correlated with other capital market returns. The weather market does not react to the currency, equity or interest rate markets so that weather assets could provide portfolio managers with diversity. The weather and climate affect the economy, not the other way around.

The mortgage market faced and solved problems similar to those that the weather market is now facing. Mortgage-backed securities (MBS) are bond-like securities built upon a portfolio of mortgages; collateralised mortgage obligations (CMO) are mortgage derivative securities with cashflow tranches built on a portfolio of MBS. The MBS (and the CMO) provides a principal and cashflow similar to other bonds but carries embedded options; one option is exercised each time the mortgage borrower prepays the mortgage thereby terminating that individual contribution to the cashflow. Pools of mortgages, MBS and CMO were difficult to analyse until pricing models became widely available to market investors. The appetite of investors for these securities with embedded options rose with the ability to estimate value and measure risk. Sales of mortgage securities to investors transferred mortgage risk to those investors with an appetite for it and returned capital to mortgage originators, who then could originate more mortgages and securities.

Securitising weather assets could make them more attractive to investors. A weather securitisation could be a bond backed by the credit of the issuing institution, and might offer tranches that assure payment of principal and would pay a coupon that would float with the weather. There were two attempts at weather securitisation in 1999–2000, and both aroused investor interest; each met with limited success. These earlier efforts are discussed in Chapter 13, where it is suggested that the lessons learned from the past could lead developing weather structures that would appeal more to the capital markets.

Securitising weather derivatives into weather-linked bonds and selling these to investors would return capital to weather asset originators, who could then originate more assets. The result would be more liquidity in the weather market and a new and manageable diversifying asset to capital market portfolios.

SPECULATORS BET ON THEIR VIEWS

Market makers/speculators take risk intending to profit from doing so: this is their role in providing liquidity in the market. New speculators may have no weather exposure until they buy, sell or swap weather derivatives. If they acquire a portfolio through trading, they hope to rebalance it as the market moves (discussed in Chapter 14). An important distinction between speculators and end-users is that an end-user begins with a natural weather exposure, and generally transacts no more than once for each perceived exposure. Speculators continuously watch market prices with an eye to capturing a mispricing and improving the risk and return factors of their portfolio. (As in all successful derivative markets there is a primary market, with transactions that involve at least one end-user. Speculators trade with each other in what is called the secondary market.)

The price of a weather derivative is an objective fact – its value to an end-user or in a speculator's portfolio, however, requires a subjective assessment. The speculators set the price as they trade with each other and compete for end-user fees. They study climate history and are attentive to changing weather forecasts to try to understand if the odds of having to payoff on a contract are changing. They also look at the current composition of their weather portfolio and its risks to evaluate their preference to buy, sell or hold individual assets. It is through the dynamic of all speculators doing this simultaneously that prices are established.

Speculators are hopeful that their evaluations of history and their divining of the future leads them to prices that are favourable to their portfolios. Clearly this is not always the outcome. Speculators trade with each other because each believes that its estimate of probabilities is superior. The payoff on each contract, however, is based strictly on the measurements of its weather index, not on anyone's estimates of the actual season.

Speculators post weather prices but Mother Nature picks the winner.

1 *US National Oceanic Atmospheric Administration (NOAA) news releases found at http://www.noaanews.noaa.gov/stories/s390.htm and http://www.noaanews.noaa.gov/stories/s549.htm.*
2 *Energy and Power Risk Management Special Report: Global Deregulation, November 1999.*
3 *Tank et al. (2002).*

4 Weather Risk Management Association 2002 survey, available at: http://www.WRMA.org.
5 Keefe (1999).
6 "A Workshop for the Weather Risk Management Industry and NOAA: Climate Forecast and Data Needs", sponsored by NOAA, the Risk Prediction Initiative, and the Weather Risk Management Association, October 2–3, 2001, Bethesda, Maryland.
8 WRMA reports the results of the survey at: http://www.WRMA.org.
7 This quote by Mark Twain can be found at http://www.freedomsnet.com/cgi-bin/q.cgi?subject=weather.
9 El Niño is an ocean event most easily recognised by anomalously warm sea surface temperatures in the Eastern Tropical Pacific Ocean. La Niña is a related but distinctly different ocean event most easily recognised by anomalously cool sea surface temperatures in the Tropical Pacific Ocean. Scientists and mariners had been aware of these oceanic events for more than a century, but public awareness increased through heightened media reports sparked by the warm winters and heavy rains in the US in a sequence that began in 1997. El Niño, La Niña and atmospheric pressure differences that disturb the tropical winds and alter sea surface elevations are all part of the phenomenon called the El Niño–Southern Oscillation (ENSO). (See Chapters 4, 6, 7 and 10 for fuller discussions.)
10 The degree-day is a legacy concept. It is a measure of energy demand that the weather market inherited from the power market, where it was used as a means to approximate the requirements for heating and cooling. It was the basis for almost all contracts in the US. Recently, the more easily understood "average temperature" has been introduced and might one day replace degree-days.
Heating degree-days, the measure of the demand for heating, are accumulated as follows:
A heating requirement term is calculated for each day in the period of interest, November–March, for example. The term is the result of subtracting the average temperature in that day from a reference temperature; if this term is negative, it is set to zero. Each day's term, positive or zero, is added to the HDD tally for the period.
Cooling degree-days, the measure of the demand for air conditioning, are accumulated as follows:
A cooling requirement term is calculated for each day in the period of interest, June–August, for example. The term is the result of subtracting the reference temperature from the average temperature in that day; if this term is negative, it is set to zero. Each day's term, positive or zero, is added to the CDD tally for the period.
Energy degree-days (EDDs) are the sum of heating degree-days and cooling degree-days in a period. In any one day, only one – either HDDs or CDDs – can be positive and add to EEDs, as the other will be zero. EDDs measure the demand for both heating and cooling over a period.
Typically, the reference temperature in the above terms is 65°F in the US, but it need not be. Europe has used 18°C and other temperatures depending on country. We also see a similar application in agriculture using growing degree-days with a reference temperature appropriate to the application.
11 Many in the market use the term weather derivative to mean a weather contract in either derivative or insurance form, the latter is sometimes called a "weather insurance product" or "weather-indexed insurance". Regardless of the name, these two formulations can be very similar contracts; perhaps the greatest qualitative differences are in the ability of the issuing party to construct and offer the requested product, and the issuing parties credit rating. We use the term weather derivative to mean either a weather derivative or a similarly constructed weather insurance product throughout this chapter.
12 USA Today (2000).

13 *WRMA reports the results of the survey at: http://www.WRMA.org.*
14 *Dischel (1999).*
15 *Dutton and Dischel (2001).*
16 *The way atmospheric temperature varies with height above the surface is often used to describe the structure of the atmosphere. The troposphere is the atmospheric layer that extends from the earth's surface to a height of about 16 kilometres, depending mostly on latitude and season, and it is distinguished from the layer above it, the stratosphere, by the transition layer of the tropopause.*
17 *Houghton, et al. (2001).*
18 *US National Oceanic Atmospheric Administration (NOAA) news release found at http://www.noaanews.noaa.gov/stories/s549.htm.*
19 *It is not yet possible to obtain pan-European data as readily as data can be obtained for the US. It is likely that the qualitative conclusions drawn from this German data fairly represents the events in Western Europe.*
20 *Dischel (2001).*
21 *Houghton et al. (2001)*

BIBLIOGRAPHY

Dischel, R., 1999, "Shaping History", *Energy and Power Risk Management Special Report: Weather Risk*, September.

Dischel, R., 2001, "Deutsches Data Duel", *Energy and Power Risk Management*, January.

Dutton, J., and Dischel, R., 2001, "Climate Predictions: Minutes to Months", *Energy and Power Risk Management Special Report: Weather Risk*, August.

Houghton, J. T., Y. Ding, D. J. Griggs, M. Noguer, P. J. van der Linden and D. Xiaosu, "Climate Change 2001: The Scientific Basis", Intergovernmental Panel on Climate Change.

Keefe, D., 1999, "Still Waiting", *Energy and Power Risk Management Special Report*: Weather Risk, September.

Tank, A. K., J. Wijngaard and A. van Engelen, 2002, "The Climate of Europe", European Climate Assessment, available at: http://www.knmi.nl/samenw/eca/index.html.

Twain, M., 1867, *From the Celebrated Jumping Frog of Calaveras County and Other Sketches*, (New York: C. H. Webb). See also http://www.twainquotes.com/18640515t.html.

West, J., 2000, "Making Money with Weather", USA Today, 15 June, available at: http://www.usatoday.com/weather/money/wxderiv.htm.

2

Financial Weather Contracts and Their Application in Risk Management

Robert S. Dischel; Pauline Barrieu

Weather Market Observer, LLC; Université de Paris VI and Doctorate HEC, France

The weather risk market exists to offer protection against the potentially adverse effects of weather and climate – it does so principally with financial contracts. Traditional insurance provides weather protection and has done so for more than a century, but the products offered in this new market are unique; weather contracts settle based on values that are keyed to a weather index, not to a measure of damage or loss – payments are fixed at the end of the contract period based only on the final index value.

An objective measure of the weather is at the centre of the contract, so finding and using appropriate weather data has become a fundamental market activity that is required for success. Mostly, this discovery is required of the active player and weather consultants already in the market; the newcomer must first understand the new risk management tools – assisting in this understanding is the objective of this chapter.

Introduction
Those accustomed to derivatives will see that the weather market has borrowed, and continues to borrow, the form and structure of weather assets from other well-established derivative markets. While some may be seduced into thinking they can simply behave in this market as they do in other derivatives market, such a perspective has been self-defeating to some experienced players who underestimated the complexity of contracts based on the capricious and uncontrolled nature of weather – it is not a trivial distinction.

Almost all human activities are exposed to weather on timescales from minutes to decades. Businesses, some of which are more weather-sensitive than others, must estimate the cost to them of weather variations and learn how to manage this unpredictable unevenness, as the weather can be favourable or adverse to their results. An evaluation of the cost of weather variability sometimes rightly leads to a decision to take no protective action or to act to mitigate risk with revised business practices; indifference to the evaluation can sometimes lead to no action by default. Each year, however, more businesses look at the weather market to see if there is opportunity to manage their climate risk with weather assets.

If the evaluation leads to a decision to hedge weather variability, then the business faces many choices in developing a meaningful weather risk programme. Unless this prospective end-user is experienced in weather and climate and is familiar with derivative concepts, the assortment of choices can be confounding and the task can be daunting – even with such familiarity it is difficult.

FINANCIAL WEATHER CONTRACTS AND THEIR APPLICATION IN RISK MANAGEMENT

As the market matures, so does the array of available products. One of the more important qualitative distinctions in weather assets is between weather derivatives and weather insurance products, as each can provide similar means for reducing financial weather exposures. And while some see weather protection as derivative hedging, and others as business insurance, for the end-user of financial weather products there is little quantitative difference – it is simply weather risk management.

Many in the market use the term "weather derivative" to mean any weather contract in either derivative or insurance form (the latter is sometimes called a weather insurance product or weather index insurance). The practice of generalising the term weather derivative to most weather assets is used throughout this chapter, as it is common in the weather market.

This chapter puts forward some of the concepts with which every participant in the weather market, newcomer or experienced player, must be aware. The first section is a general discussion of the form of financial weather contracts. Next, there is a review of some of the market's basic tools and their characteristics or structures. Finally, there is a brief anecdotal description of how one company quantified its weather exposure and entered the weather market for the first time.

Financial weather contracts

A financial weather contract is a weather contingent contract whose payoff will be in an amount of cash determined by future weather events. The settlement value of these weather events is determined from a weather index, expressed as values of a weather variable measured at a stated location during an explicit period.

A weather index is a very specific commodity. An example of a weather index could be the average temperature at Charles de Gaulle airport, France during November, December and January, as a measure of winter's severity near Paris. Others could be: the number of inches of rainfall in Sacramento, California in spring (to fill reservoirs), the number of days of freezing temperatures in Kissimmee, Florida (near the orange groves), and the number of consecutive days of temperature in excess of 95°F in August in Chicago, Illinois (that might require purchasing or producing additional power to meet supply).

WEATHER DATA AND INDEXES

Weather derivatives have potential cashflows that depend on weather events. The widespread weather market practice is that the weather data that is used for contract settlement is measured, recorded and provided by national agencies. (Obviously, in an over-the-counter (OTC) market, other arrangements are possible, if uncommon.) Weather contract originators prefer national agency data to other data for a few reasons: because it is usually more comprehensive, it is generally equally available and transparent to all parties, and it is thought to be unbiased and free of tampering and therefore free of moral hazard, the ability to manipulate data to bring about a favourable outcome. Additionally, sometimes national data is the only data.

The reader is directed to Chapter 5 of this book for a fuller discussion of the access to national data and the preparation of the data for assimilation in weather risk management. The chapter details how many nations differ on policies of access to national weather data, and that data quality, reliability and standards of measurement vary across borders. In addition, it shows why all weather data need to be analysed and (usually) treated to become useful for weather derivative analyses. Untreated and treated data are available from vendors who are specialised in their administration. This is one way that the services they offer add value.

Another data issue is best stated as a question: "How many years of a data record are needed to reliably represent past climate and to make appropriate estimates of the future climate?" Speculators in the market vary in their views but most tend to choose short records of roughly a decade. Still, there are others who prefer to use longer records, even the longest that exist. The selection of the record length and the treatment of data are inextricably intertwined with the resulting estimates of price and value.

The selection of the contract's weather index is vital to a successful hedging programme. The preference is always to select the one that best represents the weather exposure that is being hedged. Principal in the decision is the choice of the weather parameter (temperature, precipitation, wind, sunshine or other), the location where it is measured (almost always the premier level of national weather measurement site), and the period over which the information is collected (week, month, season or other).

Selecting the index depends on the risk exposure of the hedger who may be sensitive to variations in a single weather parameter such as temperature or precipitation, or some combination of parameters. Typical temperature-related indexes are average temperature, degree-days and specific temperatures such as freezing point. Typical precipitation-related indexes are rainfall, snowfall, snow depth, hail and some would add streamflow – a measure of the surface water in rivers and streams that might fill a dam, for example. Moreover, the period of the index is set to be the period over which protection is needed, or a subset of it.

There is usually a geographical difference between the weather measurement site and the environment of the weather exposure, which is often a region. The question is always asked if this difference leads to geographical basis risk because the weather and climate vary over distance. Sometimes the location difference is small and sometimes the climatic difference is small. More commonly, geographical basis risk is a concern for the hedger who must accept a compromise because no closer, more representative site is available. While this has been a deterrent to growth of this market, basis risk is a commonplace feature of most other financial markets. If the measurement site and the weather exposure area are in a region of slowly changing geographical features, geographical basis risk can be partially mitigated with quantitative relationships discovered using regression analyses.

CONTRACT FORM

In the weather market's first year or so the choices presented to participants were limited to financial derivative contracts offered initially by one of the three power marketers: Koch (now Entergy-Koch Trading), Enron and Aquila. However, very soon after, the number of originators expanded considerably. Then, as others visualised a lucrative opportunity, banks, insurance and reinsurance companies, who had been working one-on-one with clients, entered this market.

The end-user is sometimes offered weather insurance contracts that in structure and payments are almost indistinguishable from the financial derivatives against which they compete. Regardless of the name, derivative or insurance, these two formulations can be appear very similar, although there are implications that arise from the differences.

One difference is in the ability of the issuing party – a power marketer, bank, insurer, or reinsurer, for example – to construct and offer a particular requested product: banks and power marketers can offer derivatives but not insurance, whereas insurers and reinsurers can do both.

Another difference is the credit rating of the issuer; we have just witnessed the bankruptcy of the weather market's largest player, Enron, and as the dominoes begin to tilt and fall though the power traders, counterparty risk is again the important issue it should always have been.

Even though there is no requirement that it be so, the financial structures of a derivative and a weather insurance product can be virtually identical. A reason for this would be that a potential buyer of the contract, an end-user, wants the payoff characteristics of a derivative but does not want to own a derivative (either for internal policy or regulatory reasons) and chooses to buy the insurance form. If this is the case, a reinsurance company might sell the insurance contract to the end-user, and offset the acquired weather exposure by buying a virtually identical derivative contract in the secondary weather market. The reinsurance company has transformed one product into another.

In some economic sectors and in some specific firms, the risk management culture is to use insurance rather than derivatives. Sometimes this is because these companies are more comfortable with insurance contracts having used insurance for years; sometimes regulations prohibit derivative use or make it so inconvenient that is just simpler to use insurance. Different countries have different rules as to how the product will be handled for tax and accounting purposes. For example, current US rules make distinctions about derivatives and insurance, whether the instrument is used as a hedge, and whether it is an OTC instrument or it is exchange traded.

WEATHER ASSETS AND THE CAPITAL MARKETS
Alternative Risk Transfer (ART) is a means of moving risk from one entity to another and one form to another, and it has spawned an industry of innovations for the exchange of risk (and funds) between parties. The distinctions between, say, banks and insurers have not faded entirely and each is regulated differently. This has focused attention on the development of offshore entities by companies who seek to be less restricted in their ability to transform products between derivatives and insurance, as can the national reinsurers.

As the distinctions between finance and insurance markets blur, the end-user sees more variety of products that come from a wider range of originators. An example is the multiple-year weather contract that allows both the hedger and the originator to extend a season's hedging (the option position) out to a few years into the future (currently, three- to five-year contracts are the most common).

Capital market investors have so far remained spectators not speculators of weather derivatives market, probably because weather derivative pricing is opaque and weather market liquidity is limited. (There was interest from investors in the weather notes offered by both Koch and Enron in 1999 – see Chapter 13 – and in weather funds). Investor interest may rise again as new products for risk management and portfolio diversification are offered in the market. As an example of what can happen here, consider what happened with catastrophe (cat) bonds. Cat bonds are high-yield corporate bonds whose payments depend on the occurrence of a natural catastrophe, such as a hurricane or an earthquake. Because they include many different kinds of catastrophes, not only weather, they are not strictly in the weather market. Similarly, cat options are exchange-traded standardised contracts that give the buyer the right to receive a cash payment should a specified cat index equal or exceed the option strike level. These products are attractive for several reasons:

❑ The capital markets, with active trading and large investors, who collectively have US$ trillions, are a very good financing source for the risk sector. Capital market investors, if they are able to put capital to work at higher margins than they might otherwise, are willing to risk the losses that result from natural catastrophes: the expected yield of cat bonds is quite high – of the order of 500 basis points over LIBOR (the London Interbank Offered Rate).
❑ Moreover, investors see cat bonds as a way to diversify their portfolios without being exposed to additional credit risk. The occurrence of a natural catastrophe is not correlated to the financial markets and importantly it redistributes balance sheet risks through diversification.
❑ The financial structure of cat bonds is very secure.

Banks are aware that loans to customers with weather risk are more secure if the customer hedges that risk, and may therefore offer better rates to encourage them to do so. Importantly, as banks want to service an increasingly broad client base and to provide an increasingly broader service to existing clients, their attention is drawn to structured deals. This led to the creation of new securities that are tradable in financial markets. It may be time to begin thinking of a weather analogue to the historic development of the mortgage market with its mortgage-backed security/collateralised mortgage obligations (see Chapters 13 and the End Piece).

Derivative structures
A weather derivative is a contingent contract in which a payment is made depending the outcome of a weather event. The most basic derivatives are call and put options. The buyer of the option pays to enter into a contract that may require the seller of the option to pay at the end of the option period, an amount calculated from a specified measure of the weather (the weather index).

Payment is made to the buyer of a call option only if the weather index exceeds a specified level (the attachment strike) at the end of a specified period (the contract period). Payment is made to the buyer of a put option only if the weather index is lower than the attachment strike at the end of the contract period. Payments are keyed to the difference between the index and the strike level, with the exact formulation specified in the contract: the buyer of a call option might receive a payment if the index increases sufficiently and the buyer of the put option may receive a payment if the index declines sufficiently.

A digital derivative or binary derivative is one that pays if a specific event occurs, say rainfall on a particular date. But there is also a critical-day derivative that is a variation of a simple call or a put: payments are keyed to the *number of times* – not the amount – in the contract period that the index goes above or below the strike level. Swap agreements, the most actively traded contracts in the market, and collar agreements, are similar to each other, and, as we will see, they contain features of both calls and puts. Because the weather market took the design of its derivative contracts from other, more developed derivative markets, standard derivative terminology applies.

The weather market is primarily an OTC market, meaning that two parties can transact in privacy and need not make their transactions public. It is a private matter for them to construct a derivative of their own choosing (subject, of course, to regulation and, presumably, to reasonable accounting practice). These freedoms allow the buyer and seller to reach agreement on basic derivatives, as well as to negotiate structures that are more complex.

They can also choose the weather index, at various locations, that best meets the hedging need. Indexes that have been selected include average temperature, accumulated degree-days, rainfall or snowfall amount, average wind speed, accumulated sunshine and other weather variables. The weather index represents the selected weather variable at one or more locations, weighted equally or unequally, over an agreed upon period.

OPPORTUNITY IN THE EARLY MARKET
From the very beginning of the weather derivatives market, there have been two principal groups of participants separated by their interests. On one side are the end-users – those enterprises that need weather derivatives to hedge an intrinsic or natural weather risk. On the other side are the originators of weather contracts and other speculators who trade in weather risk contracts for possible gain.

An end-user must identify the weather that is advantageous or adverse to business performance to know how and what to transact; speculators often have no natural weather risk until they acquire it looking for profit. We say the weather market is made of two types of activity: transactions in the primary market involving at least one end-user who is hedging a natural exposure, and transactions between two speculators in the secondary market.

In the weather market's initial stage, end-users wanted to manage the threat of declines in consumer demand for gas and electricity for home heating and cooling; the concern was unusually cool summers and unusually warm winters, that is, mild seasons. The typical end-user of a weather derivative was an energy company that wanted to reduce the risk of not otherwise being able to manage the financial effects of climate well. They faced climate variability that appeared to be increasing at the same time they were facing the spreading deregulation of the power market. Even now, as the participation by speculators shifts from exclusively power marketers to originators in other sectors, energy end-users are still at the market's core.

The power marketers were trading in power (buying and selling electricity and gas futures, for example, as well as distributing energy in some cases) even before the birth of the weather market when they saw weather derivatives as another means to the same end – to gain from arbitrage. With the evolution of the weather market they took on weather risks hoping to earn a superior return as their reward. As they bought or sold to an end-user, they tried to unload this recently acquired weather risk by reversing their derivative position: they sold or bought the same or similar derivatives elsewhere. Usually their counterparty in this second contract was another speculator. Speculators bought and sold degree-day calls, puts, collars and swaps, all in the OTC market, and later, weather futures that trade on an exchange. (From the perspective of a speculator, a swap and a future are very much the same.)

Clearly most transactions to date have been indexed to either degree-days or average temperature and most end-users of these are energy suppliers – principally gas and electricity utilities. The volume of temperature-based contracts is still larger than all others combined, but we also see a growing trend to contracts based on other weather variables: the market share of contracts indexed to streamflow, rainfall, and snowfall is increasing (See Chapter 3 and www.WRMA.org).[1] Streamflow contracts are indexed to the above-surface flow of water in tributaries, from small streams to wide rivers, and have been offered as protection to hydroelectric generators and to water suppliers. We also see new end-users from sectors other than power. This is perhaps an early indication that weather derivatives and weather insurance products could serve a larger purpose in the global economy.

BASIC WEATHER DERIVATIVES
The remainder of this section describes the basic structures of a collar, a swap, a put, a call and a digital option. Although there are important differences between derivative structures so that sometimes one may be useful and another not, all of these derivatives share some features.

Each derivative is indexed to a weather variable, such as the volume of water flowing in a river measured at a specified location over a specified period. Each derivative contract specifies a level keyed to the index (the strike level), which, if exceeded requires the seller of the contract to pay the buyer. Payments are calculated at the contractual rate sometimes called the "tick rate". Only one side of a strike may be active, that is, only when either the index is above or below the strike will seller pay the buyer. All payments from these derivatives accumulate over the contract period, and payments are made after the contract period. Common contract periods are as short as a week or as long as a season (a few months). Some contracts include a provision to measure the index and make payments in the same season in consecutive years (multi-year contracts).

So as to avoid taking unlimited risk on the weather without adequate compensation, almost all weather derivatives have an upper bound on payments; this limit is called the "cap", and no payment exceeds the cap regardless of the final value of the weather index; seasons may be so extreme or so mild that the final calculation of the weather index is be far above or below the historical average index, but the cap would still limit the payments.

Whether a series of transactions (trading) begins with a primary or secondary market transaction, the principal contract in weather trading by dealers is the swap.

Typically, the transacting parties agree to tick rates and caps for swaps (and also for collars) that are symmetric about an agreed upon level. This means that payments to the buyer and the seller would be equally likely if the agreed upon level is also the mid-point of the distribution of the weather variable. Unfortunately, all distributions are estimates of the coming event – so the probabilities are not really known to either

---

PANEL 1

END-USER STRATEGIES

The common energy end-user hedge

If an enterprise experiences an increase in revenue or a cost saving when a weather index increases, then that enterprise is said to be "long" the index if, as the index goes up, so does the enterprise's return. If an enterprise experiences a drop in revenue or a higher cost when a weather index increases, then that enterprise is said to be "short" the index: as the index goes up, return goes down.

Energy suppliers are naturally "long" both CDDs and temperature in summer, and long HDDs but short temperature in winter. There is an exception to these long and short positions: it is that that revenue may weaken if temperatures and degree-days are extreme enough. When demand for power exceeds current supply, supplementary energy purchases might be required to meet the increased demand. Accumulating degree-days beyond this level might decrease revenue. The end-user could protect against mild seasons with low degree-day accumulations with a degree-day put, for example, or against extreme seasons with high degree-day accumulations with a degree-day call.

Because of being short mild seasons, the energy provider usually pays a premium to buy a degree-day put that goes up in value as the degree-day index goes down. In this way the short position is offset. (Buying an average temperature call in winter or an average temperature put in summer has an equivalent effect.)

Alternative energy end-user hedges

A less common transaction would be for the end-user to sell a degree-day call in winter (or sell an average temperature put in winter) to receive a non-returnable premium. As the seller of a degree-day call, the end-user earns the premium in exchange for the promise to pay the buyer should the degree-day accumulation be higher than the strike. (This payment would be out of strong revenue earned in an advantageous season.) Importantly, the end-user can "bank" the non-refundable premium as a cushion against the possible decline in revenue should a season be adversely mild. With this cushion, revenue would have to fall more than the premium before the decline would have been experienced. Selling a degree-day call, in this case, is analogous to what the institutional investor would call "selling a covered call."

In a more complex transaction, the energy supplier could acquire the protection needed by buying a degree-day put and paying for at least part of the premium by selling a degree-day call. However, the market provides a less complex way to do this: the end-user can instead sell a degree-day collar. Selling a collar provides a similar payoff to buying a put and simultaneously selling a call, and the bid–ask spread on each is built into the collar.

The gas supplier described earlier who might have bought a temperature collar (or sold an HDD collar), would find the same protection in buying the temperature call or buying the HDD put. For an end-user, the choice between the collar position and the option position is a choice of how to pay for the protection: either with cash or with a promise to pay with a share of good revenue.

party. For example, if the distribution of average temperatures of past winters is approximated well by a Gaussian distribution and the swap level is the mean of the distribution, the probability of either party paying would be 50%, with the probability of each level of payment also defined by the distribution. The estimation of these probabilities falls within the purview of valuation and pricing studies (see Chapter 8).

The symmetry about the average temperature also implies the expected or probable payoff will be equal as well (the expected payoff is the integral of the probability distribution over all possible payments). However, swaps trade with bid and offer spreads and originators "move" the averages to fit their views, sometimes using climate forecasts, and by selecting the number of years in the averaging period. The result is often a tilt in one direction, perhaps leading to asymmetric probabilities, meaning there is a shift in the probability of a payment in favour of one party. (Having an advantage implies knowing the "true" distribution, although no one does know. The end-user should, however, at least be alert to dealer practices.)

Swaps and collars usually transact with no initial exchange of money and are said to be "no cost" contracts (though rarely does one receive something for nothing). The zero-cost characteristic of a swap is sometimes attractive to speculators as they can assume risk positions, and even build a portfolio of risk, with no initial outlay of capital. Swaps can be more risky than options as downside risk can be better controlled with options.

In a swap, collar or future contract, either the buyer or seller might have to pay. Moreover, determining which party is called the buyer and which is the seller is a matter of market convention. In swap (and in a collar), the buyer is the one who benefits from the rising index; the swap buyer receives a payment from the seller only when the index is greater than the swap level, up to the cap. Conversely, the swap seller benefits from a declining index, and would receive a payment only when the index is less than the swap level, down to the limit (the floor). There is no payment due when the index is at the swap level.

Figures 1–4 are the contingent payment diagrams for a collar, a swap, a put and a call, respectively. They are based on the average temperature as measured over a winter season (the weather index), and the payoff is in Euros (€). Although these are temperature-indexed derivatives, these same structures fit almost all other weather variables for all periods. (The payoff, if both parties agree, need not be in currency but could be in any commodity, such as megawatts or chicken eggs.)

The diagrams present the range of payments that depend on the final value of the index. All of the payments in the range are possible in any one season but, of course, only one of these payments will be made and it will be keyed to the actual season. The derivatives are constructed for a temperate European location whose average winter temperature in recent decades is 10°C (50°F) and whose standard deviation of annual average temperatures is 2.5°C (4.5°F). For convenience of description, the strikes of the derivatives are set at one standard deviation from the mean temperature.

Each agreement specifies that no payment will ever exceed €1 million. Payments increase in increments of €40,000 per 0.1°C (the tick rate) up to the €1 million limit. Dividing the limit of €1 million by the tick rate reveals an active range over which payments accumulate to be 2.5°C – or one standard deviation of the historical temperature: the caps then, are at two standard deviations from the average.

A collar indexed to average temperature
In this section, we see how a collar is a general structure and a swap is a specific case: a swap is a collar whose two strikes are the same – the swap level. We can also find within the collar, the structures of a call and put when viewed from certain perspectives and with appropriate definitions: a long position in a collar is the sum of a long position in a call and a short position in a put (buying a collar is like buying the call and selling the put).

1. An average temperature collar

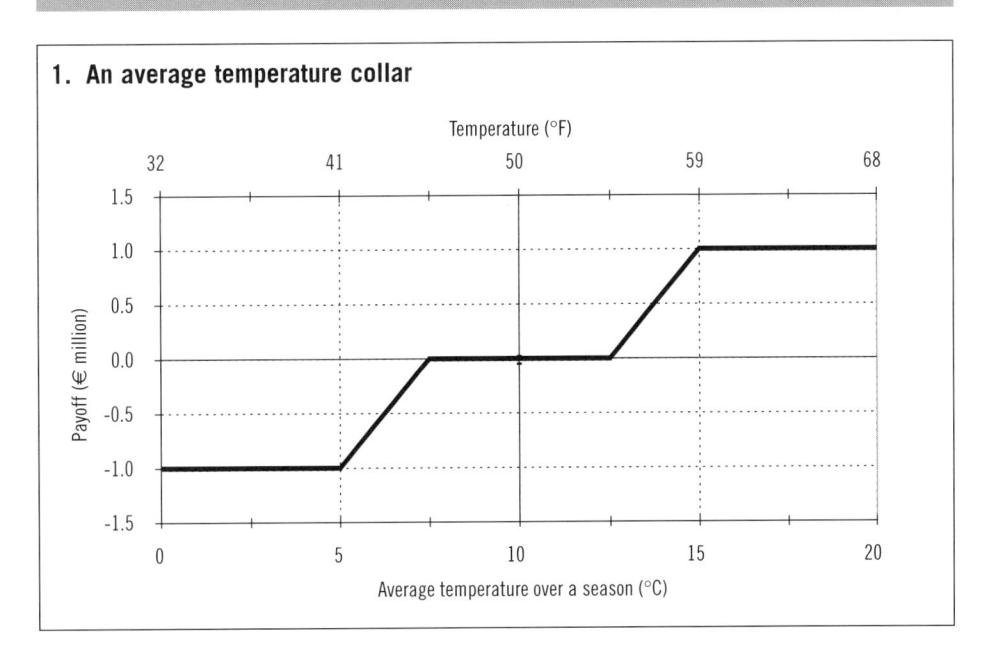

[Figure: Payoff (€ million) vs Average temperature over a season (°C), with secondary x-axis in Temperature (°F). The payoff curve is -1.0 from 0-5°C, rises linearly to 0 at 7.5°C, stays at 0 between 7.5 and 12.5°C, rises linearly to 1.0 at 15°C, and remains at 1.0 up to 20°C.]

A collar (as with a swap) is an agreement in which two parties exchange promises that one will pay the other should the index exceed (be greater than or less than) the strike on their side of the collar. The collar in Figure 1 has an upper strike set at 12.5°C, one standard deviation (2.5°C) above the average temperature (10°C). Its lower strike is set at 7.5°C, one standard deviation below the average temperature.

The possible payments in Figure 1 are seen from the buyer's perspective, that is, positive values indicate possible payments to the buyer, and negative values are payments the buyer would have to make to the seller. The buyer would receive a payment if the final average temperature is higher than 12.5°C, and will make a payment if the average temperature is below 7.5°C. No one makes a payment if the final index falls where the collar is inactive, that is, between 12.5°C and 7.5°C.

A utility end-user, such as a gas supplier, might buy this collar for a few reasons. First, the levels are selected so that no cash is paid to enter the contract, whereas protection with a call or put would require that a premium be paid. Second, the collar delivers the downside revenue protection needed against the adverse (that is, warm) season: the end-user receives a payment if the final average temperature would be more than one standard deviation above the historical average. Third, in any near average season (meaning within one standard deviation of the historical average, in this collar) the buyer keeps his revenue and makes no payment. Finally, should the buyer have to make a payment, it would be out of the strong revenues in an exceptionally good (cold) season, when the average temperature is more than one standard deviation below the average.

If the end-user had selected heating degree-days as the weather index, then an analogous transaction would be for the end-user to sell a degree-day collar. The role-reversal of buyer and seller is because HDDs increase as temperature decreases: in a mild winter, higher temperatures result in low accumulated HDDs.

An average temperature swap
A swap can be described as the special case of a collar whose upper and lower strike levels are the same. The contingent swap cashflows in Figure 2 are those that would result if the two strikes of the collar in Figure 1 were set at 10°C. The cap is kept at €1 million with the result that the active range of 5.0°C on both sides of the 10°C average, where a payment is due, is symmetric about the historical average and is closer to average than the active range of the original collar.

The only temperature at which the contingent payment would be zero is called the "swap level" (10°C in this example). Speculators set the swap level based on a few

2. An average temperature swap

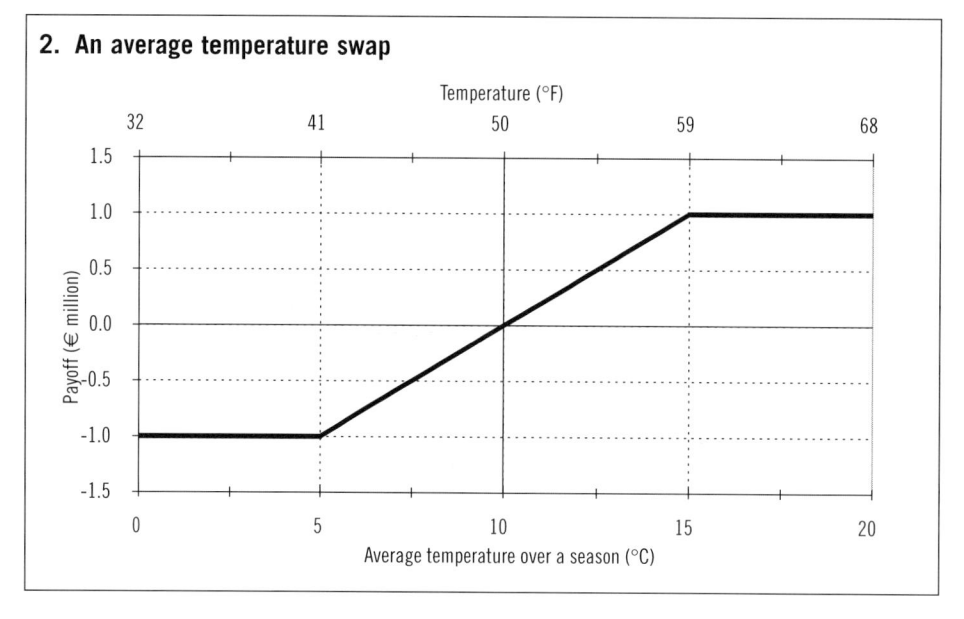

factors. These are: the historical average temperature, the forecast for the swap's weather index (average temperature in this example) and the speculator's and the market's appetite for this particular swap. Speculators continuously review the current market to see at what levels other swaps are trading, and may change levels at any time.

Swaps are a favourite derivative for speculators in the weather market as they aid in communicating views on the season's outcome. We will see later in this section that speculators can combine various derivative positions to transform one derivative structure into another and one risk position into another.

An end-user, however, generally has little appetite for a swap as it is more of a speculative tool. The swap has no inactive range between two strikes as there is in a collar and, because they would require a payment so close to the estimated average temperature, they provide unneeded protection at undesirable cost. With a swap, end-users might receive payments in winters that are slightly warmer than average even though they may not need the payment, or they might be required to make a payment from their moderate revenue in slightly cooler than average winters. That is, more of the end-users' potential upside must be promised than they would want, in exchange for more downside protection than they need.

Call and put options
The simplest and most common hedging transaction for the energy provider and many other end-users is to buy a put or a call. The decision between a put and a call depends on the index and the end-user's exposure to it.

The call option on average temperature in Figure 3 has the same cap and strike as the cap and upper strike of the collar in Figure 1. If Figure 4 were inverted, the cashflow of the put would be seen from the viewpoint of the seller and would have the same strike and floor as the collar in Figure 1. Both the put and call in the figures are viewed from the perspective of the buyer of the option. The upper line in each figure is the contracted contingent payment, or the payoff. The lower line is the net payment, that is, the payoff less the premium paid in advance.

Buyers of the call in Figure 3 pay a premium for which they will receive a payment if the index is greater than the call strike (12.5°C). If the index is below 12.5°C, they receive nothing. If it is above 15°C, they receive the cap of €1 million. If it falls between, they receive a payment at the rate of €40,000 per 0.1°C, subject to rounding.

3. An average temperature call

4. An average temperature put

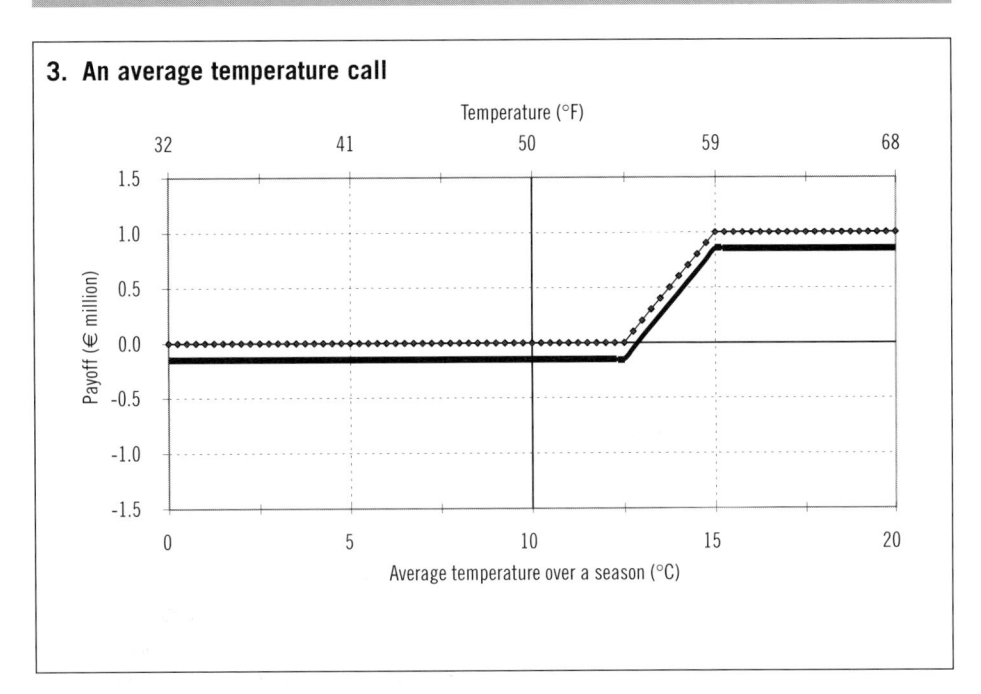

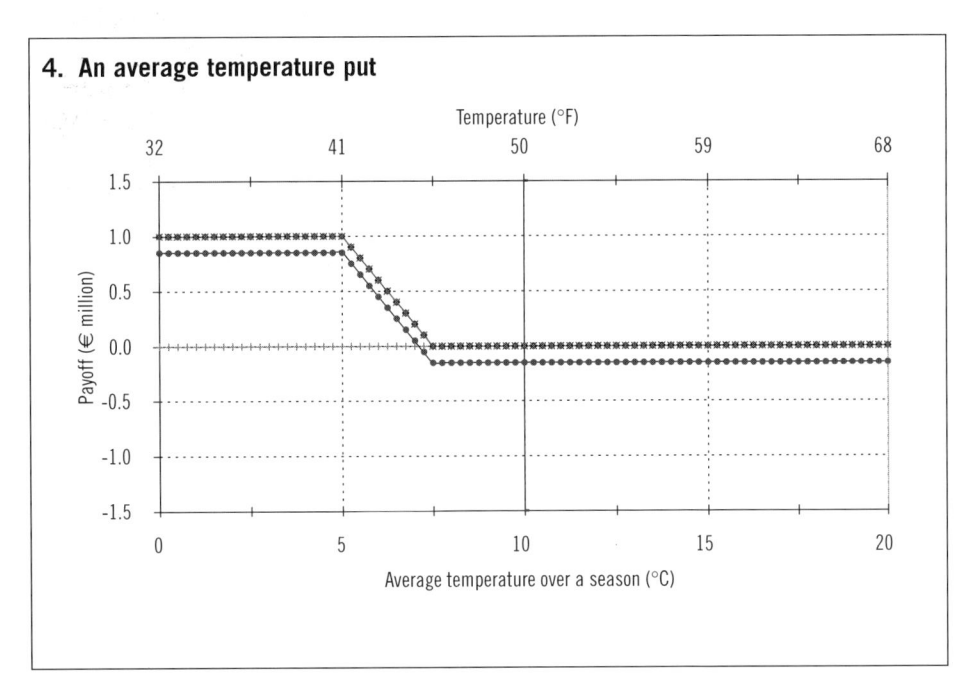

Buyers of the put in Figure 4 pay a premium for which they will receive a payment if the index falls between 5°C and 7.5°C, at the rate of €40,000 per 0.1°C, subject to rounding. If the index is above 7.5°C, they receive nothing. If it is below 5°C, they receive the cap of €1 million.

CRITICAL-EVENT CONTRACTS
Even in the market's initial stage it became clear that there were end-users who needed yet another derivative structure – one that could be used to hedge specific or critical events. These critical-event contracts pay the buyer if the weather index exceeds (or falls below) a level specified in a contract. However, unlike the payments of call and put options, the payments of digital options pay based on the number of critical events in a period, not the index value.

For example, if summer's heat becomes extreme and prolonged, buildings and even whole cities retain the unusual heat, and the need for air conditioning becomes

FINANCIAL WEATHER CONTRACTS AND THEIR APPLICATION IN RISK MANAGEMENT

> **PANEL 2**
>
> ## ALTERNATIVE DERIVATIVE STRATEGIES
>
> Conceptually, a collar can be replicated with a call and a put. Buying a call (going long the call) with the same strike as the upper strike of a collar, and selling a put (going short the put) with the same strike as the collar's lower strike, reproduces the collar's contingent payments. The corollary is that a collar can be decomposed into a call and a put with the same strikes, cap and floor.
>
> Suppose the current owner of the collar in Figure 1 would rather own only the call in Figure 3. Clearly, he/she could sell the collar and buy the call in two separate transactions. Alternatively, he/she could keep the collar and sell the put in Figure 4 in only one transaction. With this new "portfolio of derivatives," the long collar and the short put positions, the payments net to the same payments as the call. A speculator who wants to own a collar might buy a call and sell a put, perhaps but not necessarily simultaneously, and only if favourable opportunities exist.
>
> Speculators use these combinations widely as they can more easily achieve the risk position they seek. If, for example, they own a put and want to reduce their risk, they could sell the same or similar put, or failing that, they could buy a swap (or collar) and sell a call, all with the corresponding or similar strikes and caps.
>
> Swap levels change over time. Even though swap levels are based on the average temperature for the period, a speculator is always watching for a new weather or seasonal forecast and quickly includes this view in his prices (see Chapters 6, 10 and 11). Importantly, a speculator's current weather derivative portfolio determines his risk position and affects his appetite for certain geographic locations, strike levels and derivative structures – meaning the price the speculator offers or asks for can change with the changing portfolio. As swap levels change, the strike levels of related derivatives adjust to the new swap levels. The strike levels of related puts and calls must either move in synchrony with swap levels or their prices will change.

extreme. If these events are severe enough, the providers of electricity can find themselves short of power and need to "fire up" standby generators. Often these are auxiliary gas turbines, with an associated incremental cost to meet the increased demand. This digital derivative, or critical-day derivative, would pay an agreed amount for a designated number of such events, thereby alleviating some of the supplier's incremental generating cost.

A critical-day contract can be structured to pay for prolonged periods with temperatures above 95°F, below freezing point, or any other reference temperature. Critical-day contracts can offer protection to the contractor facing penalties for workdays lost to rainfall, or for the orange grower who loses a portion of his crop each time it freezes.

Entering the weather market

The following is a brief description of one company's first effort at a weather risk management programme. It is based on a real event experienced by a real company. The name of the company, the Hot Air Gas Company, is not the company's real name and name of the meteorological measurement site, Frozen Falls, replaces the name of a real location. Certain liberties were taken with weather statistics to make the exercise easily tractable.

The Hot Air Gas Company (HAGC) is a distributor of gas for home heating in a region near the Canadian border. Over most of the company's history, the normally cold winters in these northern latitudes generated strong sales for the company. However, the winters beginning in 1997, 1998 and 1999 were warmer than the 30-

year average (–0.8°C, 30.5°F), and sales in these years were weak. (See Table 1.) The winter of 2000–01 was colder than average and sales were once again strong.

The climate in this region, it appears, is more variable now than it has been over at least a century (See Figure 5 of this chapter and Figure 7 in Chapter 1). As a result, the company's revenues have become quite volatile. Mr Bavardier, the company founder and CEO, was not comfortable with a weather-exposed position that might put his company into a financially threatened situation should the winter of 2001–02 be warm as it had three out of the recent four years.

He spoke with his friend "Flash" Burns, at the Frozen Falls Fuel Company (FFFC) just across the border in the US. Flash explained that he had reduced his company's losses in recent winters by successfully hedging his company's revenue using weather derivatives as a protective action. Flash said FFFC would hedge every year regardless of the seasonal forecast and accept that in some years the hedge would not pay and expire worthless. For example, the hedge it had purchased before the winter of 2000–01 paid nothing at the end of winter because winter had been cold. That did not trouble him, as FFFC revenues had been strong. He viewed the weather hedge, he said, as a kind of business insurance that protects against bad times; he claimed it was an antidote to FFFC's inherent weather risk.

Mr Bavardier constructed the diagram of HAGC's weather contingent revenue (Figure 6) from records of sales during past winters, and adjusted it for today's market conditions (mostly the result of variations in gas prices and the changes in the company's service region). He calculated that the "cost of warm weather" was about C$4 million for each 1°C (or C$400,000 for each 0.1°C) should the winter's average temperature be greater than –2°C. Conversely, for average temperature

Table 1. Frozen Falls winter 1997–2001

November–March	°C	°F
1997–8	0.5	32.9
1998–9	1.8	35.2
1999–2000	1.0	33.7
2000–01	–0.4	29.4

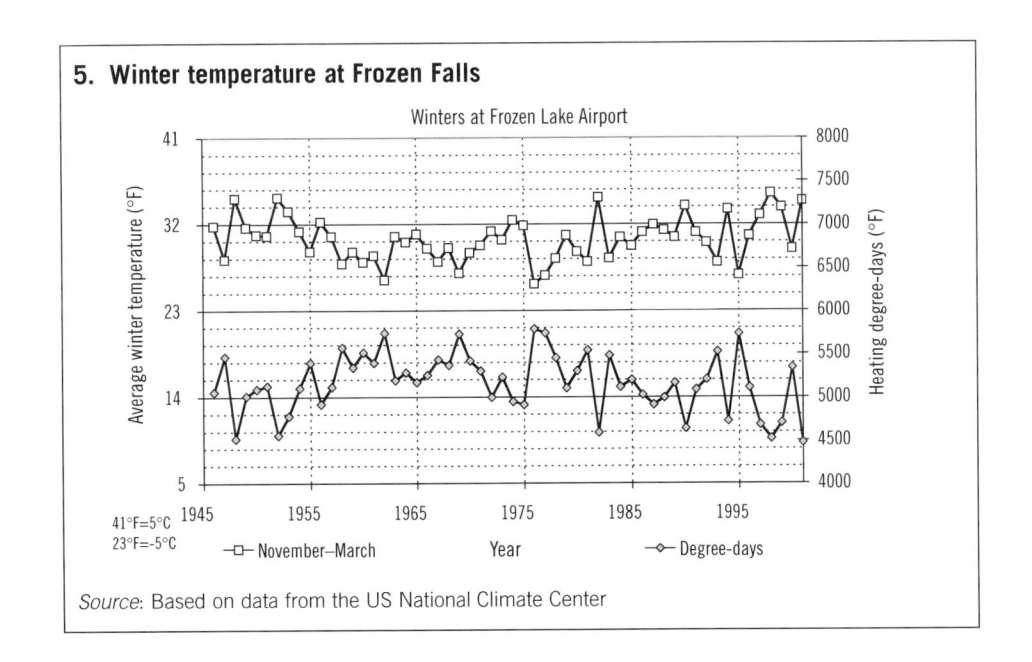

5. Winter temperature at Frozen Falls

Source: Based on data from the US National Climate Center

6. Weather contingent cashflows

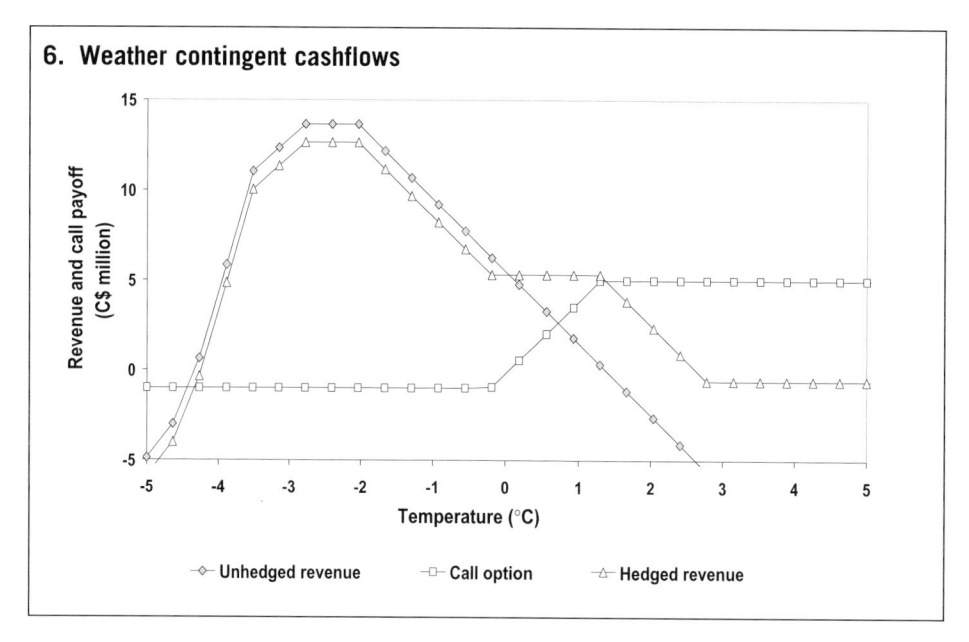

between 3°C and –2°C, the revenue could be expected to increase by C$4 million for each 1°C. (See Figures 2–4 in the Chapter 1 for supplementary information.)

He realised that because the company has weather sensitive revenue, it has been gambling on the weather: when winter was cold it won the "weather bet," when winter was warm it lost. He knew HAGC needed to take defensive action against weather variations.

With help from the consultant who provided the weather data, he was able to calculate an array of useful information. First, he prepared the histogram of winters from the temperature data used in Figure 5 and converted it into probabilities of temperature (Figure 7). The consultant suggested that he also estimate the Gaussian (normal) distribution using the average and standard deviation of these same winters because it would later give him another estimate of the value of hedging and an option's price.

Once deciding to hedge against a warm winter with a weather derivative, Mr Bavardier learned of the many decisions facing him. Among these were:

❑ *Which index to use?* The selection of the index was obvious as the Frozen Falls weather station just across the border was the nearest measurement site to the HAGC service region with a complete and audited weather history. It was important to know how well it represents his weather exposure.
❑ *What contract period to use?* He chose the November 2001–March 2002, the company's principal heating season, as the contract period.
❑ *How long should the analysed climate record be?* He did all his calculation with the recent 30-year record (average –0.8°C, standard deviation 1.3°C). The originator agreed with this as the time-series of temperatures in Figure 5 showed no obvious trend.
❑ *Which derivative was right for Hot Air Gas Company?* Although other possibilities existed, his interest was to purchase an HDD put or an average temperature call. He was not interested in selling an HDD collar or buying an average temperature collar because these transactions required a promise to give up potential revenue after a favourable season.

He chose to buy an average temperature call option because he thought temperature is easier to understand than degree-days, and one could be substituted for the other. He saw in Figure 5 that temperature and HDDs are

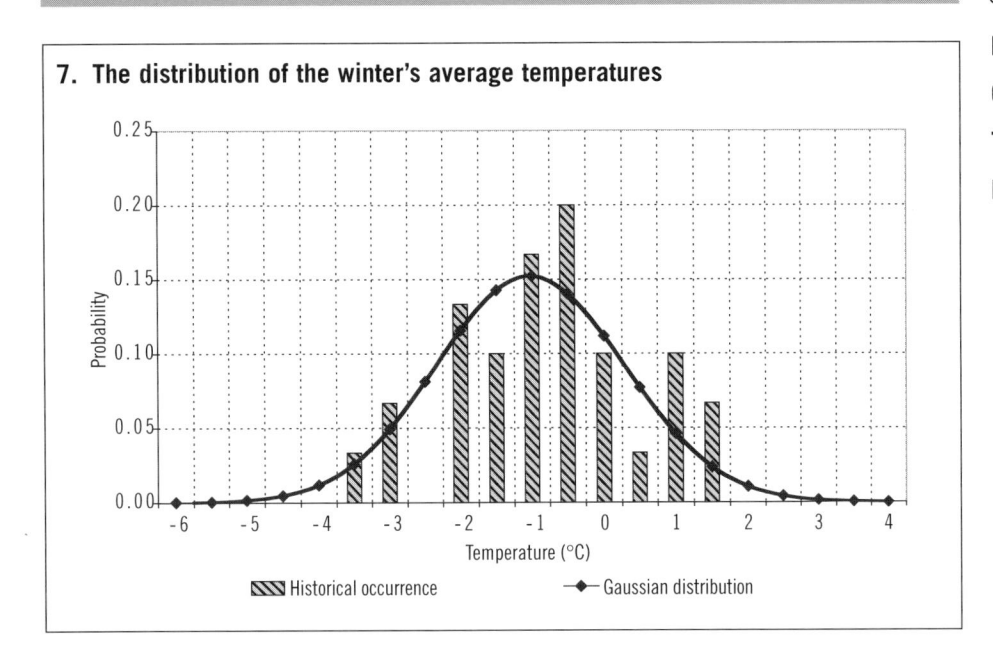

7. The distribution of the winter's average temperatures

Histogram with Gaussian distribution overlay; x-axis: Temperature (°C) from -6 to 4; y-axis: Probability from 0.00 to 0.25. Legend: Historical occurrence, Gaussian distribution.

mirror images: they are very highly and negatively correlated (correlation coefficient exceeds –0.99).

❏ *At what HDD or average temperature should the strike be set so the option would pay when his revenue fell?* He understood that setting the derivative's strike farther from the average of past winters reduced its price but also its expected payoff (the sum of the probability of each possible payoff multiplied by each payoff). He realised that the company's revenue declined to unacceptable levels at a temperature of about one-half of one standard deviation above the 30-year temperature average and so he asked that the strike be set at –0.2°C.

❏ *At what rate (the tick rate) should the option pay?* He chose the same rate as the company's "cost of warm weather" for the range of temperature he was concerned about: C$400,000 for each tenth of a degree Celsius over the strike.

❏ *How much of the option would the company need to provide adequate protection? (How much notional amount? At what HDD or temperature would the option payment be capped?)* He estimated from Figure 7 that temperatures above 1.3°C (about one and a half standard deviations above the average) occurred only a few per cent of the time, and that was good enough. If winter were that extreme, at least he would receive the option limit of C$6 million (1.5 degrees x C$4 million/degree).

Mr Bavardier had prepared Figure 6 and a similar diagram for each of the options the weather market originators offered to him. In these charts he was able to compare the company's weather contingent revenue (the line with the diamonds) with the possible cashflows from the derivative (the line with the squares), and add them to estimate the possibilities for revenue hedged with the derivative (the triangles).

He chose the call because it was the simplest transaction and he believed it offered the protection the company needed. He could see that, as the option began to pay, the hedged revenue flattened: sensitivity to the weather was neutralised as the option payments offset the revenue declines, up to a point. The hedged revenue in Figure 6 is reduced everywhere by the amount of the non-returnable premium.

Hedging with this the average temperature call option, however, placed an almost certain floor under the company's 2001–02 net revenue, as it would pay HAGC should the average temperature at Frozen Falls be more than half a standard deviation above the average of recent winters. After hedging, a loss from the coming

8. Expected revenue

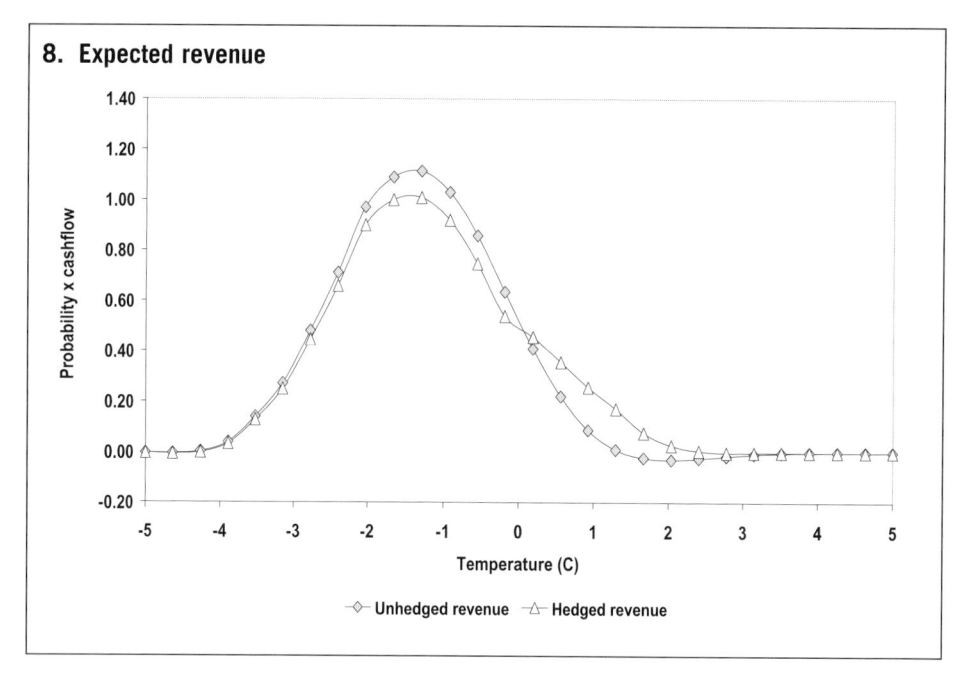

winter was now highly unlikely. The cost of the call option to HAGC was C$1 million, about 17% of the cap. (That is C$1 for up to C$6 of protection, and he could have benefited from that in three of the last four years.)

Mr Bavardier did one last calculation to confirm his choice: he charted the probability-weighted cashflows for the unhedged revenue and for the revenue when hedged with the call option (Figure 8). Probability-weighted cashflows are calculated by multiplying the cashflow expected for each winter temperature by the probability of that temperature occurring (here he used the Gaussian distribution in Figure 7).

The hedged probable revenue is higher than the unhedged probable revenue at all temperatures just above 0°C (0.05°C to be exact), the temperature at the option paid for itself. In the range of temperatures where revenue would have been unacceptably weak – that is, below C$5 million – the option "lifts" the hedged probable revenue curve above the unhedged revenue curve. Subtracting the unhedged revenue from the hedged revenue, Mr Bavardier could see that there was a trade-off between the probable cashflows of these two curves. The area between the two curves is negative below 0.05°C and positive above: they offset each other almost equally, so the price the company paid was reasonable.

Mr Bavardier bought the call a few months before the onset of the 2001–02 winter, while everyone was guessing about what kind of winter it would be. He ignored the speculation on weather forecasts because he reasoned that it was not meaningful to his situation. If winter would be warm, he needed the call option payment to supplement expected weak revenues. If winter would be cold, revenue would be strong and he would have sacrificed the premium. He decided, however, that he would never complain about paying for an option that expires worthless, anymore than he would complain about his family not collecting on his life insurance policy!

He felt strongly that there is at least one additional benefit to hedging that comes from simply being sure that the company will not lose money in the coming winter. It is that it is not necessary to set aside risk capital to protect against weather-related variable earnings, as earnings no longer depend on the weather.

The winter 2001–02 produced a November–March temperature of 1.3°C and about one and a half standard deviations and 2°C above the 30-year average on which the call option was structured. The HAGC collected C$6 million, the full cap of the call option. Mr Bavardier had neutralised the weather bet by hedging and avoided a potentially fatal financial blow to the company.

1 *The Weather Risk Management Association in its 2002 survey states that – "While temperature-related protection (for heat and cold) continues to be the most prevalent, making up over 82 percent of all contracts, rain-related contracts account for 6.9 percent of the market, snow for 2.2 percent and wind for 0.4 percent."*

3

Hedging Precipitation Risk

Thomas Ruck

Entergy-Koch Trading, LP

"Neither snow, nor rain, nor gloom of night, nor winds of change, nor a nation challenged will stay them from the swift completion of their appointed rounds." – updated US Postal Service motto.

The US Postal Service may not be concerned with precipitation in its various forms, but many other industries are not as fortunate. Individuals and corporate entities engaged in agriculture, construction, transportation, outdoor leisure activities and power generation are exposed to economic loss due to unusually high or low precipitation levels.

Introduction

This chapter will provide an overview of precipitation risk, basic strategies for risk mitigation and a description of precipitation risk transactions concluded with various customer types. It will conclude with a brief forward-looking view of potential further developments in precipitation risk management.

Identification and quantification of precipitation risk

Precipitation is a generic term used to describe various forms of water falling from the atmosphere to the earth. Precipitation can be in the form of rainfall, freezing rain, sleet, snow or hail.

Plants and animals need water to survive, but extremes of too much or too little water can be disastrous to both. Consider a corn farmer who needs adequate rainfall during the growing period, but who prefers relatively dry conditions prior to planting and at harvest. Fruit growers also prefer dry conditions at harvest; if there is too much rain, the fruit can absorb excess moisture and burst. The economic penalty is that cracked fruit sells for a discounted price. Excess rainfall during harvest can physically delay farmers from picking their crops. For vineyards, this can be disastrous, as the time window for processing grapes can be quite narrow. Too much time on the vine can result in an increase in sugar content that can significantly reduce the value of the crop. Cattle ranchers can also be affected by extreme precipitation events. Droughts can leave pastures barren of grass resulting in increased animal feeding costs. Winter rain or snowfall can create muddy feedlots that reduce animal mobility and increase animal discomfort. An uncomfortable cow does not eat and, therefore, does not gain weight. Since cattle are sold by the pound, the lower the weight, the lower the rancher's revenue. In addition, muddy cows sell for a significant discount at the slaughterhouse.

Many forms of outdoor commerce can be subject to weather-related delays. Both the Federal government[1] and the private sector[2] estimate that over US$2 trillion, roughly 25% of the US GDP, are affected by weather and climate. The construction and transportation sectors are particularly vulnerable to the weather. Building, bridge and highway construction may be disrupted during periods of heavy rain or snow. Crews

HEDGING PRECIPITATION RISK

and rented equipment sit idle. Many construction contracts include significant penalties for missed completion dates. Heavy rain or snow can also halt airplane, truck and train traffic. Flight delays alone cost the public roughly US$1 billion each year.[3] Traffic accidents increase during bad weather, as do automobile insurance claims.

Rainfall can influence individuals' leisure activity decisions. A day at the golf course, beach, amusement park or zoo can be postponed or cancelled by inclement weather. Rain or snow can also influence consumer-purchasing decisions. Umbrellas, galoshes, raincoats, sump pumps, parkas, gloves, snow shovels, rock salt and snow blowers can often be the objects of impulse buying. Even something as basic as electricity service can be disrupted when power lines are downed during ice storms.

The first step in quantifying precipitation risk is to determine the economic impact of the weather event. A round of golf can be postponed for another day so the economic impact of rainfall on a golfer is relatively small (other than in a professional context) and somewhat intangible. Nevertheless, reduced crop production and downed power lines have real and tangible economic effects: either revenue is reduced or operating or capital costs are increased. These revenues or costs may never be recoverable. It is these types of risks that the weather market seeks to efficiently mitigate.

Risk quantification can begin by answering a set of basic questions:

1. How do variations in precipitation affect business or activity?
2. What is the economic consequence?
3. What is the probability of occurrence?

If the product of the economic consequence multiplied by the probability of occurrence has a significant monetary value, then a hedge should be considered. A more rigorous statistical analysis may be required to determine the appropriate hedge to mitigate the precipitation risk, but the cost of performing this analysis may be avoided if the basic questions are answered first.

The definition of the hedge

Hedging precipitation risk is a bit more complicated than hedging temperature risk because of the intrinsic nature of the phenomenon. Temperature is typically homogeneous over a fairly wide geographical region while precipitation can be a very localised event. Most people have experienced a rain shower falling on one side of a street, but not on the other. This experience can be verified mathematically as well. For example, a comparison of the correlation coefficients for temperature – measured as cumulative cooling degree-days (CDDs) – and cumulative rainfall for Chicago O'Hare and Moline Quad City Airports (roughly 145 miles apart) reveals that the correlation for temperature is stronger than for rainfall. The correlation coefficient is a measure of the relationship between the weather at one location compared to the weather at another location. The value of the correlation coefficient can range between 1, a perfect correlation or –1, a perfect inverse correlation. The value of the correlation coefficient for CDDs at the Chicago O'Hare and Moline airports is 0.89. The square of the correlation coefficient, 0.792, is the percentage change in CDDs at Moline that can be explained by the change in CDDs at O'Hare. The rainfall correlation, however, is significantly weaker than the temperature correlation, as indicated by the lower correlation coefficient value of 0.65. Only 42% (0.65 × 0.65) of the change in rainfall at Moline can be explained by the change in rainfall at Chicago O'Hare.

One of the major challenges to hedging precipitation risk is to identify the source of the weather data that is most representative of the risk being hedged. Often, the nearest weather station is located at an airport. The difference between the weather at the hedger's location and the weather at the local airport or nearest weather station is (in this context) known as "basis risk". The decision to hedge precipitation risk or not can ultimately depend on the magnitude of the basis risk and the cost associated with mitigating it. If the cost is reasonable, the hedger will usually decide to transfer this risk to the hedge provider. If the cost is high, the hedger may decide to self-insure the basis risk and only hedge the precipitation risk at the nearest airport or alternative weather station. In the case of the Moline–O'Hare correlation example discussed above, a measure of the basis risk would be the 68% (one minus 42%) of the change in Moline precipitation that cannot be explained by the change in O'Hare precipitation. The magnitude of this risk is not insignificant and should not be ignored.

Another hedging issue is the reliability and accuracy of the weather station used to determine hedge payoffs. This is known as "measurement risk". Consider the case of a high priced precipitation hedge that uses only one station's data for settlement. A small change in the calibration of the station's sensors can have a significant financial impact on either the buyer or the seller of the weather hedge. Relying on a single station can also expose both the hedge buyer and seller to measurement tampering. One potential solution to this problem is to install additional measurement stations at or near the primary weather station. A mathematical process can be developed for eliminating station readings that deviate significantly from the other stations and the remaining station readings can be averaged to produce a final value for contract settlement. This is the weather market equivalent of the old adage, "Don't put all your eggs in one basket." The installation of additional weather stations can also address the basis risk issues previously described. A combination of stations may provide a more accurate measurement of the weather conditions that are creating weather risk than a single station. Since an automatic weather station with remote data transmission capability can be installed for about US$5,000, the cost of mitigating basis and measurement risk can be quite reasonable.

Once the weather station(s) has (have) been selected, the next step in the hedge development process is to determine the financial impact of weather on the hedger's business. This typically requires statistical analysis that is too complicated to discuss in an introductory chapter such as this.[4] However, the objective of the analysis is to determine the financial change resulting from a change in a weather variable. For example, consider the case of a wheat farmer that is interested in hedging the loss of wheat yield due to a lack of rainfall during the growing season. Regression analysis of historical yields and rainfall amounts may produce results similar to those shown in Figure 1. Notice that in this example, wheat yield decreases with decreasing

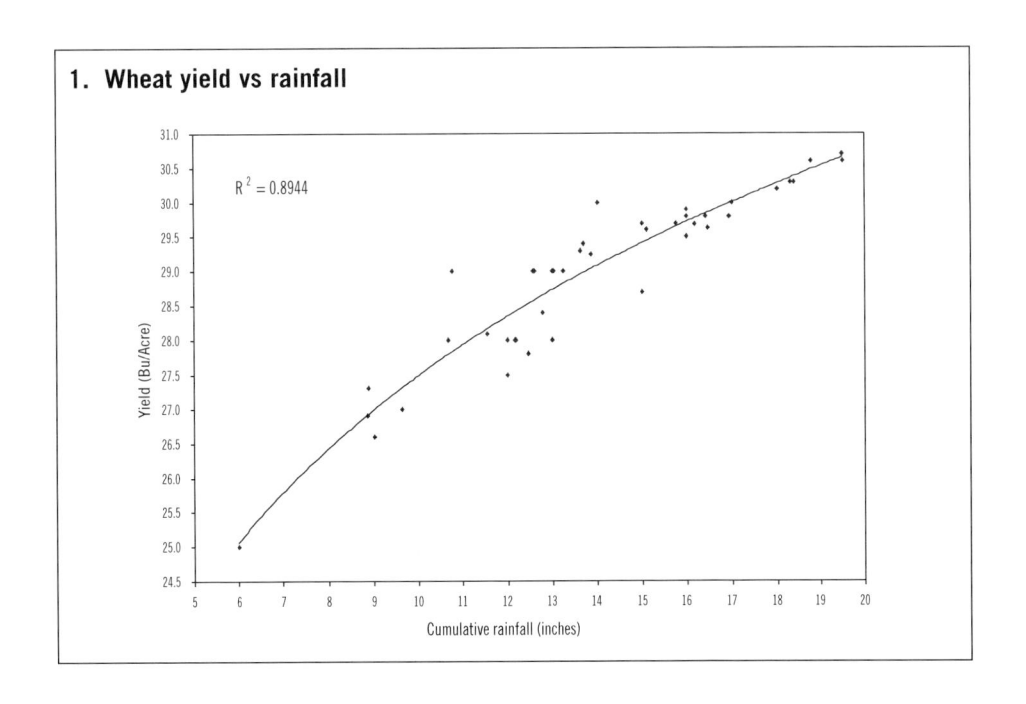

1. Wheat yield vs rainfall

$R^2 = 0.8944$

cumulative rainfall. Suppose that the farmer is concerned about a yield below 27 bushels per acre. The slope of the regression curve from 27 down to 25 bushels per acre is approximately 0.67 bushels per acre per inch. Using an average price of US$3 per bushel for the wheat and 10,000 planted acres, the farmer's weather risk can be calculated to be about US$20,000 per inch of rainfall below nine inches. This is a key number required to size the precipitation hedge. Next we can consider various hedging alternatives.

Hedging alternatives
The objective of a hedge is to provide a financial offset to the risk exposure of the hedger. In this sense, a hedge is simply a cashflow management tool. The numerical values identified during the risk quantification and hedge definition stages are the variables that can be modified during the process of evaluating hedge alternatives. In the case of the wheat farmer, the known hedge variables are:

1. The time of year (specific months over which the wheat is grown);
2. the location (the weather station nearest the farmer's land or newly installed stations adjacent to the wheat acreage); and
3. the financial impact of rain on wheat yield (US$20,000 per inch below nine inches).

The unknown variables at this point are:

1. the desired hedge;
2. the cost of that hedge; and
3. the maximum payoff amount.

Because weather hedges are a hybrid of an insurance product and a financial contract, weather hedges typically have a payoff limit similar to a coverage limit for an insurance policy.

To hedge the loss of wheat yield due to low rainfall, the farmer could consider a precipitation swap contract. A precipitation swap would provide the farmer with a financial payment if rainfall were below a predetermined amount. However, the farmer would also have to make a financial payment to the hedge provider if the rainfall were above the swap level. Typically, the financial payments are the same for both the hedger and the hedge provider, but they can also be varied to create asymmetric payoffs. The swap level is typically determined through negotiation with the hedge provider. Assuming that nine inches was the agreed level for the swap and that the payoffs were symmetric at US$20,000 per inch of rainfall, the hedger would receive US$20,000 per inch of rainfall below nine inches. The hedge provider would receive US$20,000 per inch of rainfall above nine inches. Because the payoff amounts are a constant per unit of rainfall, these payoffs are called "linear payoffs". In either case, maximum payoff amounts could be limited to an agreed value such as US$100,000. The market calls these maximum payment amounts the "notional amount".

Figure 2 provides a payoff diagram for this swap example. By using a swap, the farmer guarantees a wheat yield of 26.8 bushels per acre (equivalent to nine inches of rainfall) but foregoes any increases in yield due to higher rainfall amounts. The farmer would also need to enter into a forward sale contract or a futures hedge to lock in a wheat price to guarantee revenues for that crop. A swap has no upfront cost, but requires the hedger to exchange upside gain for downside risk protection. Most producers of commodities are unwilling to give up any upside in revenues, so a swap may not be an attractive hedging alternative. The next alternative, a floor contract, may be more desirable.

A precipitation floor contract provides protection against a minimum amount of

2. Precipitation swap contract payoff diagram

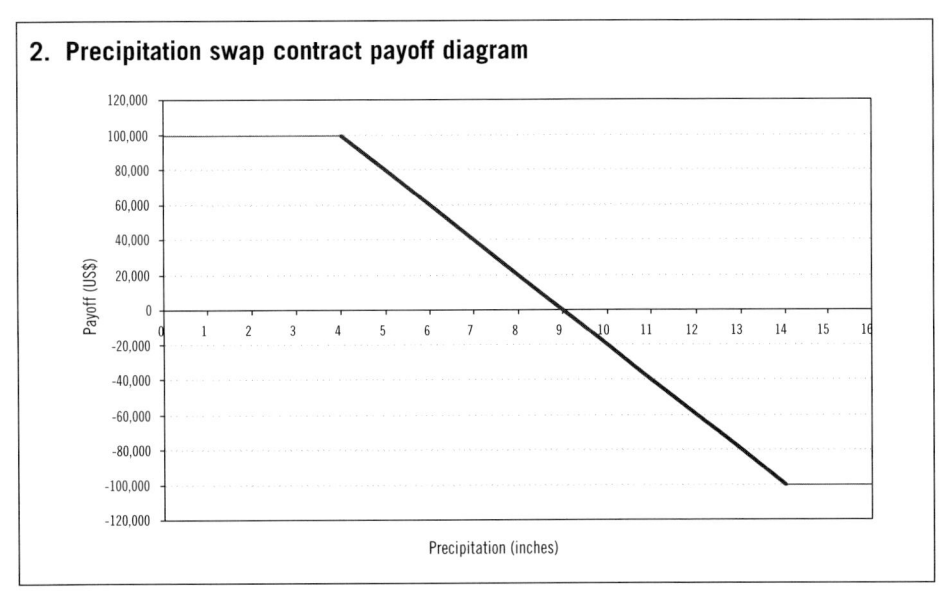

rainfall, but allows the hedger to retain all of the upside benefits of increased rainfall. A floor is essentially a put option. As with the purchase of any type of option, a floor contract requires the hedger to pay an upfront premium, but after paying this premium, there are no additional payments made by the hedger to the hedge provider. If the wheat farmer desired a floor at nine inches of rainfall (the strike level for the floor contract), the hedge provider may require a premium of US$20,000 for this hedge. The actual price quoted by the hedge provider will be a function of the historical value of the hedge and the risk transfer premium required by the hedge provider to assume the weather risk from the hedger. In return for purchasing the floor contract, the hedger would receive US$20,000 for each inch of rainfall below nine inches up to a limit of four inches (assuming a US$100,000 maximum payoff amount), but would also retain all of the benefits of increased rainfall above nine inches. Figure 3 illustrates the cashflows for this example floor contract. A rainfall amount of eight inches would represent the breakeven rainfall level for the hedger because at that rainfall amount, the hedger will recover the upfront premium payment.

3. Precipitation floor contract payoff diagram

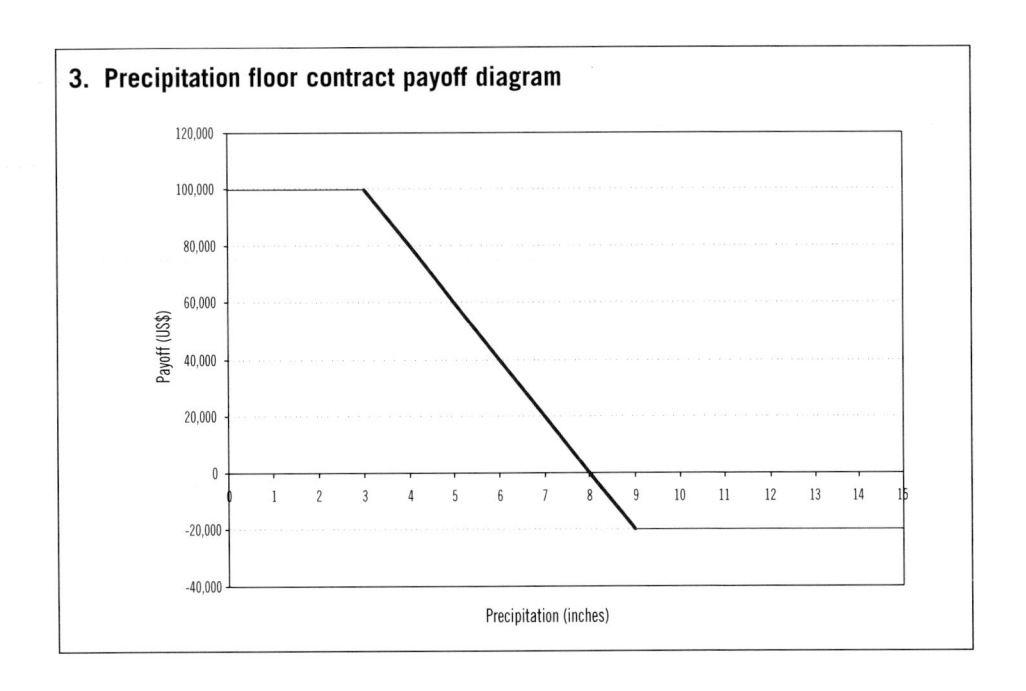

HEDGING PRECIPITATION RISK

At times, a downside risk hedger may view the floor contract premium to be too expensive for the amount of financial protection provided. In this scenario, there are two possible solutions:

1. Lower the strike level of the floor contract. This reduces the level of financial protection provided by the hedge, but should also reduce the premium payment required. By lowering the strike level, the hedger exchanges some of the upfront premium for a portion of reduced future cashflow if the precipitation amount is below the critical level but above the floor contract strike level.
2. Consider exchanging some of the upside benefit for downside protection. In the case of the swap, the hedger was required to forego all of the upside. By selling a precipitation cap contract, a hedger can partially or entirely offset the cost of the floor contract, but still retain some of the upside benefit of higher precipitation.

A cap contract is the opposite of a floor contract and is essentially a call option. A precipitation cap contract provides a financial payoff if the rainfall amount exceeds the cap strike level. The hedger could define a precipitation level above which they would be comfortable foregoing any upside cashflow gains and offer to sell this contract to the floor hedge provider. If the premium value of the cap is exactly the same as the premium of the floor, then the combination of the two contracts will be costless to the hedger. This type of hedge is known as a "costless collar" or a "zero cost collar". Figure 4 shows the cashflows for a costless collar with a floor strike of nine inches and a cap strike of 12 inches of rainfall.

Notice that the payoff diagrams for a swap and a collar are similar. A swap is simply a combination of a floor and a cap with the same strike levels. A collar can be thought of as a "imprecise swap" because the collar has a middle range between the strike levels in which the hedger's cashflows will fluctuate. A swap narrows this range to a single value. Collars are especially useful structures when the hedger desires cashflows to be stabilised within an acceptable range.

Hedging alternatives for non-standard precipitation risks

The example of hedging the theoretical precipitation risk of a wheat farmer described above is typical of the basic types of precipitation hedges that are commonly structured. In some cases, however, the precipitation risk is not linear. The cashflow at risk may be a lump sum amount that needs to be hedged with a digital payoff contract. Taking the example of an amusement park, a light rain even throughout the

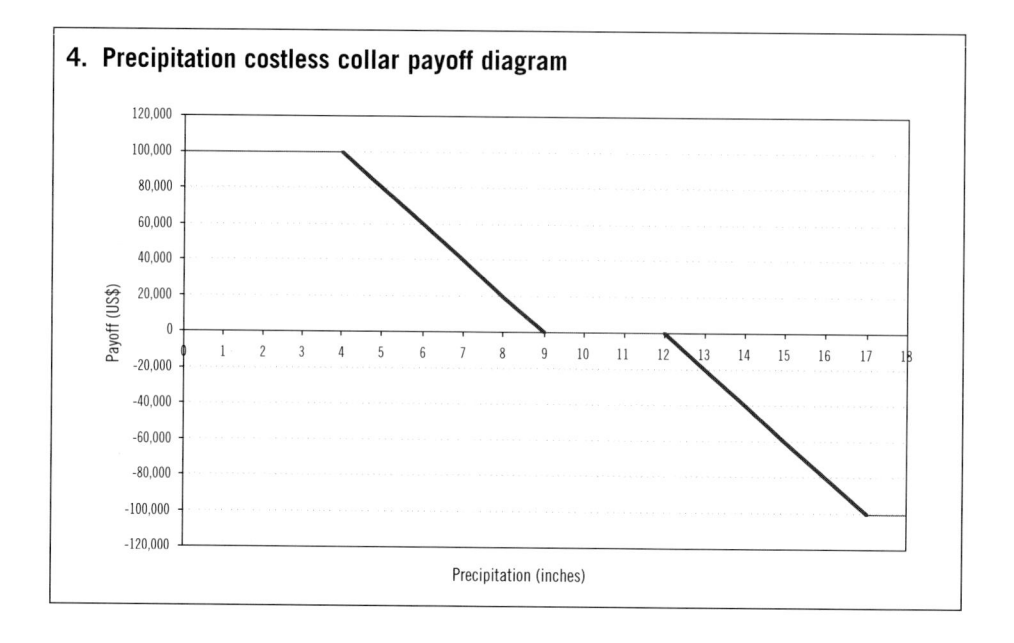

4. Precipitation costless collar payoff diagram

PANEL 1

MEASURING PRECIPITATION – AN IMPRECISE SCIENCE

For any weather derivative, the measurement of the weather index is of critical importance. Particularly, the measurement of precipitation and snow can be imprecise and inconsistent. It is necessary to consider the source of the weather data and its accuracy.

Precipitation in its various forms is measured at ground-level weather stations that can either be manual or automated. In the US, automation of weather stations began in the 1970s. In the 1990s, electronic monitoring devices with computer links began to be installed to allow more frequent weather observations to be recorded and stored in the National Weather Service (NWS) database. Prior to the introduction of automatic stations, human observers logged the weather observation data. The accuracy of these data is a function of the skill and experience of the weather observers.

A metal cylinder with an opening of eight inches in diameter is the standard gauge used to record precipitation. It is used with an internal tube, a funnel and a calibrated ruler used to determine the amount of precipitation collected. A support collar that has legs fixed into the ground holds the entire apparatus. In the winter, only the external cylinder is used to collect precipitation for measurement. If frozen precipitation has been collected, the observer adds a known quantity of warm water to the measuring tube to melt the frozen water. The observer then subtracts the known amount of water from the total reading to determine the precipitation amount to be recorded.

Snow measurements include three different readings: precipitation amount, new snowfall and total depth of snow on the ground. A fourth measurement is often taken, but is not required by the NWS: the water content of the snow. It is an important variable for hydrologic forecasts and for determining the weight of snow on man-made structures. A rough rule of thumb is to convert one inch of snowfall into 10 inches of rainfall. The definitions of the climatological components of snow observations are:[5]

- *Precipitation*: the accumulated depth of rain or drizzle and also the melted water content of snow or other forms of frozen precipitation, including hail, that have fallen in the past 24 hours or since the previous observation.
- *Snowfall*: the depth of new snow that has fallen and accumulated in the past 24 hours or since the last observation.
- *Snow depth*: the combined total depth, at the time of observation, of both old and new snow on the ground from a representative location.
- *Water equivalent of snow*: the amount or depth of water obtained by melting a representative core of snow on the ground.

In addition to human error, inaccuracies in manual measurements of snowfall can be caused by:

- windy conditions – especially a factor with light, dry snow;
- wet snow – can clog the opening of the measurement cylinder; and
- poor location for the gauge – too close to nearby structures or vegetation.

Clearly, the manual measurement of precipitation and snowfall is not an exact science. Automatic weather stations have been designed to eliminate some of the inconsistencies in weather observations, but they too are subject to mechanical problems. For example, many automatic gauges utilise tipping buckets to measure precipitation. The buckets are designed to measure precipitation in increments of 1/100th of an inch. When 0.01 inches of precipitation accumulates in the bucket, the bucket tips over and an electric switch closes to record the event. Problems can occur when the heater located in the upper portion of the precipitation collector fails resulting in icing of the bucket apparatus. Birds sometimes nest inside the collector and cause the apparatus to jam.

Regardless of whether the readings are taken by humans or automated machines, precipitation data is subject to a fair amount of variability that should be considered by both weather hedgers and speculators alike.

entire day may not deter some people from enjoying the rides and attractions, but a heavy downpour in the morning or a thunderstorm in the afternoon may keep people at home or cause others to leave early. Gate revenues and *per capita* expenditures in the park are at risk, but it is unlikely that they will be correlated with total daily rainfall amounts. A lump sum payoff tied to some other weather variable would seem to be more appropriate. One type of digital payoff weather contract that can be used in this scenario is called a "critical-day contract". A critical-day contract accumulates the number of times a weather event meets certain criteria and provides a financial payoff if the number of events exceeds a predetermined level. The number of events becomes the index against which the contract is settled rather than the actual weather itself. For example, assume that an analysis of an amusement park's historical attendance records reveals that attendance drops significantly whenever daily rainfall exceeds 1.5 inches between 08.00 and 12.00 hours from May to August. Also, assume that the economic impact of this rain event is calculated to be US$100,000 per event and that on average the event occurs five times a year. A critical-day cap contract could be structured to provide the amusement park with a cash payment of US$100,000 per event (the event is 1.5 inches of cumulative rainfall between 08.00 and 12.00 hours). If the cap strike level was set to zero (ie, a payment is made for each event), the premium cost could be quite high since the average expected payoff each year would be US$500,000. To reduce the premium of the critical-day cap contract, the amusement park management might decide to raise the cap strike level to five events or higher to protect against a year in which there are more rains events than normal.

Digital payoff contracts can also be structured as "all or nothing payoffs". In this case, if the weather event occurs, the full payment amount is provided to the hedger. If the weather event does not occur, then no payment is made. The key variables that determine the premium for these types of contracts are the payment amount and the probability of payoff.

Precipitation contract applications
While the number of executed precipitation contracts is still quite small relative to the number of executed temperature contracts, the demand for precipitation hedging is clearly growing. The potential for precipitation hedging actually exceeds that of temperature hedging because of the lack of correlation previously discussed. Current estimates of the relative size of the precipitation hedging market to the temperature hedging market vary between 10% and 25% based on the notional amounts for all transacted contracts. This is somewhat impressive given that precipitation hedging is relatively new.

In the year 2000, the Sacramento Municipal Utility District (SMUD) executed the first well-publicised US precipitation contract. According to the press releases, SMUD was concerned about the loss of hydroelectric production from their American River run-of-the-river dams during periods of low rainfall.[6] They were interested in a costless precipitation collar that provided a payoff when rainfall was below the strike level during the October 1–September 30 water year. Because SMUD benefited from excess hydroelectric production during periods of high rainfall, they were willing to offset the purchase of a precipitation floor with the sale of a precipitation cap. Several hedge providers entered in various contracts with SMUD for varying amounts of precipitation risk. While no total monetary figures were ever made public, the estimated notional value of all contracts was in the order of US$100 million.

One of the most interesting aspects of the SMUD deal was the weather station used to settle the contracts. The station was a manual co-operative observation station located in the Sierra Nevada Mountains east of Sacramento, California. The station was in the watershed for the American River and rainfall at that site was determined by SMUD to be highly correlated to their hydroelectric production. The station had a good history of data and the observations were deemed to be accurate and reliable. To mitigate measurement risk, the hedge providers installed several new automatic weather stations adjacent to the existing manual station. To further protect against station tampering, a video camera was installed to monitor activity at the stations. In the first year covered by the hedge transactions, the cumulative rainfall was well below normal and SMUD received a payoff from the hedge providers – the hedge worked. SMUD was a pioneer in precipitation risk management.

Following the SMUD transaction, many of the hydroelectric producers in the Pacific Northwest were contacted and offered similar precipitation hedge contracts. Entergy-Koch was able to define a basket of weather stations whose cumulative rainfall was highly correlated to the flow rate of the Columbia River, but the hydroelectric producers viewed the basis risk between this basket of stations and their hydroelectric production to be too great. They preferred to hedge the stream flow of the Columbia River itself. In early 2001, after a few months of modelling the impacts of snow melt, reservoirs and fish runs on the Columbia River system flow rates, Entergy-Koch began offering stream flow hedges based on the flow rate measured at The Dalles dam. The Dalles is one of the last dams on the Columbia River before it empties into the Pacific Ocean and is commonly monitored by hydroelectric producers. By the time the Dalles stream flow hedges were available, snowfall in the Northwest was already well below normal and the flow of the Columbia River was about half its normal rate. As a result, the hydroelectric producers viewed the strike levels for the stream flow hedges that were offered to be too low and no transactions were concluded. In hindsight, even the strike levels that were offered would have been somewhat valuable because the Columbia River flow rates dropped even lower. Nevertheless, the development of the Columbia River stream flow hedge demonstrates how a new weather product can be created through a dialogue with potential customers.

In 2001, AON Re Canada and Agriculture Financial Services Corp (AFSC) concluded perhaps the first agriculture-related precipitation contract.[7] The US$46 million precipitation contract covered lack of rainfall for cattle ranching on about 2 million acres of native pastureland. Infrared images taken by the US National Oceanic and Atmospheric Administration (NOAA) satellites were used to create a vegetative index that was a measure of the "greenness" of the pastureland. The vegetative index was then correlated to rainfall amounts over the covered region. Nineteen Environment Canada weather stations were selected to settle the rainfall contract. The precipitation strike level was set at 75% of the May–July historical average. The maximum payoff for the contract was US$3 per acre and payments were varied on a sliding scale as a function of lower rainfall amounts.

Also in 2001, a small precipitation hedge was concluded between Entergy-Koch and a corn production company in Nebraska. The company owner is a progressive farmer who had previously transacted a temperature-related growing degree-day hedge and therefore was relatively familiar with weather hedging mechanics. The precipitation risk that was hedged was low rainfall in July and August that could negatively affect crop yields. Within a few days, the strike levels and premiums were agreed and the transaction was closed. The only the minor complication in the transaction was the selection of the weather stations used to settle the transaction.

After some analysis, a basket of five weather stations was created. These stations bounded the corn acreage being hedged and were viewed to be a good representation of the microclimate that could affect the farmer's yields. The contract did not pay out since rainfall was above the strike level but the hedger was very satisfied with the results since they were not required to use expensive well water for irrigation of their corn crop.

The first European rainfall-indexed weather contract was executed in 2001. In this transaction, a German electric utility, Elektrizitatswerk Dahlenburg, purchased a weather hedge which provided a financial payoff if summer rainfall was excessive. The exposure that was being hedged was farmer demand for power: many farmers who purchase electricity from Elektrizitatswerk Dahlenburg use the power to drive irrigation pumps and when rainfall is above normal, the farmers turn off their pumps and demand for electricity drops. Elektrizitatswerk Dahlenburg was astute in identifying the risk and in uncovering a cost effective way to hedge it.

In addition to the precipitation hedges described above, there have been a small number of snowfall-related transactions concluded with ski resort operators. It is likely that a number of other precipitation contracts have been concluded in the US and Europe, but they may not have been publicised due to confidentiality issues.

The future of precipitation contracts
Given the relative novelty of precipitation contracts, what further developments in precipitation hedging might be anticipated? One potential improvement in hedging precipitation risk would be to create hourly precipitation indexes. Most of the precipitation contracts that have been transacted to date have been based on cumulative rain or snowfall over a relatively long period such as one or more months or an entire year. For some potential hedgers like retail establishments, airports and amusement parks, rainfall exposure occurs over a much shorter time duration. For these and other businesses, critical hour contracts may be the more appropriate hedging tools.

Another potential development for precipitation hedging is to combine the precipitation hedge with one or more other weather factors. For example, ski resorts might find their revenues to be better correlated with a "skier-day" index that incorporates snow depth, temperature, wind speed and sunshine as elements for the hedge payoff. Other businesses may also discover multiple factor indexes that correlate well to their weather-related cashflow risks. Hedge providers will be challenged to develop cost-effective, highly customised products to mitigate these risks.

Conclusion
In whatever form they are transacted, precipitation contracts have been established as viable risk transfer instruments for various industries. Based on recent favourable experiences with these financial hedging products, the use of precipitation contracts by non-energy business sectors is expected to grow over the next several years. By providing additional education to potential hedgers and by creating additional customised products, hedge providers can enhance the popularity of precipitation hedging products. Today, some people are not just talking about the weather, they are acting on it.

1 *National Research Council, The Atmospheric Sciences: Entering the Twenty-First Century, Washington, DC: National Academy Press, 1998. Preface, Contents, and Summary available on the Internet (24 January 2001).*
2 *Chicago Mercantile Exchange (1999).*
3 *National Research Council, (1994).*
4 *The reader wishing to follow this line of thought is directed to Chapters 12 and Doesken and Judson (1996) 13 of Devore (1991).*

5 *The Snow Booklet – A Guide to the Science, Climatology and Measurement of Snow in the United States*, Nolan J. Doesken and Arthur Judson, Colorado Climate Center, Department of Atmospheric Science, Colorado State University, Fort Collins, CO 80523, 2nd Edition.
6 *SMUD (2000).*
7 *WRMA website, URL: http//:www.wrma.org.*

BIBLIOGRAPHY

Chicago Mercantile Exchange, 1999, "Weather Futures & Options", Press Release, August 23, URL: http://www.cme.com/products/index/weather/products_index_weather_about.cfm.

Devore, J. L., 1991, *Probability and Statistics for Engineering and the Sciences*, Third Edition, (Pacific Grove, CA: Brooks/Cole).

Doesken, N. J., and Judson, A., 1996, *The Snow Booklet: A Guide to the Science, Climatology, and Measurement of Snow in the United States*, Second Edition, (Colorado Climate Center).

National Research Council, 1994, *Toward a New National Weather Service – Weather for Those Who Fly*, (Washington, DC: National Academy Press).

National Research Council, 1998, *The Atmospheric Sciences: Entering the Twenty-First Century*, (Washington, DC: National Academy Press).

SMUD, 2000, Annual Report, available at: http://www.smud.org/info/reports.html.

4

Weather and Climate – Measurements and Variability

Steve Smith[1]
ACE Tempest Reinsurance Ltd

*Blow winds, and crack your cheeks; rage, blow
You cataracts, and hurricanes spout,
Till you have drench'd our steeples, drown the cocks.*
King Lear, Act 3, Scene II
William Shakespeare

Weather derivatives are essentially financial contracts based on weather observations made at one or many weather stations over some time period. The question of what drives the weather we observe and experience from day to day is, therefore, of great interest; (re)insurers try to answer this question because it is the weather over the period of interest that determines the outcome of the financial contract.

Local geographic features influence local weather everywhere. The mountains on three sides of Los Angeles, for example, obstruct the flow of air and make the formation of photochemical smog more likely. This affects both the day to day weather and the climate of the Los Angeles Basin (Barry and Chorley, 1998). However, we also see that many times that large geographic regions experience similar weather conditions over long periods. As an example, the winter of 1999 was warmer than average across much of the continental US.[2] So we must also ask the question "What drives the climate?" We see that, while weather can be thought of as a 'random' set of events, there is some underlying order that we can examine. This is the study of the climate.

'Climate' is differentiated from 'weather' simply in terms of timescale. The state of the atmosphere we observe over the 1- to 10-day timescale is considered weather. Over the course of a season, such as winter, the general state of the atmosphere is regarded as climate. It is perhaps better to think of this using a statistical concept: the climate is the mean state of the atmosphere; the weather is the noise about that mean. This raises an interesting linguistic question. The majority of weather derivative contracts deal with the weather experienced over a season, whether it will be warmer or colder than expected. Thus, are weather derivatives then more accurately described as 'climate derivatives'?

INTRODUCTION
Many different and competing atmosphere and ocean interactions affect the climate: orbital dynamics and the sun, El Niño Southern Oscillation (ENSO), the North Atlantic Oscillation (NAO), Pacific Decadal Oscillation (PDO), Quasi-Biennial

Oscillation (QBO) and global climate change – these are all covered later in this chapter. First, though, this chapter considers how we make measurements of weather, what variables we can determine and what variables are useful.

Weather measurements

TEMPERATURE

When we talk about measurements of temperature, we explicitly mean air temperature. Typically, we refer to the temperature at some height above the ground (usually two metres for most observing systems). Air temperature is a measure of the average kinetic energy of atmospheric gas molecules. The Zeroth Law of Thermodynamics tells us that two objects in thermal equilibrium have the same temperature (Feynmann, Leighton and Sands, 1989). So, a measuring device in thermal equilibrium with the air will be at the same temperature as the air. To be clear: when we 'measure the temperature' we are actually determining the temperature of the sensor in contact with the air.

A temperature sensor is a device that has sensitivity to changes in temperature. The most commonly thought of device is the mercury thermometer – as the temperature changes, the mercury confined in a tube expands or contracts to a degree observable by the naked eye. However, in most meteorological applications these types of manual thermometers have been replaced by electrical sensors enabling easy automation. There are three basic types of electrical thermometers in common usage:

Thermocouples: These are wires made of two types of metal, which generate a voltage when opposite ends of the wire are at different temperatures. These are not found often in meteorological applications.
Resistance temperature devices: These devices have stable and linear changes in electrical resistance as temperature changes. Typically, platinum wires are used as resistance temperature devices in meteorological applications.
Thermistors: These are similar to resistance temperature devices but, rather than being made of platinum wires, are made from semiconductors. These usually have non-linear responses to temperature so sensor devices are typically made from several different thermistors that respond to different temperatures to obtain a near-linear response over a broad range of temperatures.

It is not enough to place a sensor in the open air and expect a representative air temperature measurement. Sensors have to be shielded from the sun, which can anomalously warm the sensor, from wind, rain and snow, which can anomalously cool the sensor, and from other influences, such as interference from animals, birds and insects. Thus, temperature sensors are mounted in shields or enclosures that can be passive, where ambient air passes freely, or active, where air is captured outside the shield and blown over the sensor.

Care must also be taken to ensure that the siting of the temperature sensor is in line with the environment that is required to be measured. For example, the Federal Aviation Administration (FAA) requires that airport weather stations measure weather variables where they will have the most direct impact on aircraft, ie, at take-off and landing. Therefore, in the mid- to late-1990s, as part of the Automated Surface Observing System (ASOS) upgrade programme, many airport weather stations were relocated to the end of runways (see the FAA's ASOS website: www.faa.gov.asos). The result was that after the relocations many weather stations appeared to be measuring colder temperatures than before (D'Zurko and Robinson, 1999). This was not due to the airport cooling, but instead that weather stations once sited closer to airport buildings, where it is warmer, were now placed in the cooler outfield. As another example, many city weather stations are located on rooftops. In the sun, these rooves radiate heat, making temperatures appear warmer than they actually are.

Temperature is measured using instrumentation, at a known height and in known units – degrees Celsius (°C) or Fahrenheit (°F). These measurements are used as a basis for a weather derivative in the following ways:

Average temperature: The published maximum and minimum temperature on a given day are averaged to give the average temperature for that day. The average temperatures for the length of the contract are totalled, or averaged again, to generate an index for the contract.

Maximum or minimum temperature: Some contracts are based on just the maximum or minimum temperature. Often, it is the occurrences of maximum or minimum temperatures above or below a given threshold that is important. For example, we may wish to write a contract based on the number of days in a given period where the minimum temperature is below 32°F.

Degree-days: This is the most commonly used temperature variable for weather derivatives. First, define a temperature threshold. For a cooling-type degree-day, how much the average temperature exceeded the threshold is recorded. If the average temperature on a day does not exceed the threshold, a zero is recorded. For a heating-type degree-day, we are concerned about whether the average temperature is below the threshold. The name "cooling-type degree-day" derives from the fact that when days are warmer than the threshold, cooling is required; very hot days require much cooling. Likewise, a heating-type degree-day, records when the temperature is colder than the threshold, requiring heating. The genesis of these indexes is from the energy industry. Further, in the US weather derivatives market, the standard heating degree-day (HDD) and cooling degree-day (CDD) indexes are used. In the US, the temperatures are measured in degrees Fahrenheit and the temperature threshold is set at 65°F. It has been found that 65°F is a good psychological indicator: below this temperature people generally turn on their heating, above this temperature they turn their air-conditioning systems on. So, mathematically we define the CDD and the HDD indexes for a given time period as:

$$CDD = \sum_{i=days} \max(T_i - 65, 0)$$

$$HDD = \sum_{i=days} \max(65 - T_i, 0)$$

RAINFALL

Rain is still a surprisingly ill-defined area of study. The mechanisms that produce rain continue to remain, to some extent, unknown. Clouds form as air is uplifted, either by heating or by being pushed up by orography. As the air rises, water starts to condense into small water and ice particles and clouds are formed. Eventually, the water and ice particles can no longer be held up by convection and they fall as rain. However, the physics of the formation of water and ice particles into droplets large enough to fall as rain is not well known. There are two competing theories: the Bergeron–Findeisen theory (Bergeron, 1960) (which states that large ice particles grow by deposition of water onto the surface of the particle) and coalescence theories (which state that water and ice droplets form as particles collide and 'sweep up' smaller particles, as can be seen in Barry and Chorley, 1998).

Rainfall is measured as the depth of water fallen per unit area. Typically, weather reports state that some number of inches of rain fell (neglecting to mention that this is a value per unit area). Rainfall is almost universally measured using a tipping rain gauge. The rain gauge contains an open top, which typically has a large diameter (around one foot). The sensor allows rain to fall into a heated collector shaped like a funnel; the collector is heated to melt any frozen precipitation, such as snow or hail, for collection. Collected water is funnelled to a tipping bucket, which incrementally measures the accumulation of the water and causes the momentary closure of a switch for each increment. The tipping bucket is typically designed to measure in increments of 0.01 inch of rain. As water is collected, the tipping bucket fills to the point where it tips over. This action empties the bucket in preparation for additional measurements and signals another 0.01 inch of precipitation.

Weather derivatives using rain as an index will usually refer to the accumulated rainfall over the period of a contract, eg, the total number of inches of rainfall measured over the life of the contract.

SNOWFALL[3]

Snow is simply ice particles that, unlike rain, reached the ground without melting. It should be noted that, unlike rain, fallen snow contains a large fraction of air meaning that 12 inches of snow does not contain the same amount of water as 12 inches of rain. In fact, even the densest snowfall is only 40% water – four inches of water from 10 inches of snow – while most snow is considerably less dense.

Unlike rain, snow is very difficult to measure. Also, it is necessary to take different types of snow measurement: snow depth (useful for owners of ski resorts, for example) and water content of snow (a prime factor in determining if a river will flood when the snow thaws). The majority of snow measurements are highly manual, relying on a skilled observer; the most reliable method of determining snow depth is still with a flat board and a ruler – the observer places the ruler in the snow down to the depth of the board to read the depth, then clears the board ready for the next day's measurement. Water content is determined using a rain gauge. However, rain gauges need to be modified with windshield or baffle to capture more snow as, unlike rain, wind-driven snow seldom falls vertically. Other techniques are also used, from acoustic or optical methods (sound or light signals are bounced off the surface of the snow to determine depth) to satellite imagery (to map the extent of snow cover). One interesting technique is gamma ray detection. Snow cover attenuates the gamma radiation that the Earth continually emits in relation to the snow's water content. Thus, examining the degree of attenuation can determine the water content of the snow. Gamma ray detectors are most usually deployed on aircraft to help estimate the flood potential from snowmelts.

Few derivatives have been written based on snow measurements though we could imagine that the natural indexes would be snow depth accumulated over a season or the water content of snow accumulated.

WIND

Air flows from high pressure to low pressure: we experience this as wind. Wind should flow down the pressure gradient, however, the motion of a rotating atmosphere is not quite as simple. The force exerted by a pressure gradient is opposed by the Coriolis force, the force due to air parcels moving in a rotating frame of reference (ie, the force due to the Earth's rotation). In the absence of friction at mid-latitudes, the pressure gradient and Coriolis force balance and air parcels flow along paths of equal pressure – this is known as "geostrophic balance" and is why wind flows around a low-pressure system and not simply towards the low-pressure centre. Also, low-pressure weather systems rotate clockwise in the southern hemisphere and anti-clockwise in the northern hemisphere, meaning that the Coriolis force acts in an opposite sense within each hemisphere. Near the ground, the effect of friction affects geostrophic balance, causing air to be pulled in towards a low-pressure centre (see Houghton, 1986).

When measuring wind, it should be remembered that wind is a vector quantity: it has a magnitude and a direction. Any measuring system should measure both these quantities. The most commonly used measuring techniques are mechanical: a vane measures the direction of the wind and a cup or propeller measures the magnitude. The cup or propeller is calibrated such that a given number of rotations correspond to a certain windspeed. It should be noted that both cups and propellers have a threshold (due to mechanical friction) below which they cannot measure. This is usually not such a large issue for weather derivative contracts or insurance products as they are typically concerned with high (damaging) winds.

There are non-mechanical ways to measure wind but these are not widely used. The primary non-mechanical method is using sonic detection and ranging (SODAR). A SODAR sends out a series of high-pitched "chirps" from its speakers at regular intervals, each being broadcast in a slightly different direction (to capture the full three-dimensional direction vector). As they travel away from the speakers, the signals are reflected back to the device by the moving air molecules. The frequency of the returning sound is dependent on the air's movement. By measuring the frequency change the direction and speed of the wind can be calculated.

STREAMFLOW

Large amounts of precipitation, rain or snow must go somewhere. Frequently, rivers flood due to extreme precipitation events or rapid spring thaws. Monitoring rivers is, therefore, of crucial importance in determining what measures might be needed to mitigate floodwaters. Many areas with enhanced flood risk have countermeasures that can be employed, ranging from causeways to divert waters around urban centres, to dikes and dams.

Stream discharge, or streamflow, is usually measured as a volume of water passing a point on the river in a unit time. Typically, the measurement of choice is cubic feet per second. Streamflow is generally estimated from measurements of the stream stage height, ie, the depth of the river. The deeper the river, the faster it flows. This is, of course, based on the development of a stage height/discharge function. These functions are calculated using flow metres (propeller-type measuring devices) and knowledge of the shape of the river channel at the point of interest. The stage height itself is measured by a variety of means, the most robust of which is a simple ruler or staff.

OTHER WEATHER VARIABLES

Weather contracts can use many other variables as the basis for an index; if a meteorological quantity can be measured reliably and consistently then a contract can be based upon it. For example, as relative humidity is measured by most weather stations, it is conceivable that one may wish to write a contract based on relative humidity (technically, relative humidity is a derived quantity typically based on a pair of dry-bulb and wet-bulb thermometer readings). Contracts can also be written on pairs of variables: for example, we may wish to protect ourselves against freezing rain or ice, in which case we would structure a contract based on precipitation and occurrence of temperature below 32°F.

There are other indexes that should be mentioned. Various 'misery' indexes are published by meteorological agencies that we may wish to use as the basis for a contract. The Wind Chill Factor is an attempt to quantify how cold it feels in the presence of wind. The Heat Index attempts to quantify how hot it feels when the weather it both hot and humid. It should be noted that these indexes are attempts to quantify how the weather feels to an average person. They are subject to many approximations and may not be entirely applicable in many situations.[4]

Catastrophic weather

While not the subject of weather derivatives *per se*, catastrophic weather events produce significant financial risks. For example, second to the World Trade Center event of September 11, 2001, the largest insured loss was produced by Hurricane Andrew in 1992 (see the Insurance Services Office's website: www.iso.com/docs/stud006.html). While these catastrophic events are most usually covered by traditional insurance and reinsurance programmes, several products have appeared in the market over the last 10 years which have a derivative-type nature to them:

Catastrophe bonds: These are traditional bonds, which pay a coupon, but where the principal of the bond, or the value of the coupon may depend on the occurrence (or not) of certain types of catastrophe.

Industry loss warrants: These are essentially standardised insurance contracts where the payoff is based not upon a particular company's loss but a measure of the loss caused to the insurance industry as a whole. Since these are standardised they can be (and are) regularly traded and used by insurers and reinsurers as a risk management tool.

TROPICAL CYCLONES

Tropical cyclones form where there is a large expanse of warm water (typically above 80°F) and only small amounts of vertical shear (change in wind direction and/or magnitude with height). As water evaporates from the warm ocean surface, it rises and creates and area of low pressure. Winds rush in towards the low-pressure centre and, due to the Coriolis force, begin to rotate. Air is then drawn into the system at the surface, is heated by the ocean, rises through the centre of the system and is expelled at high altitudes away from the centre. This creates a self-sustaining heat engine: the warmer the water, the faster the air rises in the centre of the low-pressure system; the faster the air rises, the lower the pressure at the centre of the system becomes; the lower the pressure, the greater the winds.

Tropical cyclones, as the name suggests, can only be formed in the tropics, ie, between five and 20 degrees of latitude; at higher latitudes, ocean water is too cold for tropical cyclones to form; closer to the equator there is insufficient Coriolis force to generate circular motion. Also, water temperatures are typically only high enough to allow tropical cyclones to form during boreal late summer and autumn (in the northern hemisphere, August and September are the most active months). In the Atlantic and eastern Pacific basins, strong tropical cyclones are referred to as hurricanes, in the western Pacific as typhoons. Weak tropical storms (with maximum winds less than 74 miles per hour) are classified as tropical storms. Strong tropical cyclones are usually classified according to the Saffir–Simpson scale, as seen in Table 1:

Table 1. The Saffir–Simpson scale

Saffir–Simpson category	Maximum sustained windspeed (mph)
1	74–95
2	96–110
3	111–130
4	131–155
5	>156

Tropical cyclones dissipate once they are isolated from their energy source; they rapidly lose their power once they leave the warm waters of the tropics or make landfall.

In any one year, approximately 80 tropical cyclones will form, of which over half will rate on the Saffir–Simpson scale.[5] Fortunately, few of these will make landfall. For example, in an average year, 3.2 tropical cyclones will affect the continental US, 2.9 will make landfall and only 1.7 will be of hurricane strength. Nevertheless, just one hurricane can be devastating; Hurricane Andrew, which made landfall near Homestead AFB, Florida on August 24, 1992 as a strong category four hurricane, caused (directly and indirectly) 40 deaths and estimates of insured loss are around US$18 billion (in 1992 dollars).

Recent studies (see Goldenberg *et al.*, 2001) have shown that the years 1995–2000 have seen the largest recorded hurricane activity in the Atlantic basin.

Goldenberg *et al.*, 2001 postulate that this is due to increases in the Atlantic sea-surface temperature and a decrease in the vertical windshear. They also warn that, due to the persistence of these changes on a multi-decadal timescale, this increased activity period could last for 10–40 years.

EXTRA-TROPICAL WINDSTORMS

Extra-tropical cyclones (ETCs) are simply low-pressure frontal systems, of the type normally experienced in northern Europe, which have intensified to the extent that winds of near tropical cyclone strength are produced. Cyclogenesis, the process by which low-pressure systems form, occurs throughout the extra-tropical region, ie, above 20–30 degrees of latitude. Warm and cold airmasses collide in an unstable atmosphere. The most favourable regions for cyclogenesis are in the Pacific near the Asian coastline and in the Atlantic between North America and Greenland. Unlike tropical cyclones that have a simple annular structure, ETCs are complex, involving multiple low- and high-pressure regions. Also, again unlike tropical cyclones, ETCs can intensify once making landfall. While rarely producing winds as high as tropical cyclones, ETCs can affect large geographic areas subjecting them to high winds for several days.

In recent memory, several ETCs have caused serious damage to parts of northern Europe. The UK windstorm 87J which hit the south of England on October 15–16, 1987 (a few days before the 1987 stock market crash) formed in the Bay of Biscay so fast that forecasters were caught by surprise. 87J caused close to US$2 billion (in 1987 dollars) in insured loss (as derived from Swiss Re's "Sigma" reports). The end of 1999 saw two winter storms cause havoc in France and Germany; on December 26, 1999, ETC Lothar crossed northern France, southern Germany and Switzerland in the space of a few hours. Windspeeds of the order of 110 miles per hour were recorded in the Paris area. The following day, ETC Martin crossed France further to the south. Martin was slightly weaker than Lothar but windspeeds over land close to 100 miles per hour were recorded. Between them, Lothar and Martin caused over 80 deaths and created an insured loss in excess of US$8 billion (in 1999 dollars) (as derived from Swiss Re's "Sigma" reports).

While by far the main damage caused ETCs is that owing to their intense winds, they can lead to flooding, particularly along the east coast of the UK. First, the direct wind interaction of ETCs with the sea produces a flux of water. Second, the low barometric pressure causes the water level to rise beneath the ETC. The combined effect is to cause water to amass, which, when combined with natural tides, can wreak havoc, as seen around the coastal regions of the UK. Other effects of ETCs, such as increased rainfall, snowfall or abnormal temperatures are usually not significant, from the standpoint of insured loss.

It should be noted that, given the right conditions, any 'normal' weather could develop into a storm that will produce some insured loss. Examples of these are the Nor'easter storms that hit the US East Coast during winter and storms that hit the Pacific Northwest of the US (so-called "Pineapple Express" storms that follow a storm-track originating near Hawaii). While these can frequently cause inconvenience to those in the storm path (Nor'easters frequently deposit large snow amounts), they very seldom produce significant insured losses.

SEVERE THUNDERSTORMS

Severe thunderstorms are large-scale (around 50km in extent) events that may produce hail, tornadoes and straight-line windstorms (downbursts). The Great Plains states of the US (those between the Rocky and Appalachian Mountains) are at the greatest risk from severe thunderstorms as the cold air coming off the Rockies meets warm moist air from the Gulf of Mexico. The interaction of these two airmasses over the relatively flat Midwest can generate large systems (mesocyclones – systems which can range in size up to 1,000km) that can give rise to intense tornadoes, hail storms and downbursts.

Tornadoes are small areas of intense low pressure around which winds can reach in excess of 280 miles per hour. The diameter of tornadoes ranges from 300 to 2,000 feet. As tornadoes are local events, for which actual wind measurements are scarce, they are typically classified on the "Fujita" scale, which relates an approximate windspeed to a subjective analysis of the damage caused by the tornado. Hailstorms are simply intense thunderstorms that have enough internal convection for water and ice particles to grow to diameters upwards of an inch. Such hailstones, especially when accompanied by strong winds, can cause significant property and crop damage.

OTHER CATASTROPHIC WEATHER
Beyond these basic types, many extreme weather events have the potential to cause catastrophe. One example is the 1998 ice storm experienced by Canada and the north eastern United States (Jones and Mulherin, 1998). Warm moist air from the Gulf of Mexico met cold Arctic air creating freezing rain during the week of January 5, 1998. Many places in southeast Canada and northern New England received over three inches of freezing rain, with over an inch accreting on exposed surfaces. Trees and powerlines collapsed. In total, an economic loss over US$4 billion (in 1998 dollars) was recorded.

Climate

So far, this chapter has discussed weather – measuring it and giving some extreme examples of it. This, as mentioned at the beginning of the chapter, is the 'noise' in the atmospheric system. We now turn our attention to the 'mean' of the atmospheric system: the climate. Specifically, we will discuss those effects that drive the climate.

ORBITAL DYNAMICS AND THE SUN
Almost forgotten are the primary drivers of the ocean-atmosphere system – the Sun and the Earth's orbit around it. The Sun's radiation is the primary driver of the ocean-atmosphere system providing all the energy to drive the Earth's atmosphere, and it is the Earth's orbital inclination (23.5° to the plane of the ecliptic) that induces the seasons (see Barry and Chorley, 1998, Chapter 1 for a more in-depth appraisal of the Earth's climate system).

The key to understanding the effect of the Sun's radiation upon the Earth's climate system is the concept of differential heating. This exhibits itself over differing timescales. First, at all timescales, the Equator is heated by the sun to a greater extent than other parts of the Earth since the Sun's radiation falls perpendicularly on the Equator thus the heating per unit area at the Equator is greater than at the poles. Further, radiation must travel a greater distance through the atmosphere to get to the poles than it must to the Equator so more of the radiation is attenuated at the poles. The simple fact that the Equator is warmer than the poles drives the motions of Earth's climate system. The First Law of Thermodynamics states that, in the absence of work being done, heat will travel from a warmer to a cooler body so heat must be transported from the Equator to the poles, via the oceans and the atmosphere.

For seasonal timescales, the orbital inclination of the Earth plays a role. During the northern hemisphere's winter, less radiation falls on the northern hemisphere as the attenuation of the Sun's radiation is increased. Hence, the northern hemisphere is colder than the southern hemisphere. While this is all very well known, it is worth keeping in mind that seasonality is by far the largest observable climate effect.

The final, often neglected, effect is the Earth's rotation. Throughout the day, the amount of radiation received at a point in the Earth's surface varies sinusoidally from a maximum at midday to a minimum at midnight. We experience this most markedly in temperature: night is colder than day. Also, differential heating plays a different role to create sea breezes. During the day, land surfaces heat more rapidly than the oceans (since water has a high specific heat capacity – it can store more heat but it takes more energy to warm it up). So, air rises over the land and falls over the ocean. This creates a wind that blows off the ocean to the land. During the night, this process is reversed as the ocean (thanks to the heat it has stored) is warmer than the land. Thus, wind blows from the land to the ocean. There is also a seasonality to sea breezes. Since they are created by differential heating, they are greatest during the summer when the absolute amount of heating is greatest. A final consequence of the diurnal (day–night) cycle is the effect on atmospheric chemistry. For example, the diurnal cycle moderates the production and destruction of ozone in the stratosphere (Crutzen, 1970 and Smith, 1996).

THE EL NIÑO–SOUTHERN OSCILLATION

The El Niño–Southern Oscillation (ENSO) is a cycle in tropical atmosphere and ocean events. It is most evident in the anomalously warm and anomalously cold ocean temperatures in the central and eastern equatorial Pacific Ocean. The anomalously warm state is referred to as El Niño and La Niña is the name for the corresponding anomalously cold state.[6] As well as changes in the ocean temperature, pressure gradients in the atmosphere change: since the atmosphere and ocean are a coupled system, changes in one are reflected in the other. Taken together, El Niño and La Niña events and accompanying changes in atmospheric pressure are known as the El Niño–Southern Oscillation.

H. H. Hildebrandsson first suggested ENSO in 1897, using 10 years worth of pressure measurements noting relationships between different time-series (Hildebrandsson, 1897). The first major breakthrough was through the efforts of Sir Gilbert Walker, who defined the "Southern Oscillation" as "the tendency of pressure at stations in the Pacific ... to increase, while pressure in the region of the Indian Ocean decreases."[7] In fact, the Southern Oscillation Index (SOI), the pressure differential between Darwin and Tahiti, is still used as a diagnostic for ENSO (indexes based on ocean temperature measurements are the current standard).

In fact, the SOI can be used as a leading indicator for ENSO: changes in the SOI appear to precede changes in the ocean surface temperature. We now know the pressure signals that Walker and others identified to be the atmosphere's response to an ENSO cycle. In fact, a consequence of the ENSO cycle is quite clearly seen in the "Walker circulation". The Walker circulation is a wind circulation where warm trade winds approach Indonesia, are warmed by the ocean, rise, travel eastwards and then sink over the eastern Pacific Ocean. The rising air over Indonesia is warm and wet (large amounts of rain are produced) and is associated with a low-pressure system. The falling air produced by this circulation is dry and cold and is associated with a high-pressure system. The ENSO cycle acts to moderate this circulation. During El Niño (the warm phase of ENSO), the Walker circulation appears to weaken. In fact, it is still present but slips eastward. During the ENSO cold phase (La Niña), the Walker circulation intensifies (Raymond, 1994).

Warm and cold events happen irregularly every two to seven years, typically lasting from 12–18 months (Green *et al.*, 1997). Figure 1 shows a time-series of the most commonly used ENSO index, the Niño 3.4 anomaly index. Niño 3.4 is the average sea-surface temperature of the region of the Pacific Ocean bounded by 120°W–170°W and 5°S–5°N. ENSO events are classified as a Niño 3.4 anomaly (deviation from average) of ±0.4°C (this is noted in Figure 1 as a set of dotted lines). Anomalies greater than 0.4°C are warm events (El Niño events), less than –0.4°C are cold events (La Niña events). The strong warm events in 1982 and 1997 can clearly be observed in Figure 1. Likewise, the cold events in the 1970s and the strong cold event in 1998–1999 can be seen.

What causes ENSO? There is no clear answer but in general all the current theories describe ENSO as an interaction between modes of variability in the ocean–atmosphere system (Hirst, 1988, Suarez and Schopf, 1988, Battisti and Hirst, 1989, Penland and Magorian, 1993 and Penland and Sardeshmukh, 1995). One can think

WEATHER AND CLIMATE – MEASUREMENTS AND VARIABILITY

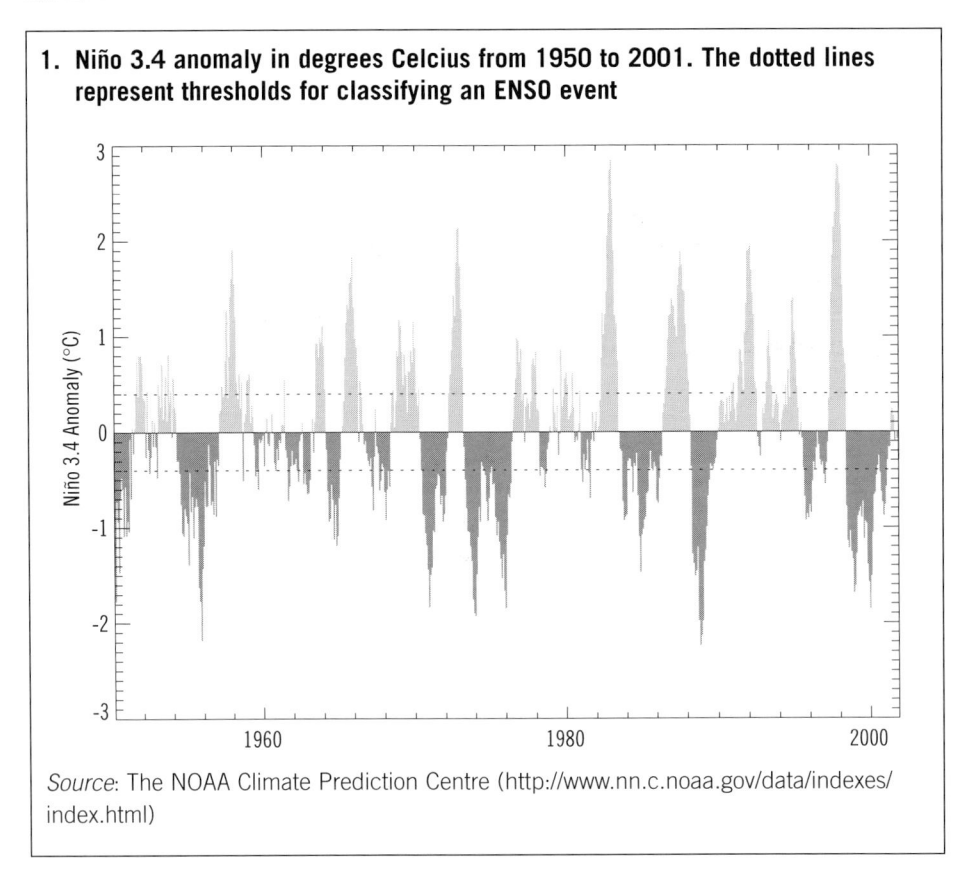

1. Niño 3.4 anomaly in degrees Celcius from 1950 to 2001. The dotted lines represent thresholds for classifying an ENSO event

Source: The NOAA Climate Prediction Centre (http://www.nn.c.noaa.gov/data/indexes/index.html)

of the ocean–atmosphere system as consisting of many different ways of oscillating or moving. These are the "modes" of the system. They are caused by several factors such as the shape of the Earth (being a sphere creates the possibility of certain modes) and the shape of the oceans (some modes are created, or excited, by the shapes of ocean basins). In particular, ENSO is thought to be caused by the interaction of some of the modes that can be created or sustained within the Pacific Ocean basin. The exact method of the interaction of these modes is beyond the scope of this chapter but the interested reader is directed to Hirst, 1988, Suarez and Schopf, 1988, Battisti and Hirst, 1989, Penland and Magorian, 1993 and Penland and Sardeshmukh, 1995).

The strong warm ENSO year of 1997 and cold ENSO years of 1998 and 1999 have been attributed as the cause of much extreme or unusual weather over the entire globe. What then are the effects of ENSO? First, globally: during El Niño the abnormally warm waters in the equatorial central and eastern Pacific give rise to enhanced cloudiness and rainfall in that region. Rainfall is reduced over Indonesia, Malaysia and northern Australia. Within the tropics, the eastward shift of thunderstorm activity from Indonesia into the central Pacific during El Niño results in abnormally dry conditions over northern Australia, Indonesia and the Philippines in both seasons. Drier than normal conditions are also observed over southeast Africa and north Brazil, during the northern winter season. During the northern summer season, Indian monsoon rainfall tends to be less than normal, especially in northwest India. Wetter than normal conditions during El Niño are observed along the west coast of tropical South America, and at subtropical latitudes of North America (Gulf Coast) and South America (southern Brazil to central Argentina). During an El Niño winter, mid-latitude low-pressure systems tend to be more vigorous than normal in the region of the eastern North Pacific. These systems push abnormally warm air into western Canada, Alaska and the extreme northern portion of the contiguous US. Storms also tend to be more vigorous in the Gulf of Mexico and along the southeast coast of the US resulting in wetter than normal conditions in that region. During La Niña, these effects are reversed – it is drier where it was wetter under El Niño, warmer where it was colder, etc. It should be noted (as in Mason and Goddard, 2001) that the impact of ENSO on precipitation should not be over-stated – only 20%–30% of land area experiences increased probabilities of abnormal precipitation during an ENSO event. Also, there are, where precipitation is concerned, definite asymmetries between warm and cold ENSO events.[8]

Focusing on the US, we can quantify ENSO effects more specifically.[9] During winter, the temperature across the entire south of the US shows a strong negative correlation with ENSO. So, when there is a warm (El Niño) event, the southern US will be colder than normal in winter. Alternately, the northern US, particularly the northern Midwest and New England, has a strong positive correlation with ENSO in the winter – warmer than normal during a warm event and colder than normal during a cold event. During summer months, temperatures across the US show less strong correlations with ENSO, having a slight negative correlation for all areas east of the Rockies. Winter precipitation shows very strong positive correlations with ENSO for all areas west of the Rockies (particularly California), and for many of the southern states (Florida, South Carolina and Texas) – these places will be wetter than normal during a warm event, and drier than normal during a cold event. Summer precipitation again shows weaker correlations – the mid-Atlantic coastal states and New England show slightly negative correlations with ENSO (drier during warm events) while the Great Plains show positive correlations. As for snowfall in the US, Smith and O'Brien (2001) found increased snowfall during an ENSO cold event early during the winter season, less snowfall during a cold event in northeast US, and less snowfall during both warm and cold events in the Ohio Valley and Midwest US.

ENSO also strongly influences the occurrence and path of tropical cyclones. Kimberlain and Landsea (2001) present an exhaustive report on ENSO's influence on tropical cyclones. They found that:

❏ Tropical cyclone activity in the Atlantic basin is reduced dramatically during a warm event.
❏ During warm events, tropical cyclones in the Atlantic basin are weaker and shorter-lived while in the Pacific basin they are stronger and longer-lived.
❏ Tropical cyclone tracks in the Atlantic shift northward, in the Pacific basin they shift towards the centre of the basin during warm events.
❏ During warm events, western north Pacific tropical cyclones show enhanced recurvature (the northward and eastward curving of tropical cyclone tracks when they leave the tropics).

Cold events show some opposite effects on cyclones. For example, during a cold event, the Atlantic basin experiences more and stronger tropical cyclones. It should be noted, however, that this is the average behaviour. Hurricane Andrew occurred during a warm ENSO event, when fewer, weaker tropical cyclones would be expected in the Atlantic basin.

Finally, how well is the ENSO cycle predicted? Typically, because of the persistence and relatively slow movement in the ocean, ENSO can be predicted with reasonable accuracy over 0–3 months. Further, as noted before, the SOI can be used as a leading predictor of the ENSO cycle. However, longer timescales are troublesome; although the climate science community made great claims of accuracy of forecasts for the 1997 warm ENSO event, Landsea and Knaff (2000) critically evaluated the forecasts of the 1997 event from various models and found that the skill of these models in predicting the onset, magnitude and end of the event was not as great as claimed. In many cases, these models performed worse than a simple statistical model (the ENSO-CLIPER model, Knaff and Landsea, 1997). It is hoped that the lessons learned from this exercise will enable more skilful forecasts of ENSO in the future.

THE NORTH ATLANTIC OSCILLATION[10]

The North Atlantic Oscillation (NAO) is the dominant mode of winter climate variability in the North Atlantic region ranging from central North America to Europe and into northern Asia (Hurrell, 1995). Typically, the pressure patterns over the North Atlantic show a low-pressure area near Iceland, and a high-pressure area near the Azores. The NAO is a cyclic displacement in this pressure pattern. It is measured by comparing the observed pressure near the centres of action near Iceland and the Azores. Figure 2 shows an NAO index calculated by using data from Stykkisholmur in Iceland and Gibraltar in the Mediterranean. In Figure 2, the data shown are normalised winter averages (where winter is defined as December–March) of the difference of atmospheric pressure at these two stations. Also shown is a five-year running average of this data. It is seen that the NAO has been in a positive phase (greater than zero) for much of the last 20 years.

What causes the NAO? This is still the subject of many debates. Some questions have been resolved. For example, the links between the NAO and ENSO have been shown to be weak (Hurrell *et al.*, 2001). It has been speculated that the increase in the positiveness of the NAO over the last 20–30 years is a sign of global climate change (Visbeck *et al.*, 2001), but this is far from being resolved. Many researchers think that the NAO is a projection of the Arctic Oscillation (AO) – a 'wobbling' in the annular pattern of pressure seen over the Arctic at high altitudes (Hurrell, 2001). However, Ambaum *et al.* (2001) have cast doubt on this, speculating that any correlation between the NAO and the AO can be interpreted as a statistical artefact.

Regardless of the cause, the NAO is a very real phenomenon that affects the weather across a large part of the northern hemisphere. The main effect of the NAO on northern Europe is to guide winter storm tracks; during the positive phase of the NAO storms are guided towards the UK, northern France and Scandinavia; during the

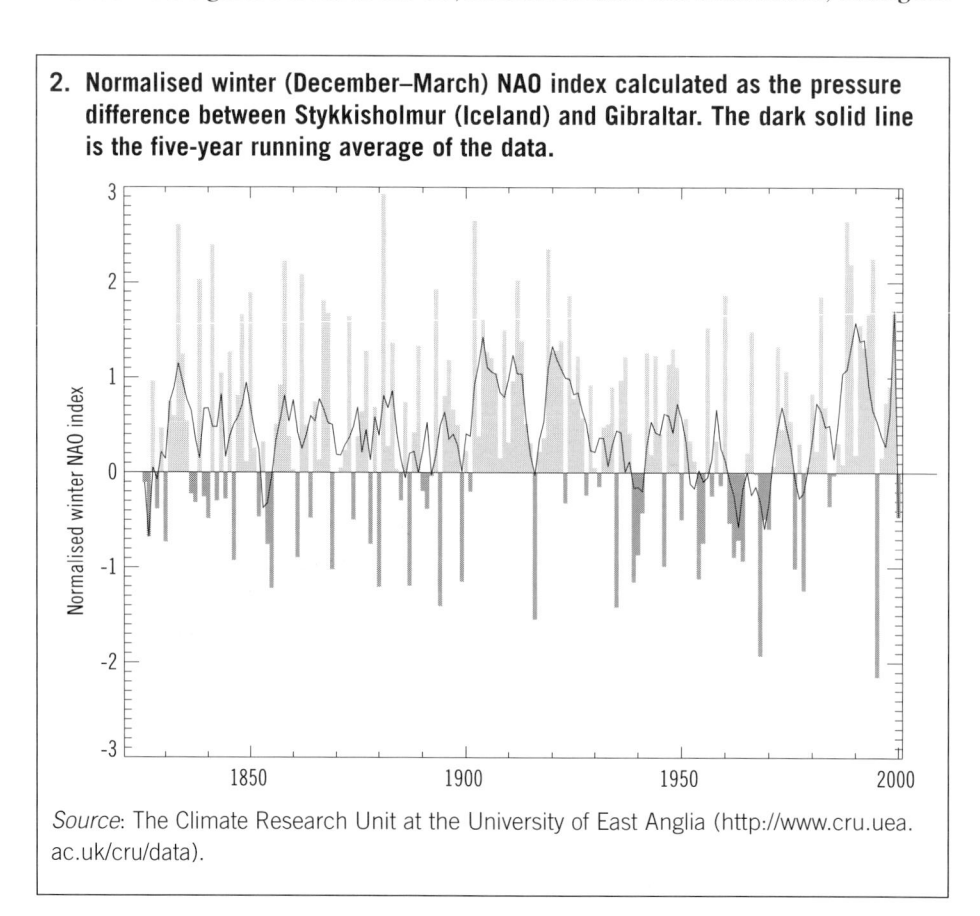

2. Normalised winter (December–March) NAO index calculated as the pressure difference between Stykkisholmur (Iceland) and Gibraltar. The dark solid line is the five-year running average of the data.

Source: The Climate Research Unit at the University of East Anglia (http://www.cru.uea.ac.uk/cru/data).

negative phase storms are guided along more southerly tracks towards the Iberian Peninsula. Consequently, during the positive phase northern Europe is warmer, wetter and windier than average. During the negative phase, northern Europe is colder, drier and less windy on average. The NAO also affects the US East Coast – during the positive phase the East Coast is milder and wetter than average, whereas during the negative phase it is colder and snowier (see David Stephenson's NAO thematic website at the Department of Meteorology, University of Reading. URL: www.met.rdg.ac.uk/cag/nao/index.html).

OTHER CLIMATE OSCILLATIONS

The ocean–atmosphere system has many different oscillations (or modes), the main ones have already been noted above. There are others, however:

❏ The Pacific Decadal Oscillation (PDO) is a cyclic change in the atmosphere over the Pacific Ocean, and has a pattern very similar to ENSO but with a much longer timescale: Whereas ENSO events last 12–18 months, PDO events last 20–30 years.

PANEL 1

GLOBAL CLIMATE CHANGE

One of the most widely discussed topics of the last 10 years has been the environment and humanity's impact upon it. It is clear that the activities of humanity in the industrial age have changed the Earth's climate. The global average surface temperature has increased over the twentieth century by 0.6°C (Houghton *et al.*, 2001).

Globally, the 1990s were the warmest decade on record (Houghton *et al.*, 2001), with 1998 being the warmest year on record. Snow cover and ice cover have declined while the global average sea level has risen. The Intergovernmental Panel on Climate Change (IPCC) has reported that over the next century the global average surface temperature will increase by 1.4°C–5.6°C and the sea level will rise by 0.09m–0.88m. Snow and ice cover will continue to decrease and it is likely that we will see tropical cyclones of increased strength, more precipitation and more days of extreme temperature (see Houghton *et al.*, 2001).

What has humanity done to create this? By burning fossil fuels and related activities, human society has introduced more carbon dioxide and methane into the atmosphere to increase the 'greenhouse effect'. The Earth's temperature is controlled by a balance between the incoming radiation from the Sun and the amount of radiation that the Earth emits back to space. The greenhouse gases trap more of the outgoing radiation, altering the balance between the heat received from the sun and heat re-radiated to space. The more greenhouse gases we put into the atmosphere, the more this balance will change and the hotter it will become. As a point of reference, the carbon dioxide in the atmosphere has doubled over the last 100 years. So far, efforts to reduce greenhouse gas emission have been minimal.

Further, other gases that human industry has produced have altered the chemical balance of the atmosphere. The most striking example of this is the Antarctic ozone hole, first discovered in 1985 (see Farman *et al.*, 1985). Here, the production of chlorofluorocarbons (CFCs) has led to the destruction of a large part of the stratospheric ozone layer over the Antarctic. The ozone layer is important in preserving the temperature balance in the stratosphere and in protecting the surface from harmful ultraviolet radiation. There have been major efforts to reduce CFC emissions and it is hoped that the Antarctic ozone hole will begin to repair within the next 50 years.

❑ Also the PDO is more prominent over the North Pacific whereas ENSO is typically confined to the tropics. The causes of the PDO are, at present, unknown. The effects of the PDO are also very similar to ENSO although the PDO signal is hard to separate from the ENSO signal. The PDO seems, to some extent, to moderate the ENSO cycle.[11]

❑ The Quasi-Biennial Oscillation (QBO) is a very regular oscillation of the zonal winds (the winds that blow around latitude circles) in the lower stratosphere. The QBO has a weak correlation with ENSO and its main effect is to moderate the production of tropical cyclones by moderating or enhancing vertical windshear. More hurricanes are produced in the Atlantic basin during the westerly phase of the QBO as vertical shear is reduced.[12]

CONCLUSION

This chapter explained how weather measurements are made, how extreme weather is created and how weather is influenced by larger timescale variations – the climate. While the Earth's climate system is one of almost infinite complexity, the basic principles, as outlined in this chapter, are fairly straight-forward; the Sun's radiation produces a differential heating effect; the climate system attempts to move heat from the Equator to the poles, via the oceans and the atmosphere; the climate system can move the heat in many different ways, or modes; the interaction of these modes produces climate signals, such as the El Niño–Southern Oscillation. So, this answers a question originally posed at the beginning of the chapter: what drives the climate? The climate is driven by the physical imperative to move heat from the Equator to the poles. The way that the climate system performs this function creates a mean state upon which various oscillations are imposed. This leads us to answer the second question from the introduction: what drives the weather? The weather is driven by the climate and by the various oscillations inherent to it. However, it should be noted that the weather, on a day to day basis has a large amount of intrinsic randomness.

Given all we know about the mechanisms that govern the Earth's climate system, there is still much we do not know. The accurate prediction of weather on timescales greater than two weeks is, at present, almost impossible. The prediction of climate state (the "mean weather") has little skill beyond a few months. This introduces risk. However, this is a risk that business now has the tools to mitigate. An energy company in the northeast US cannot know whether it will be colder or warmer than normal next winter. We can, perhaps, say that if a warm ENSO event happens it will be warmer than normal. Even so, the energy company can use weather derivatives to hedge, at least partly, the risk from this unknown weather. A large corporation in Florida may not know if or when a hurricane may damage its buildings, but it can purchase insurance to reduce its risk. Finally, the prices at which weather derivatives or insurance (and reinsurance) are offered to these companies will depend on the view of the climate that the weather derivative market maker or the insurer takes. An understanding of the climate, its causes and effects, is therefore crucial to allow an efficient transfer of risk to take place.

1 *The author would like to thank Dr Manoj Joshi, Jemma Levy, Sloane Miller and Brock Webel.*
2 *For the winter November 1998–March 1999, apart for the states of California, Oregon and Washington, and climate divisions in the north of Nevada, all climate divisions in the continental US were at least 1°F warmer than the 1961–1990 November to March longterm average. Some climate divisions in the Midwest were up to 7°F warmer than the 1961–1990 mean (see the NOAA CDC mapping website http://www.cdc.noaa.gov/USclimate/USclimdivs.html).*
3 *Readers are pointed towards Doesken and Judson (1997) for a very comprehensive discussion of snow and how to measure it.*

4 *For example, the National Weather Service in the US recently reformulated its calculation of wind chill, so care should be taken if comparing their wind chill numbers before and after the reformulation.*
5 *Based on data from 1968–1989, there are globally on average 83.7 systems with a windspeed of $17ms^{-1}$ or greater (tropical storm and higher), with a global average of 44.9 systems with a windspeed of $33\ ms^{-1}$ or greater (Saffir–Simpson Category 1 and higher) (see Chris Landsea's Tropical Cyclone FAQ http://www.aoml.noaa.gov/hrd/tcfaq/tcfaqHED.html).*
6 *The name "El Niño" means "the Christ child" or "the boy child" in Spanish and originates in Peru, as a warming event typically first shows its presence as a warming in the waters off the South American coastline during the Christmas season. La Niña means "the girl child" or, less commonly, "El Viejo" – "the old man".*
7 *See Walker (1928).*
8 *The interested reader is advised to consult the comprehensive and annotated list of web-based ENSO resources at the Association of College and Research Libraries (www.ala.org/acrl/research98.html).*
9 *See US regional websites www.mcc.sws.uiuc.edu/elnino.html for the Midwest and www.wrcc.sage.dri.edu/enso/enso.html for the West Coast.*
10 *The interested reader is directed towards David Stephenson's NAO thematic website at the Department of Meteorology, University of Reading (http://www.met.rdg.ac.uk/cag/NAO/index.html) for a more complete survey of the NAO.*
11 *For further reading on PDO's the reader is directed to Zhang et al. (1997) and Trenberth and Hurrell (1994).*
12 *For further reading on QBO's the reader is directed to Andrews et al. (1987).*

BIBLIOGRAPHY

Ambaum, M. H. P., B. J. Hoskins and D. B. Stephenson, 2001, "Arctic Oscillation or North Atlantic Oscillation?", *Journal of Climate*, 14, pp. 3495–507.

Andrews, D. G., J. R. Holton and C. B. Leovy, 1987, *Middle Atmosphere Dynamics*, First Edition, (London: Academic Press).

Barry, R. G., and R. J. Chorley, 1998, *Atmosphere, Weather and Climate*, Seventh Edition (London: Routledge).

Battisti, D. S., and A. C. Hirst, 1989, "Interannual Variability in the Tropical Atmosphere-Ocean System: Influence of the Basic State, Ocean Geometry, and Nonlinearity, *Journal of Atmospheric Science*, 46, pp.1687–712.

Bergeron, T., 1960, "Problems and Methods of Rainfall Investigation," *The Physics of Precipitation*, Geophysical Monograph 5 (Washington, DC: American Meteorological Society).

Cruzten, P. J., 1970, "The Influence of Nitrogen Oxides on the Atmospheric Ozone Content", *Quarterly Journal of the Royal Meteorological Society*, 96, pp. 320–5.

Doesken, N. J., and A. Judson, 1997, *The Snow Booklet – A Guide To The Science, Climatology, And Measurement Of Snow In The United States*, Second Edition (Colorado State University).

D'Zurko, D., and D. Robinson, 1999, "A Study of the Effects of ASOS on Temperature Data Continuity for New York State Sites", NYGAS Technology Brief (99-721-1), New York Gas Group.

Farman, J. C., B. G. Gardiner and J. D. Shanklin, 1985, "Large Losses of Total Ozone in Antarctica Reveal Seasonal ClOx/NOx Interaction", *Nature*, 315, pp. 207–10.

Feynman, R. P., R. B. Leighton, and M. L. Sands, 1989, *The Feynman Lectures on Physics: Commemorative Issue*, First Edition (Boston: Addison Wesley).

Goldenberg, S. B., C. W. Landsea, A. M. Mestas-Nuñez and W. M. Gray, 2001, "The Recent Increase in Atlantic Hurricane Activity: Causes and Implications", *Science*, 293(5529), pp. 474–9.

Green, P. M., D. M. Legler, C. J. Miranda V and J. J. O'Brien, 1997, "The North American Climate Patterns Associated With The El Niño-Southern Oscillation", COAPS Project Report Series 97-1 (Tallahassee: The Florida State University).

Hildebrandsson, H. H., 1897, "Quelques Recherches sur les Centres d'Action de l'Atmosphere I–V", *Handl. Suenska Vetensk Ak*, 29.

Hirst, A.C., 1988, "Slow Instabilities in Tropical Ocean Basin-Global Atmosphere Models", *Journal of Atmospheric Science*, 45, pp. 830–52.

Houghton, J. T, Y. Ding, D. J. Griggs, M. Noguer, P. J. Van der Linden and D. Xiaosu, 2001, *Climate Change 2001: The Scientific Basis – Contribution of Working Group I to the Third Assessment Report of the Intergovernmental Panel on Climate Change (IPCC)* (Cambridge University Press).

Houghton, J. T., 1986, *The Physics of Atmospheres*, Second Edition (Cambridge University Press).

Hurrell, J. W., 1995, "Decadal Trends in the North Atlantic Oscillation: Regional Temperatures and Precipitation", *Science*, 269, pp. 676–9.

Hurrell, J. W., 2001, "Climate: North Atlantic and Arctic Oscillation (NAO/AO)", *Encyclopedia of Atmospheric Sciences*, (eds) J. Holton, J. Pyle, and J. Curry, (San Diego: Academic Press).

Hurrell, J. W., C. Deser, C. K. Folland and D. P. Rowell, 2001, "The Relationship between Tropical Atlantic Rainfall and the Summer Circulation over the North Atlantic", CLIVAR Atlantic Boulder Conference.

Jones, K. F., and N. D. Mulherin, 1998, *An Evaluation of the Severity of the January 1998 Ice Storm in Northern New England – Report for FEMA Region 1*, US Army Corps of Engineers, Cold Regions Research and Engineering Laboratory.

Kimberlain, T .B., and C. W. Landsea, 2001, *The Effect Of ENSO on North Pacific and North Atlantic Tropical Cyclone Activity*, Final Report for the Risk Prediction Initiative, Bermuda Biological Station for Research.

Knaff, J. A., and C. W. Landsea, 1997, "An El Niño-Southern Oscillation Climatology and Persistence (CLIPER) Forecasting Scheme", *Weather Forecasting*, 12, pp. 633–52.

Landsea, C. W., and J. A. Knaff, 2000, "How Much Skill was there in Forecasting the Very Strong 1997–1998 El Niño?", *Bulletin of the American Meteorological Society* 81, pp. 2107–19.

Mason, S. J., and L. Goddard, 2001, "Probabilistic Precipitation Anomalies Associated with ENSO", *Bulletin of the American Meteorological Society*, 82, pp. 619–38.

Penland, C., and P. D. Sardeshmukh, 1995, "The Optimal Growth of Tropical Sea Surface Temperature Anomalies", *Journal of Climate*, 8, pp. 1999–2024.

Penland, C., and T. Magorian, 1993, "Prediction of Niño 3 Sea Surface Temperatures using Linear Inverse Modeling", *Journal of Climate*, 6, pp. 107–76.

Shakespeare, W., 1993, *King Lear*, (Boston: Longman).

Smith, S. E., 1996, "Satellite Measurements of Dinitrogen Pentoxide in the Stratosphere", DPhil Thesis, University of Oxford.

Smith, S. R., and J. J. O'Brien, 2001, "Regional Snowfall Distributions Associated with ENSO: Implications for Seasonal Forecasting", *Bulletin of the American Meteorological Society*, 82, pp. 1179–91.

Suarez, M. J., and P. S. Schopf, 1988, "A Delayed Action Oscillator for ENSO", *Journal of Atmospheric Science*, 45, pp. 3283–7.

Trenberth, K. E., and J. W. Hurrell, 1994, "Decadal Atmosphere-Ocean Variations in the Pacific", *Climate Dynamics*, 9, pp. 303–19.

Visbeck, M., J. W. Hurrell, L. Polvani and H. M. Cullen, 2001, "The North Atlantic Oscillation: Past, Present and Future" *Proceedings at the 12th Annual Symposium on Frontiers of Science*, 98, pp. 12876–7.

Walker, G. T., 1928, "World Weather", *Monthly Weather Review*, 56, pp. 167–70.

Zhang, Y., J. M. Wallace, and D. S. Battisti, 1997, "ENSO-like Interdecadal Variability: 1900–93", *Journal of Climate*, 10, pp. 1004–20.

5

Weather Data: Cleaning and Enhancement

Auguste C. Boissonnade; Lawrence J. Heitkemper and David Whitehead

Risk Management Solutions; Earth Satellite Corporation

High quality weather data is used for pricing weather trades, for managing the risk associated to these trades, and for settlement. Forecast weather data is used for evaluating the weather risk up to the time where forecast data ceases to show significant skill. After this period, a combination of seasonal forecasts and historical data is used in pricing weather trades. In addition, near-real time weather data provides the market participants with discrete weather data values to mark the market model for daily risk management purposes and, ultimately, for settlement purposes at the expiration of the weather trades.

The weather market requires data that meets key criteria for use in valuation. This chapter identifies and discusses these criteria, major weather data issues and means for resolving them.

Introduction

The main source for weather data in each country is the national meteorological service (NMS). Each NMS operates independently in setting its standards for weather data collection, archival and distribution.

Weather data are generally classified as either synoptic data or climate data. Synoptic data is the real time data provided for use in aviation safety and forecast modelling. Climate data is the official data record, usually provided after some quality control is performed on the data. Special networks also exist in many countries that may be used in some cases to provide supplementary climate data. These topics and related issues will be further discussed in the next section.

Weather data released by the NMSs often must be "cleaned", ie, replacing missing and erroneous data before the weather market can use it. The types of errors to be expected and methodologies for treating them are discussed in the "Data Cleaning" section.

After weather data are cleaned, problems might still exist that need to be corrected before the data can be used for modelling weather risk. These problems are known as "inhomogeneities" or "discontinuities" caused by station relocations, changes in instrumentation or changes in the surrounding environment near the station site. Treatment of inhomogeneities in the weather data time-series is generally referred to as "homogenisation" or "homogeneity adjustment". This chapter provides a thorough discussion on data enhancement, which is data adjustment resulting from discrete changes in a station's weather data time-series caused by station changes. Non-discrete inhomogeneities such as urbanisation effects are treated as trends.

There are many applications for weather data in weather risk modelling, including the potential use of remotely-sensed data for settlement. Conclusions on weather data use in risk management are presented in the final section.

WEATHER DATA: CLEANING AND ENHANCEMENT

Weather data

Ideally, the market needs timely and accurate weather data. In order to achieve this, data should be continuously recorded from stations that are properly identified, manned by trained staff or automated with regular maintenance, in good working order and secure from tampering. The stations should also have a long history and not be prone to relocation.

Unfortunately, the main charter of the NMSs in most, if not all countries is "protection of life and property". The emphasis of these offices is focused on accurate diagnosis, tracking and forecasting of severe weather events. The collection and archiving of weather data is important because it provides an economic benefit but the local/national economic needs are not as dependent on high data quality as is the weather risk market. Thus, most government weather service offices do not emphasise data quality issues.

The World Meteorological Organization (WMO) is the international association of NMS members but has no enforcement authority.[1] While various organisations have standardised products, it is apparent that each organisation stands alone in its practices in setting its standard for data collection, archival and distribution. For example, although the WMO has established a standard schedule for the reporting of data at specific times of each day, it has no ability to acquire what data is reported at these times. Daily minimum and maximum temperatures are reported from midnight to midnight of the same calendar day in the US and Germany. However, other European countries report these values in different ways (for example, the United Kingdom Meteorological Office (UKMO) reports the daily minimum temperature from 09.00 Coordinated Universal Time (UTC) of the previous day to 09.00 UTC of the current day, and the current day's maximum temperature from 09.00 UTC of that day to 09.00 UTC of the following day).[2] These individual standards and reporting practices create difficulties in dealing with weather data on a global scale and necessitate "country-dependent" quality control procedures. Knowledge of these standardisation issues and other issues, including the maintenance of historical station information (metadata), maintenance of observation networks, data availability, data usage rights and data pricing, are extremely important to the weather market; without understanding how weather data is collected and processed and without knowledge of the station history, it is very difficult to assess the reliability of these data and to use them. As the weather derivative industry continues to expand globally, these issues are being resolved by the private weather industry. Agreements are made with NMSs for redistribution of raw data and for accessing information on local weather data collection practices and station history.

METEOROLOGICAL NETWORKS

Weather data are recorded from several meteorological networks. The primary purpose of weather networks is to provide synoptic data and climate data. Generally, synoptic data are data obtained simultaneously over a large area of the atmosphere. SYNOP data (data found at stations reporting in accordance with the WMO synoptic reporting standards) undergoes very little quality control and is exchanged among NMSs in order to produce forecasts. Climate data, which are sometimes derived from synoptic data, are data generally used for climatological studies. This data is subjected to quality control procedures and most often forms the official record of historical weather and climate. Climate and synoptic weather data sets may or may not be the same. The same stations may be used for collecting both data sets. However, more quality control checks will be implemented with climate data. In other instances, physically separate observation instruments are used for recording synoptic and climate data.

There are various conventions for identifying stations established by various organisations. The WMO has established a five-digit numbering system for identifying stations for international reporting of synoptic data from these stations. There is also a naming convention for hourly METAR (the acronym roughly translates from the French as Aviation Routine Weather Report) reports consisting of four characters. In the US, the WBAN (Weather Bureau Army Navy) five-digit convention is often used in writing contracts to identify the station. The individual country meteorological networks are identified using the convention selected by the individual country. The lack of a comprehensive international identification standard sometimes causes confusion for the market.

United States
The official source of weather data in the United States is the National Climatic Data Center (NCDC[3]), an agency of the National Oceanic and Atmospheric Administration (NOAA[4]). The US does not significantly differentiate between synoptic and climate networks. US network classification is based upon a tier system with the highest quality stations called "first-order" stations (primary stations). The differences in the class of observing stations are in the variables measured, the frequency of observations being reported, generally the length of data history, and the type of operators manning these stations. The weather market most often selects the first-order stations for pricing and management of US weather risks. These stations operate 24 hours a day and are maintained by National Weather Service (NWS) trained and certified staff.[5] These first-order stations are responsible for observing official climate data as well as providing synoptic data for forecasting purposes. Observations made at these stations span a wide number of variables including temperature, precipitation, surface pressure, humidity, wind speed and direction, cloud cover, snow depth, visibility and solar radiation. These observations are taken and distributed frequently, either on an hourly basis or at specific times of the day. These observations are also cleaned by NWS and reported each month in Local Climatological Data (LCD) reports.

Trained staff that are supervised by NWS personnel also maintain "second-order" stations. These stations record temperature, precipitation and many, but not necessarily all, of the other variables measured at first order stations. These stations provide unofficial climate data as well as some synoptic data. However, no official monthly LCD reports are provided for these stations.

Europe
Similarly, in Europe, there are multiple meteorological networks. The UK has also classified its stations in a tier system based on weather observations quality. The UK system is comparable to the first-order and second-order US system classification, but it is not officially classified in such a way. Many of these stations provide both climate and synoptic data. Stations reporting climate data are located at a combination of commercial airports, Royal Air Force bases and observatories.

In Germany, there are several networks, each one identified by the type of weather observations (for example, rainfall, climate, synoptic, etc).[6] The rainfall network contains up to 4,000 stations while the climate network is comprised of approximately 200 stations.

Japan and Australia
In Japan, there are two types of network, the AMEDAS (Automated Meteorological Data Acquisition System) and SYNOP.[7] Although there are more AMEDAS stations than SYNOP stations (more than 1,000 compared to less than 150), the weather market prefers using SYNOP stations because most of these stations are manned (about 120) and daily temperatures are obtained from continuous readings. Additional supplemental data is available through provincial offices, but the data quality can be questionable. Data reports are sometimes missing and the data is not strictly quality-controlled. Also, the stations may not provide continuous 24-hour reporting.

In Australia, both climate and synoptic data are observed at many of the same stations within the different meteorological networks.[8] The stations used by the weather market are mostly comprised of stations located at airports and manned by government officials.

The Climate Reference Network has been established in Australia to provide data for climate change research. These stations are currently of little use to the weather industry due to their remote locations and limited history. In addition, cooperative volunteers, usually residing in agricultural areas of the country, provide a major supplemental rainfall network of approximately 7,000 stations. This rainfall network may have quality problems, as it is not very well controlled.

Special networks
As well as climate and synoptic networks, NMSs sponsor an array of specialised observation networks. These networks serve the purpose of providing greater spatial coverage and/or providing specialised observations. These networks vary significantly in accuracy, reliability and availability. The US NWS operates a network of cooperative observers.[9] This network of approximately 12,000 active stations serves to provide a greater spatial coverage of meteorological observations. A subset of these stations with a few first- and second-order stations are part of the 1,221-station US Historical Climatology Network (HCN), a database compiled by NCDC and used in analysing US climate (Easterling *et al.*, 1999). Many of these stations record only the maximum and minimum temperatures and precipitation, usually on a daily basis. Data for these stations should be used with caution; there are several known problems with the cooperative networks. The observers are volunteers and not professionals, the equipment is basic, data is often missing or erroneous, real-time data availability is poor and the times at which observations are made may vary ("time of observation bias").

PANEL 1

SYNOPTIC VS CLIMATE DATA

One of the issues encountered by the weather market is the definition of climate and synoptic data. In order to ease confusion and set a standard, the Weather Risk Management Association (WRMA[10]) has defined synoptic data – as "data that are collected in real-time at various stations around the globe and provided through the Global Telecommunication System (GTS[11]). The data, minimum and maximum temperature and precipitation – are normally provided four times daily at 00.00 Greenwich Mean Time (GMT), 06.00 GMT, 12.00 GMT, and 18.00 GMT. There is usually a 12-hour minimum and 12-hour maximum temperature that is recorded, but the time which the 12-hour values represent depends on the local time in the country of measurement". Climate data is defined as, "data that are quality controlled by the respective NMSs where the data is collected. The 'climate data' are the 'official' station data of that country". In a review of these two types of data, the WRMA has recommended that climate data is the most appropriate data to be used for the weather derivatives industry.

Historical synoptic data tends to have a shorter record than climate data; the longest history of synoptic data only extends back to 1974. Synoptic records tend to be less accurate because all temperatures were rounded to whole degrees Celsius before 1982 and quality control procedures may not have been thorough. Synoptic data may also contain a large percentage of missing values.

Climate data is the official historical data for a country. This status dictates a dataset with a long period of record (typically extending back to at least 1961), substantial quality control, fewer missing values, greater resolution and a clearly defined definition

Differences between synoptic and climate data, Berlin Templehof, January 1, 1983 to December 31, 2000		
Differences in the range	No. of Tmin differences (%)	No. of Tmax differences (%)
>10°C	0.02	0.97
5.0 – 10°C	0.41	1.31
2.0 – 5.0°C	3.45	3.11
1.0 – 2.0°C	4.78	5.74
0.5 – 1.0°C	4.50	6.39
0.0 – 0.5°C	6.96	9.42
any difference	20.12	26.95

Reprinted from: Jewson, S., and D. Whitehead, 2001, "In Praise of Climate Data", *Environmental Finance*, 3:2, pp. 22–3

of the data. The climate of a given day, defined by the relevant NMS, will typically be based upon the calendar day (ie, the average of what the climate has been on that date in the past), whereas the synoptic observations are reported according to WMO schedules. WMO observation schedules are created to provide data for running forecast models, while climate data serves the purpose of providing climatology for a given location.

Differences between ongoing synoptic data and climate data are most apparent in the terms of availability, accuracy and reliability. By its nature, synoptic data is available within minutes to hours of an observation in order to provide data for forecasts, whereas climate data can be delayed by hours to months. These differences in delivery time are the result of varying levels of quality control (QC). Without substantial QC, a high percentage of missing, as well as possibly incorrect data values may be passed to the user. Synoptic data compromises accuracy for availability.

Due to differences in observation time, when analysing the actual time-series, the differences between synoptic data and climate data can be quite large. For example, the table illustrates these differences between the two data sets for Berlin Templehof from January 1, 1984–December 31, 2001 (synoptic data before 1984 was not available). German synoptic data is observed daily in two 12-hour blocks; maximum temperature is observed from 06.00 GMT to 18.00 GMT, while minimum temperature is observed from 18.00 GMT to 06.00 GMT. German climate data is observed over a period of midnight–midnight. The differences between synoptic data and climate data are greater than 1°C occurring approximately 5% of the time.

As the weather derivatives market continues to mature, NMSs are realising the need to provide climate data in a more timely manner. In order to meet this need many NMSs are now producing preliminary climate data from which the final official climate values are derived. Preliminary climate data is composed of observations that undergo a limited QC procedure, but are not thoroughly edited.

STATION HISTORY AND METADATA
Clear advantages exist for organisations that conduct a thorough station history search prior to cleaning and enhancing data. The advantages of acquiring a good quality station history are that it provides the analyst with precise knowledge of when potential discontinuities occurred and what the physical causes of the discontinuities are. Historical information gathered on stations is generally referred to as "metadata".

The availability of future (ongoing) metadata is also very critical for making informed decisions concerning the selection of stations for weather contracts. The historical pricing of a trade will be irrelevant if a station moves during the period of the contract and a sudden shift in daily observations occurs. For the most part, NMSs do not have systems in place to inform users of future station changes. This task requires knowledge of the organisation of the individual meteorological offices, a working relationship with individuals responsible for station changes, and the ability to overcome translation difficulties. Private vendors have compiled and are providing continuously updated information on station history to users.

Many discontinuities do not result from a change in the instrument or relocation of the instrument itself, but from physical changes to the instrument's surrounding environment. These changes can be as big as the construction of buildings nearby to the instrument or installation of a new parking lot, or as seemingly insignificant as the removal of grass from under the sensor. The reporting processes set up by the various NMSs do not specify these surrounding environment changes to be documented; this creates shortcomings in the metadata recording. The private sector has also developed products to overcome this shortcoming. This is accomplished by employing a rigorous metadata research approach including in-depth interviews with airport managers and local meteorological office personnel.

US metadata
The US currently has the most complete metadata in the world available from NCDC via the Internet; information concerning station movement (latitude, longitude and elevation), observers (NOAA, NWS, contractors, etc) and administrative issues are available.[12] However, NCDC metadata is not complete and significant metadata may be missing from the NCDC archives due to local offices not reporting all of the changes and errors inputting the data in the archive.[13]

International metadata
Official metadata is less available in other countries than it is in the US. Metadata is often available in limited quantities, in hard copy and difficult to access. For example, metadata is available for the UK but it is time consuming to access because governmental pre-authorisations are required. Although a modernisation programme is underway by the UK Met Office, currently accessing the full metadata in the UK requires visiting and searching of hard copy materials to one of three locations (depending upon the station in question): Bracknell, Edinburgh or Belfast, in order to obtain the information desired. A similar process is required for acquiring Dutch metadata. A limited amount of German metadata is available in electronic format, but a search of hard copy data is required to obtain all German metadata.

AVAILABILITY AND COST
The unavailability and cost of meteorological data was one of the greatest limiting factors in the global expansion of the weather derivatives industry. Although weather data in the US are inexpensive and very easy to access, many national meteorological services are poorly equipped to meet the needs of the industry and pricing was often set at levels that make the use of the data uneconomical for weather risk valuation.

As the importance of weather data beyond forecasting has become increasingly evident, NMSs have reacted in diverging ways. Some countries, such as the Netherlands,[14] have moved to upgrade systems and implement a free data policy, while countries such as Finland[15] have acted to assert greater control over their data by limiting access. Data usage rights vary from restrictive end-user agreements to free distribution of all products. Data costs can vary from insignificant amounts to thousands of Euros per station for historical data.

Data usage rights and data cost have become increasingly heated issues among the various NMSs. In order to handle these issues in Europe, ECOMET (an economic weather group comprised of representatives of 20 European member countries) was established in 1995.[16] ECOMET's stated objective is "to preserve the free and unrestricted exchange of meteorological information between the NMSs for their operational functions within the framework of WMO regulations and to ensure the widest availability of basic meteorological data and products for commercial applications". In theory, any ECOMET member should act as a one-stop shop for European meteorological data. In practice, this is simply not the case; globally, in order to obtain historical and ongoing climate data, it is necessary to contact each country individually or to work with a private sector entity to secure the data. Each country places different restrictions pertaining to data usage, redistribution and the cost of data. Table 1 illustrates current data availability for selected countries.

Data cleaning

Unfortunately, "raw" weather data obtained from NMSs may be missing or be incorrectly reported. The sources of missing weather data may be that the instrument was broken and the data was never recorded, that there was a break in the transmission of the weather data or that the weather data was recorded and archived, but subsequently lost. Errors in weather data frequently are caused by poorly calibrated instrumentation, but also may be caused by errors in recording the data or while digitising older hard-copy records. In either case, this data must be "cleaned" before accurate valuation analyses can be performed.

ERRORS IN WEATHER DATA REPORTS

Weather data is reported in various formats, METAR and SYNOPTIC, being two common formats.[17] The METAR format is the standard international format for

Table 1. Weather data availability for selected countries

Country	Meteorological Organisation	Historical Data (availability)	Ongoing Data (availability)	Metadata (availability)	Ease of Access
Australia	Australian, Bureau of Meteorology	Yes	Yes	Limited	Easy
Belgium	Royal Meteorological Institute of Belgium	Yes	Yes	Limited	Very difficult
Brazil	Instituto Nacional de Meteorologica (INMET)	Yes	Yes	Limited	Difficult
Canada	Environment Canada	Yes	Yes	Limited	Easy
Finland	Finnish Meteorological Institute	Yes	Yes	No	Very difficult
France	Météo-France	Yes	Yes	Limited	Difficult
Germany	Deutscher Wetterdienst	Yes	Yes	Limited	Moderate
Italy	Italian Air Force*	Yes	Limited	No	Very difficult
Japan	Japan Meteorological Agency	Yes	Yes	Yes	Moderate
Mexico	Servicio Meterológico Nacional	Yes	Yes	Limited	Difficult
Netherlands	Royal Netherlands Meteorological Institute	Yes	Yes	Limited	Moderate
Norway	Norwegian Meteorological Institute	Yes	Yes	Limited	Moderate
Spain	Instituto Nacional De Meteorologica (INM)	Yes	Limited	Limited	Moderate
United Kingdom	The Met Office	Yes	Yes	Limited	Moderate
United States	National Weather Service	Yes	Yes	Yes	Very easy

* The Italian Air Force is the recognised National Meteorological Service of Italy by the World Meteorological Organization (WMO)

WEATHER DATA: CLEANING AND ENHANCEMENT

reporting hourly values in metric units. Supplementary daily maximum temperature and minimum temperature values are reported at midnight Local Standard Time in the US, providing the earliest information to the weather derivatives market of the previous day values. SYNOPTIC format is the international standard format for reporting values in metric units at standard synoptic hours as discussed in the previous section. A typical reporting error would be to leave out a character in recording the data. Subsequently, all values would be misinterpreted.

QC procedures applied to the data itself will detect other types of errors. Errors include a daily minimum temperature greater than the daily maximum temperature and other unreasonable values in a given day, conflicting values from different sources and inconsistencies in consecutive day values.

The frequency, type and magnitude of the errors are dependent on the country, historical time period, network and type of weather data. For example, China and Japan have a rigorous QC system for their real-time synoptic reporting system and few errors are found, while data from other countries could contain missing and erroneous values. Error frequencies in the US and European countries are low, typically less than 1%, with a wide range from almost no errors to more than 2%, but some large errors do occur.

CLEANING OF WEATHER DATA

Weather data cleaning consists of two processes: the replacement of missing values and the replacement of erroneous values. These processes should be performed simultaneously to obtain the best result.

The replacement of one missing daily value is fairly easy. However, the problem becomes much more complicated if there are blocks of daily missing values. Such cases are not uncommon, particularly several decades ago. The problem of data cleaning then becomes a problem of replacing values by interpolations between observations across several stations (spatial interpolation) and interpolations between observations over time (temporal interpolation).

Spatial interpolation

There are many potential approaches to spatial interpolation of random meteorological inputs. Some of the first work in interpolation in the meteorological field was by Cressman (1959). Cressman's interpolation method corrected a grid-point value linearly combining the residual values between a predicted and an observed value. Barnes (1964) expanded on the work of Cressman, using a linear combination of the observations themselves to generate a "first-guess" field. This field was then used as the input to a second pass of the data. The Barnes approach is still used today in the creation of the objective analysis fields for inputs into numerical meteorological models. The Cressman and Barnes techniques are designed for interpolating a set of grid-point values from a random set of inputs, but can be adapted to interpolate any non-standard grid-point value and to incorporate additional variables, such as point-to-point correlations.

With the advent of Geographic Information System (GIS) technology, there has been an increase in the investigation of interpolation methods to create non-linear surfaces, from which any geo-referenced data point can be derived. This technology is applicable to the problem of replacing missing weather data because the spatial surface of many weather variables approximates non-linear surfaces.[18]

Theoretically, a functional fit, such as the Kriging method, is ideal for interpolation because the weights calculated are chosen such that the function derived provides optimal interpolation. The Kriging approach also has the advantage that it makes it possible to determine the distance at which a value no longer makes a useful contribution to the interpolation. Unfortunately, when the Kriging approach is employed, weather data often does not provide a recognisable optimal functional fit, especially for variables such as rainfall, where the point-to-distance correlation declines rapidly.

One of the simplest forms of interpolation when distance is the only consideration is an inverse-distance or inverse-distance squared weighted averaging. This method works well for fairly homogeneous data such as daily temperature when there is a fairly uniform distribution of data around the target location. The method does not work well for more complex solutions, such as daily rainfall.

Multiple linear regression approaches can also be applied when there is more than one input determining the shape of the data field. For example, such an approach is preferred to inverse-distance weighted methods when interpolating rainfall as a function of distance, elevation and month of the year because this approach often provides a better "fit" of the data and is easier to calculate.

New methods of spatial interpolation of maximum and minimum temperature values employ artificial neural network (ANN) techniques (Snell *et al.*, 2000). While these approaches show superior statistical results compared to simple techniques in "downscaling" gridded forecast temperature values to specific locations, they have not been thoroughly tested for the application of unequally spaced input data nor have they been thoroughly tested for weather variables other than temperature.

Temporal interpolation
Temporal interpolation methods using statistical probability distribution functions have been employed. Gullet *et al.* (1992) and Mekis and Hogg (1999) employ such techniques in rehabilitating daily Canadian rainfall time-series. Hierarchical polynomial regression techniques have also been employed for filling in missing temperature and wind data for agricultural simulation models (Acock and Pachepsky, 2000). Such methodologies may be satisfactory for historical cleaning of small gaps of missing data, but provide little value in real-time data cleaning required for the near-term settlement of weather derivatives. Spatial interpolation methods are thus preferred for real-time cleaning.

COMPLICATIONS IN WEATHER DATA CLEANING
The actual application of data cleaning techniques is complicated by the fact that countries report with different daily timing conventions. For example, in the US, a calendar day is midnight–midnight Local Standard Time, the Canadian day is 06.00–06.00 GMT and the UK reports on a 09.00–09.00 GMT day, with the minimum temperature reported actually being for the previous day. The reporting times are also sometimes different for different parameters. For example, the US reports daily maximum/minimum temperatures at midnight local time and reports 24-hour rainfall at 12.00 GMT. These inconsistencies in reporting require that specific data cleaning techniques be developed for each weather variable and each country.

Another complicating factor in cleaning weather data is that it can be unwise to assume that the closest points correlate higher than ones further away. Microclimate issues often produce inadequate spatial correlation results when using standard geographic spatial cleaning techniques. Optimal techniques, such as a modified Barnes technique, evaluate historical correlation and distances between stations when developing a spatial cleaning approach.

A further complicating factor is that each weather variable has its own unique spatial correlation characteristics. For example, rainfall correlation with distance is uniquely different to temperature or humidity's correlation with distance. The distance correlations may also change with season.

Given all of these requirements for interpolation of weather data, a weather data interpolation model should address the problem of estimating values at non-standard spaced points, using both distance and weather variable correlation to derive the weights.

WEATHER DATA: CLEANING AND ENHANCEMENT

Impact of instrumentation changes/relocations

Unbiased weather data is crucial in developing models for use in pricing weather risk trades. The private sector undertook projects to address discontinuities in historical temperature data in the US and other countries. Climatologists have long recognised that most climate datasets contain inhomogeneities or discontinuities introduced by non-climatic factors, such as changes of instrumentation, changes of station physical location, changes in surrounding, changes in operation procedure, human errors and changes of operators. Because these non-climate "signals" are "noises" for climate research, climatologists have long attempted to correct these inhomogeneities (Conrad and Pollak, 1950). The task is complicated by occasional malfunctions of instruments (a fan breaking causing rapid excessive warming, for example) or "drifts" out of calibration (for example, dirt and grime can cause a slow warming/cooling trend or a "calibration drift" of the instrument). In many cases, these incidents last for a few weeks or a few months until the next maintenance. These malfunctions are usually not critical but they create additional "noise" in the detection of discontinuities.

As an illustration, Figure 1 shows the annual HDD indexes for a hypothetical situation where a cooling discontinuity is assumed to occur in 1995, causing an upward shift in the observed HDDs prior to 1995. The bottom series is calculated from the cleaned data and the upper series is estimated after discontinuities have been accounted for; both the cleaned and enhanced means have been superimposed. The purpose of enhancing data is to remove the discontinuities and to adjust the weather series to present day observations. Estimates of the index mean and standard deviation will be biased if the discontinuities are not removed. As a result, the trending analysis will also be biased.

The impact of discontinuities on valuation of weather contracts can be large. For example, a 10-year burn rate for a November–March HDD weather contract can change by as much as 60 degree-days if a discontinuity of 0.8°F is present in the data in 1995 (0.8°F is the average magnitude of discontinuities found by Schrumpf and McKee, 1996, when analysing temperature differences between observations using the recently installed Automated Surface Observing Systems (ASOS) with the previous thermometers).[19] This bias can lead to a large spread in valuation assuming the typical nominal value to be US$5,000 per degree-day.

UNITED STATES
EarthSat and RMS found in a study of 200 first-order US stations (Clemmons and VanderMarck, 2000) that:

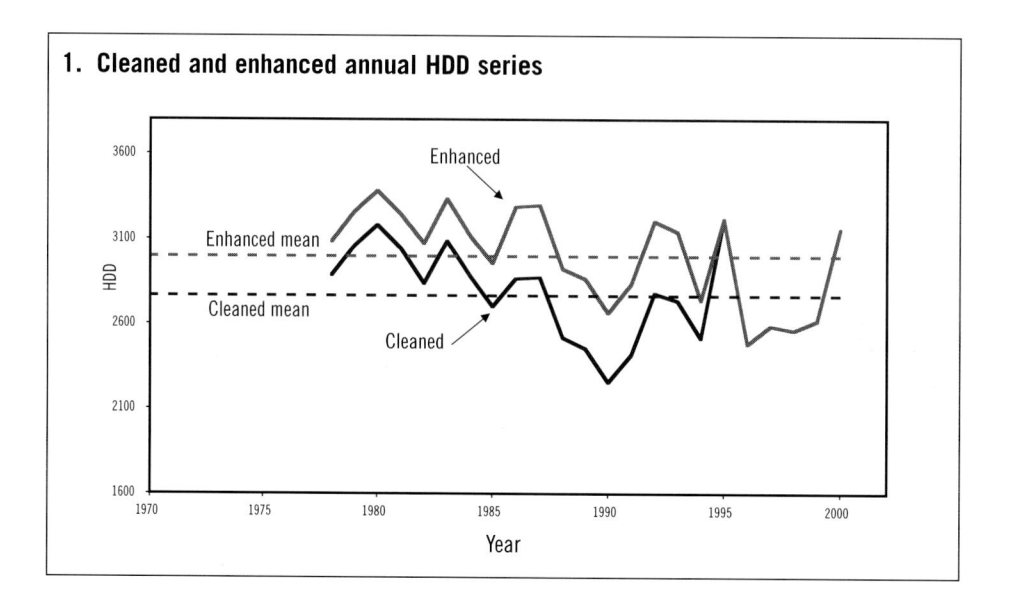

1. Cleaned and enhanced annual HDD series

1. almost all the stations have at least one discontinuity over the past 50 years;
2. more than 50% of these stations have at least one discontinuity over the past 10 years; and
3. some stations have discontinuities larger than 2.5°F in either the minimum or maximum temperature time-series.

These conclusions are not surprising upon reviewing the history of these stations. First order US stations experienced an average of 13 reported events since 1950 that were potential causes for discontinuities. Equally important is that the rates of such events did not decrease much over the past 10 years. An average of three reported events per station were found to be causes for potential discontinuities since 1992.

Several national changes of instrumentation occurred in the US over the past 40 years. The HO63 hygrothermometer temperature readings gradually replaced maximum/minimum thermometers as the official maximum/minimum recorder between the mid 1960s and the early 1980s, although the HO63s were often initially used only for recording hourly temperatures. A new generation of instruments, the HO83, installed between 1982 and 1986 is generally regarded to have a warm bias. McKee *et al.* (2000) reported that the HO83s had, on average, a warm bias of +0.57°F compared to a calibrated field standard.

A large programme took place for installing the ASOS type of instrumentation in the 1990s for approximately 850 stations (both first order NWS and second-order Airway stations). At many airports, coincident with the installation of ASOS, the instruments were generally relocated from areas near the Weather Service Office to the end of runway, generally in grassy areas. For example, the instrument in Charlotte, North Carolina was relocated in 1998 from an area near concrete between the Weather Service Office and Butler Aviation Terminal to a low grassy area about 500 feet perpendicular to the end of one of the main runways. An extreme discontinuity, greater than 2°F, was caused by this relocation, which is clearly seen in Figures 2 and 3, showing the monthly temperature difference series between Charlotte and two of its neighbouring stations.[20]

Studies have demonstrated that the installation of the ASOS and relocation of the instrumentation resulted in a cooling of the records in many locations (Schrumpf and McKee, 1996 and McKee *et al.*, 2000). Schrumpf and McKee report, in a sample study of 76 stations (not all first-order stations), that the average temperature difference between ASOS and HO83 instruments is a cooling of –0.79°F with a range of –2.56°F to +0.61°F.

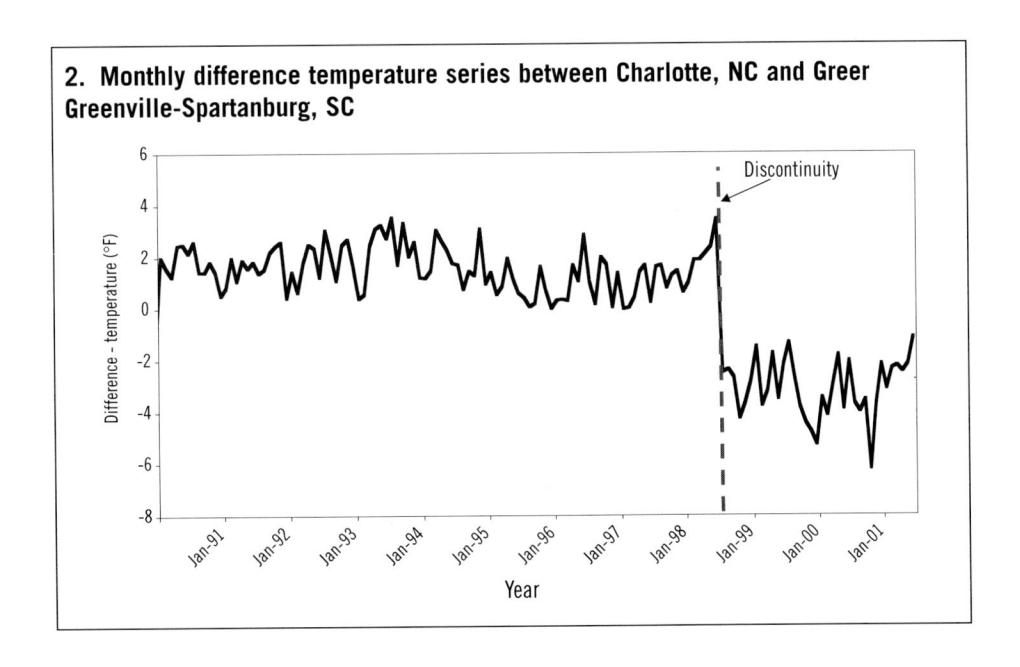

2. Monthly difference temperature series between Charlotte, NC and Greer Greenville-Spartanburg, SC

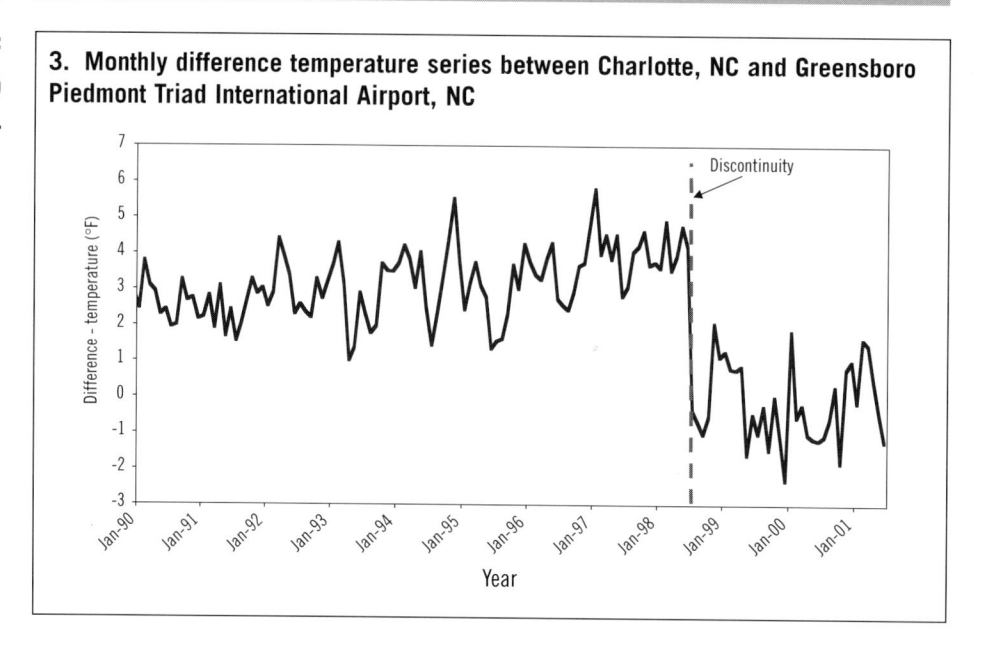

3. Monthly difference temperature series between Charlotte, NC and Greensboro Piedmont Triad International Airport, NC

A recent survey among ASOS managers on 200 first-order stations shows that, between February 2001 and February 2002, 10 stations experienced events that have potential for discontinuities (mostly station relocations due to airport construction and expansion) and that the rate of pending station changes over the next 12 months will be only slightly less than the rate of experienced events over the past 12 months.

OTHER COUNTRIES

Studies by international researchers (Peterson *et al.*, 1998 and Tuomenvirta, 1998) have also reported the existence of discontinuities in international weather data. Discontinuities are also caused by events such as instrument changes and station enclosure changes (for example, the recent installation of the semi-automatic meteorological observation stations, SAMOS in the UK).

In Japan, a review of metadata indicates that very few relocations occurred until the early 1990s, after which time the number of station moves has been increasing.

In Australia, the general policy of the Bureau of Meteorology is to build a new site, usually at an airport. Then the old site, usually a city site, is discontinued. New identification numbers are assigned at the new sites. If there is a need to provide a continuous record for the two sites, a comparison of rainfall and temperature history is performed in an attempt to remove any in homogeneity caused by the move. In addition, occasional instrumentation relocations at airports do occur, mainly to accommodate airport construction.

Methods for treating discontinuities

There is a potential that any changes in instrumentation, surroundings and relocation of instrumentation may create a serious discontinuity in the data series. An ideal treatment of weather data series containing discontinuities would be to reconstruct the historical observations as if the present instrumentation and surrounding stations conditions were uniform over the historical observation period. Such a task is very difficult because the impact of discontinuities on measured weather variables is "diluted" by the impact of climate changes. A good solution is to use the best features of several methods for identification and quantification of discontinuities in weather series.

Methods should be applied to each weather variable independently. For example, climatic and non-climatic factors affect the minimum and maximum temperature time-series differently. Three questions need to be answered when preparing to remove discontinuities from series:

1. What are the potential dates of discontinuity?
2. Did a discontinuity actually occur?
3. What is the magnitude of the discontinuity?

The first question may be answered by information available on the station (metadata) or by performing supplementary statistical tests on the series if the metadata is incomplete, as is the norm. The second question may be answered by a combination of subjective and objective methodologies. The third question is best answered with mathematical and statistical models.

POTENTIAL DATES OF DISCONTINUITIES
Metadata are invaluable in helping determine the date of discontinuities. Two general classes of events are reported in the metadata: events associated with a station move or instrumentation change (called "true" events) and those described as administrative changes. An example of an administrative change would be the renaming of an airport. Although administrative changes should not affect weather data, there are a limited number of cases where true events occurred coincidentally at these dates. Consequently, methods for identification and quantification of discontinuities first analyse data behaviour before and after true events. A follow-up analysis of data behaviour before and after administrative events is then conducted with stricter acceptance criteria than for true events.

In an ideal situation, true and administrative events reported in the metadata will explain all discontinuities. Unfortunately, this is not the case. Not all events are reported and tests need to be performed for dates other than those listed in the metadata (called "blind dates"). One approach is to perform tests at each month for identification of potential discontinuities using more stringent acceptance criteria than those used for true or administrative events.

DETECTING DISCONTINUITIES
Several papers summarise methods for treating discontinuities (for example, Vincent and Gullet, 1999, Mekis and Hogg, 1999, Alexandersson and Moberg, 1997, Peterson *et al.*, 1998 and Sneyers, 1990). These methods either try to detect and/or to assess the magnitude of the discontinuities by using only data from the station itself (single station) or by developing a reference series from neighbouring stations (multiple stations) thereby removing the effects of regional climate signals.

Detecting discontinuities using one station
The series itself can be used for detecting large discontinuities after processing the data. Techniques such as those described by Zurbenko *et al.* (1996) and Rhoades and Alinger (1993), where filters are applied to the series, or those reported by DeGaetano and Allen (1999), where quantiles of the distribution of the series are plotted over different periods of time, can be used to detect large discontinuities greater than 2°F. For example, Figure 4 shows a monthly temperature series for Charlotte, North Carolina, around 1998, the year in which a large discontinuity was detected. Although the discontinuity is not easily detected by visual inspection of the monthly series, it can be more easily seen on the superimposed 12-month moving average.

Detecting discontinuities using multiple stations
The preferred solution for detecting non-climatic signals in a climatological series (the "target") is to develop a second series (the "reference"). The reference is generally a weighted aggregate of time-series from neighbouring stations of the target station. The difference (temperature) or the ratio (precipitation) between the target and the reference series attempts to reduce, if not to eliminate, most of the climatological signals.

The use of a reference in the detection of discontinuities in a target assumes that

WEATHER DATA: CLEANING AND ENHANCEMENT

4. Monthly temperature series for Charlotte, NC

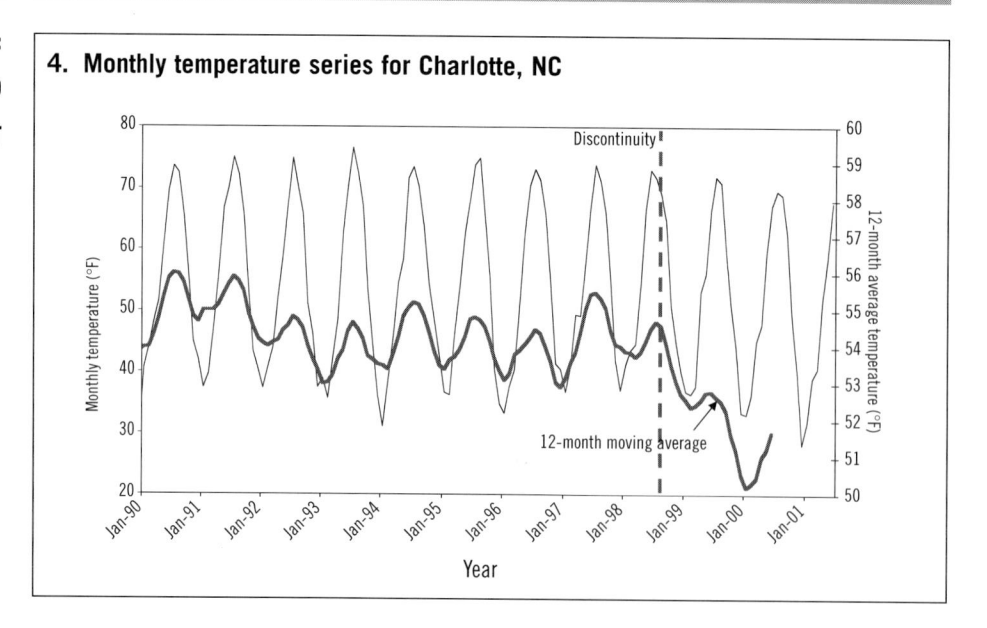

the reference series include the regional climate trends and fluctuations present in the data of the target, and that the reference series do not contain discontinuities themselves during the time period of analysis of the discontinuity in the target.

In practice, it is extremely rare to find neighbouring stations that have exactly the same regional and local climatological signals as the target station and, at the same time, do not contain any discontinuities. Enough stations are needed so that a major discontinuity in one of the neighbouring stations' time-series is not significant in the aggregated reference series and will not prevent the identification of discontinuities in the target station. However, if too many distant (or less correlated) neighbouring stations are used, the resulting reference may not reflect adequately the true climatic fluctuations of the target station.

A simple distance test is not sufficient because the closest stations might correlate poorly while further stations might be subject to similar climate conditions as the target station. Therefore, the first step is to select stations that are strongly correlated with the target station. However, two stations with a strong correlation may mean that the correlation is due to remote teleconnections. Consequently, the analyst also should use some distance function or regional climate check in the selection of the reference stations.

For the above reasons, references are developed by weighting values from a set of neighbouring stations. Each weight is a function of a distance and of a correlation factor.

Once the reference is developed, the major issues in detecting and quantifying discontinuities are:

❑ How can a signal be distinguished from the noise with confidence?
❑ What time periods should be tested before and after a suspected discontinuity?
❑ What if the apparent signal varies over time?
❑ What are the trends in the underlying data and how should they be addressed?
❑ What are the smallest values that can be detected for a discontinuity?
❑ What should be done to detect discontinuities that are close to each other in time?
❑ What should be done to handle discontinuities in the neighbouring stations?
❑ What should be done to address the variability in the shifts over different time periods?

METHODS FOR DETECTING DISCONTINUITIES AND ADJUSTING WEATHER DATA

The above issues are addressed by using both subjective and objective methods; subjective methods have a quantitative component but are characterised by a final judgment by an experienced meteorologist, whereas objective methods are fully automated procedures without human intervention.

Subjective methods
Subjective judgement by experienced meteorologists is important for modifying the weights given to various neighbouring stations based on many factors that cannot all be systematically programmed. For example, a visual review of a difference series between a target station and its neighbouring stations may give an indication of the quality of neighbouring stations' data and of the quality of metadata. This knowledge can lead to a subjective change in parameters of the tests to use in determining and quantifying discontinuities found with objective methods.

An advantage of subjective techniques is that they provide the meteorologist with simple visual tools for the quick identification of periods of time during which one or more discontinuities could exist. Acceptance or rejection of these discontinuities is decided after reviewing the results obtained from objective analyses. One of the first subjective techniques developed was the "double mass analysis" (Kohler, 1949). It plots cumulative difference temperature values over time between a target and its reference. For example, Figure 5 shows a difference time-series between Charlotte and one of its neighbouring stations and, superimposed, the cumulative differences. If the observed temperatures behaved similarly in both stations, then the cumulative difference between the two stations increases uniformly over time. Graphically, the cumulative temperature differences over time would be almost linear, which is observed up to the date of the discontinuity.

Graphical analysis of the residuals between observed values of the target series and predicted values of the same target series, obtained from linear regression analyses between the target and its references, is useful for detecting discontinuities because the graph of the residuals against time will show an abrupt change if such discontinuities exist. Also, a common assumption is that the residuals are independent random variables with a normal distribution of zero mean and constant standard deviation. The quantile–quantile (q–q) plot is one graphical technique for determining if the residual data sets come from a normal distribution.

Discontinuities can also be visually detected by plotting regression lines on the series before and after a given date. Figure 6 shows the difference series between Charlotte and one of its neighbouring stations and, superimposed, the two regression lines before and after the 1998 discontinuity and the regression line over both periods. Should there have been no discontinuities, the two regression lines would have theoretically been the same as the regression line over the entire series.

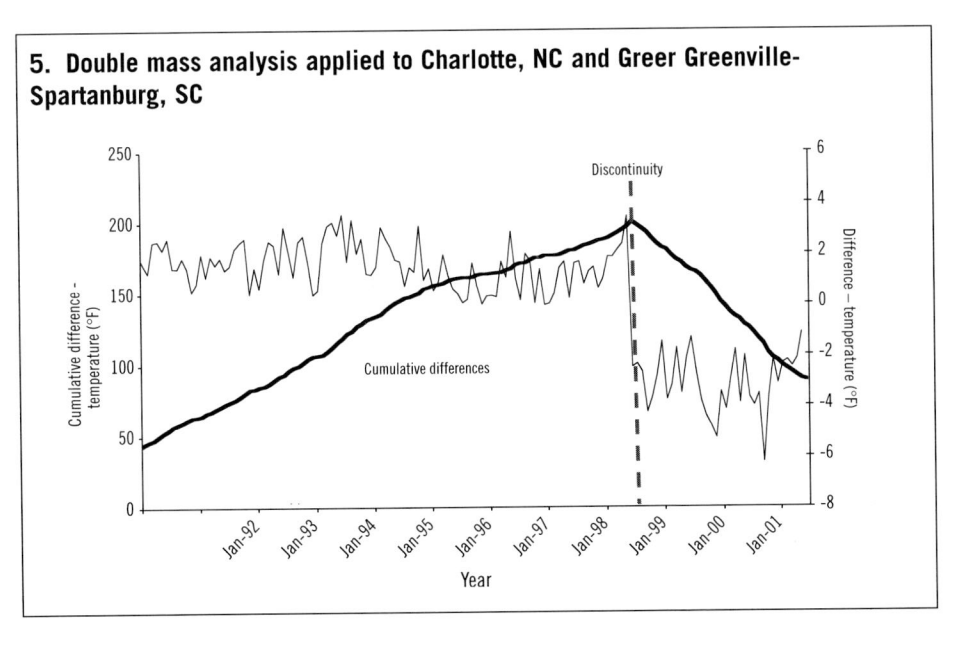

5. Double mass analysis applied to Charlotte, NC and Greer Greenville-Spartanburg, SC

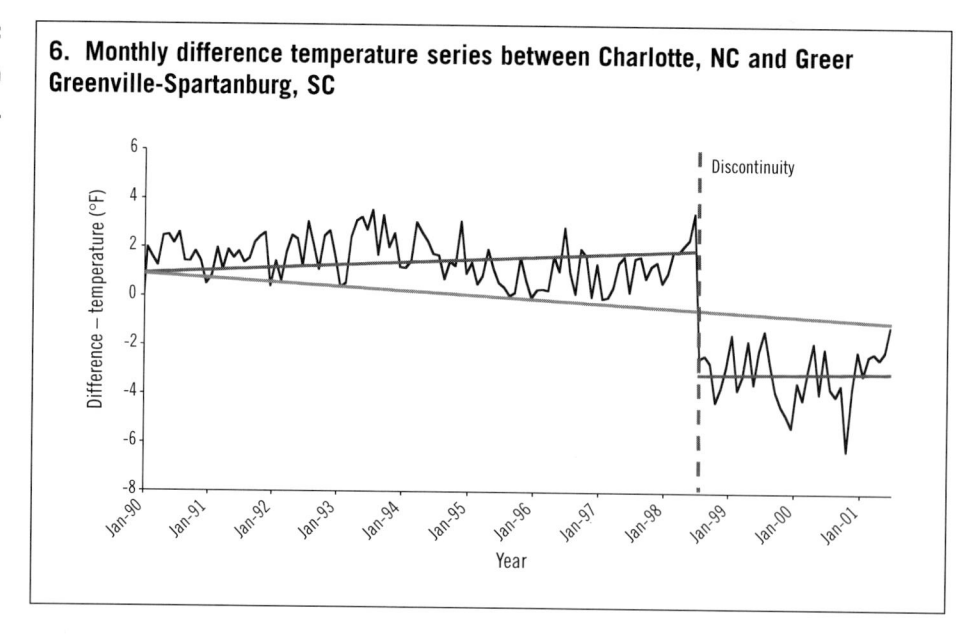

6. Monthly difference temperature series between Charlotte, NC and Greer Greenville-Spartanburg, SC

Visual inspection of these plots can usually identify with success large discontinuities (greater than 1.5°F) within a series. Visual inspection also provides "hints" for much smaller discontinuities, which will be further analysed using objective methods.

Objective methods
Most 'stand-alone' objective methods can detect single large discontinuities (1.5°F or larger) with more than 95% (99% for the best methods) accuracy in climatological series. The success rate decreases to about 50% for detecting discontinuities of the order of 0.5°F (Easterling and Peterson, 1995). It is therefore important not to rely solely upon objective methods when detecting discontinuities.

Typical objective methods are mathematical tools that detect the time and magnitude of discontinuities by testing the assumed statistical properties of the difference (or ratio) of assumed truly homogeneous target and reference series. Most of the statistical tests are formulated to automatically detect a discontinuity in the difference (or ratio) series at a given date because the parameters of the distribution of the series (either in the time or frequency domains) have significantly changed statistically before and after this date. Tests are based on the null hypothesis that the series is homogeneous (for example no changes in the means of the difference series before and after a date) and on the alternative hypothesis that the series becomes non-homogeneous at a given date. These types of tests are used in methods developed by Potter (1981), Karl and Williams (1987), Alexandersson (1986) and DeGaetano and Allen (1999).

Another common group of tests using the same approaches have developed statistical significance tests based on either the residuals of linear regression analyses, the parameters of the regressions or the autocorrelation of the residuals (Bois, 1970, Vincent, 1998, Solow, 1987, Easterling and Peterson, 1995).

Other tests use test statistics based on the significance of the rate of changes of the filtered difference (or ratio) series (for example, Zurbenko *et al.*, 1996, Rhoades and Alinger, 1993) and those based on non-parametric tests (Pettitt, 1979 and Sneyers, 1990).

More recent methods (Caussinus and Mestre, 1996 and Szentimrey, 2001) simultaneously generate reference series and detect discontinuities. These methods first identify multiple dates where discontinuities are possible for a set of series in the same climate region. Each of these series can then be used as a homogeneous reference to the other series for the time intervals where no discontinuities are found. When a discontinuity in a series is consistently found with the references derived from the other series, the discontinuity is attributed to this series.

Some discontinuities can be easily detected because they produce a large permanent change in the difference (or ratio) series but other apparent discontinuities are short-term shifts resulting from either changing microclimate conditions or shifts resulting from temporary instrument malfunctions. For these reasons, series need to be analysed over a sufficient time period before and after a potential discontinuity to ascertain a real discontinuity. Typically, several years are required for reliable quantification of discontinuities. However, it is possible to detect and to make initial magnitude estimates of discontinuities within 15 months of the discontinuity date, and even earlier if large discontinuities are present and reviewed by trained meteorologists.

Ideally, no discontinuities should be present in neighbouring stations around the time where a discontinuity has been found in the target stations. If this happens, discontinuities in the target will not be detected or will be falsely detected. Also, true discontinuities will be "smoothed" across several stations. These problems can be prevented by analysing difference (or ratio) series of the target stations with individual neighbouring stations and by estimating the magnitude of discontinuities without using these neighbouring stations.

Because it is important to detect with a high rate of success all discontinuities for pricing a weather transaction, any homogenisation or enhancement methodology must rely on a combination of both subjective and objective methods. An optimal method is to select a sample of found discontinuities of all magnitudes corroborated with metadata and feedback from station operators. This sample can then be used for developing statistical criteria for objective methods and criteria for best applying subjective methods. This process is time-consuming and departs from usual homogenisation methods employed by meteorologists studying climate data but it is needed for enhancing weather data used in valuation of weather transaction.

Quantification of discontinuities
Once the dates of discontinuities have been identified, it would be ideal if "laboratory" experiments were available for assessing the magnitude (shift) of these discontinuities. For example, comparison of overlapping time-series between the old and new instrumentation could detect discontinuities when an instrument change or relocation occurs. Unfortunately, these comparisons are only performed at a very limited number of stations and usually over a very short period of time.

The magnitude of the discontinuity is generally calculated as the differences of the means of the difference (or ratio) series before and after the dates at which discontinuities have been identified. The means are estimated over several years because the difference or ratio series are not really stationary in time (impacts of residuals of climatic and non-climatic signals still corrupt the series). Effects of discontinuities on temperature series might be season-dependent. However, historical detailed meteorological conditions at any time and the physical reasons for the existence of discontinuities are often unknown. Thus, it is often best to apply a constant shift through the year at the risk of creating biases otherwise.

After the discontinuities have been applied, it is appropriate to rerun the methodology to see if any discontinuities may have not been totally resolved in the first iteration of the methodology.

VALIDATION OF METHODOLOGIES FOR DETECTING AND QUANTIFYING DISCONTINUITIES

An excellent way to test the validation of the methods is to perform pair-wise tests, using the same methodology used in detecting discontinuities to systematically compare the target station with several of its neighbouring stations. Both subjective and objective tests need to be performed individually on the target series and the most correlated neighbouring stations to insure the discontinuities are true and that their magnitudes are not corrupted. When possible, interviews with station operators should be conducted for verification of the discontinuities, especially within the past 10 years. It may, however, be less possible to get such feedback for discontinuities older than 10 years.

Finally, homogenisation of any given target station should not drastically change the regional climatology. A rank test is usually performed on the monthly station climatology to make certain the station climatology is not being changed. If the station climatology is being changed, eg, if new monthly temperatures extremes are being created by the homogenisation, extra tests are then needed for validation.

Conclusion

A robust weather derivatives market requires timely, inexpensive, high quality weather data. The main provider of data for each country is the NMS, who has the ultimate responsibility for providing the weather data. Weather data are generally classified as either "climate" or "synoptic". Few countries, such as Australia, do not distinguish between climate data and synoptic data. In most countries, however, there is a significant distinction between these two data sets. Climate data is the official data provided by the NMSs. Synoptic data is the data provided at regular reporting times for input into global weather models.

Weather data provided by the NMSs often has missing values and errors, and is not adjusted for changes in station locations during the period of record. The private weather industry plays a vital role in the weather risk market by providing methodologies for cleaning and enhancing weather data. Weather data cleaning is performed to remove missing values and to replace erroneous values. Weather data enhancement is performed to adjust the weather data time-series for changes in station location.

There are many methods for data cleaning and data enhancement. Data cleaning is achieved by performing a series of quality control checks on the data to identify missing values and to "flag" suspect values. These suspect or missing values can then be replaced using spatial or temporal interpolation techniques. Either method may be used for historical data cleaning but spatial interpolation techniques are preferred for real-time weather data cleaning because they do not require data before and after the replacement date for the interpolation. There are several methods of spatial data interpolation. The model that should solve the problem of estimating non-standard spacing of points would use both distance and point-to-point correlations to derive the interpolation weights.

Data enhancement is performed to determine when discontinuities occurred and to determine the magnitude of the discontinuities. Metadata is very important in evaluating the potential dates of discontinuities. Either single station methodologies or multiple station methodologies may be employed to enhance the time-series. Multiple station methodologies using a reference time-series are often preferred because they remove the effect of regional climatic trends. Validation of any methodology should be conducted using pair-wise tests and operator interviews.

A new technology area yet to be fully explored by the weather derivative industry for weather data is remote sensing technologies. Current radar and satellite technologies can diagnose surface temperature using infrared sensors and rainfall using microwave sensors after proper calibrations (Brown *et al.*, 2001). This could be a solution for areas without instrumentation or where there are reliability issues in the data or moral hazard problems.

1 *The WMO's website can be found at: http://www.wmo.ch/index-en.html.*
2 *The UKMO's website con be found at: http://www.met-office.gov.uk.*

3 The NCDC's website can be found at: http://lwf.ncdc.noaa.gov/oa/ncdc.html. Weather data can also be obtained from Regional Climate Centers, see http://www.noaa.gov/regions.html.
4 The NOAA's website can be found at: http://www.noaa.gov.
5 The NWS's website can be found at: http://www.nws.noaa.gov.
6 More information can be found on the Deutscher Wetterdienst website. URL: http://www.dwd.de.
7 More information can be found on the Japan Meteorological Agency's webiste. URL: http://www.kishou.go.jp/english/index.html.
8 More information can be found on the Bureau of Meteorology website. URL: http://www.bom.gov.au.
9 http://www.nws.noaa.gov/om/coop/what-is-coop.html.
10 The WRMA website can be found at: http://www.wrma.org.
11 Information on Global Telecommunication Systems can be found on the WMO's website. URL: http://www.wmo.ch/web/www/TEM/gts.html.
12 http://lwf.ncdc.noaa.gov/oa/climate/stationlocator.html.
13 For example, when Earth Satellite Corporation (EarthSat) and Risk Management Solutions (RMS) expanded the set of metadata entries currently available in NCDC website log, this research yielded 60% more entries than are available in the NCDC website logs.
14 More information can be found on the KNMI website. URL: http://www.knmi.nl.
15 More information can be found on the Finnish Meteorological Institute website. URL: http://www.fmi.fi.
16 The ECOMET website can be found at: http://www.meteo.be/ECOMET.
17 For example, the National Weather Service METAR/TAF information can be accessed at: http://205.156.54.206/oso/oso1/oso12/metar.htm. European METAR information can be accessed at: http://blinder.lfv.se/met/metar.europe.htm.
18 Davis (1973) performed some of the original work in this area, but his application was directed at geological formations. Burroughs et al. (1998) was one of the original textbooks written, in which the use of interpolation methods was discussed in a GIS format. Burroughs discusses some of the key methods of "optimal" interpolation, including the Kriging method attributed to the original work of mathematicians Krige (1966) and Matheron (1967).
19 As part of a recent modernisation programme, NWS has been installing a standardised instrument package called the Automated Surface Observing Systems (ASOS). Information on these weather stations is available at: http://www.nws.gov/asos/.
20 Note that similar comparisons were made with other neighboring stations confirming the discontinuity belongs to Charlotte.

BIBLIOGRAPHY

Acock, M. C., and Y. A. Pachepsky, 2000, "Estimating Missing Weather Data for Agricultural Simulations using Group Method of Data Handling", *Journal of Applied Meteorology*, 7, pp.1176–84.

Alexandersson, H., 1986, "A Homogeneity Test Applied to Precipitation Data", *International Journal of Climatology*, 6, pp. 661–75.

Alexandersson, H., and A. Moberg, 1997, "Homogenization of Swedish Temperature Data. Part I: Homogeneity Test for Linear Trends", *International Journal of Climatology*, 17, pp. 25–34.

Barnes, S. L., 1964, "A Technique for Maximizing Details in Numerical Weather Map Analysis", *Journal of Applied Meteorology*, 3, pp. 396–409.

Bois, P., 1970, Une methode de controle de series chronologiques utilisees en climatologie et en hydrologie, Laboratory of Fluid Mechanics, University of Grenoble I, p. 49.

Brown, P. E., P. J. Diggle, M. E. Lord and P. C. Young, 2001, "Space-Time Calibration of Radar-Rainfall Data", *Journal of the Royal Statistical Society*, 50, pp. 221–42.

Burroughs, P. A., and R. A. McDonnell, 1998, *Principles of Geographical Information Systems for Land Resource Assessment (Spatial Information Systems and Geostatistics)*, Oxford Press.

Caussinus, H., and O. Mestre, 1996, "New Mathematical Tools and Methodologies for Relative Homogeneity Testing", Proceedings of the First Seminar for Homogenization of Surface Climatological Data, Budapest, pp. 63–82.

Clemmons, L. and P. VanderMarck, 2000, "Raining Stats and Logs", Future and Options World, December 2000, pp. 34–7.

Conrad, V., and C. Pollak, 1950, *Methods in Climatology*, Second Edition, (Harvard University Press) p. 459.

Cressman, G. P., 1959, "An Operational Objective Analysis System", *Mondial Weather Review*, 87, pp. 367–74.

Davis, J. C., 1973, *Statistics and Data Analysis in Geology*, (John Wiley & Sons).

DeGaetano, A. T., and R. J. Allen, 1999, "Single-Station Test to Adjust in Homogeneities in Daily Extreme Temperature Series", Eleventh Conference on Applied Climatology, Dallas, Texas, American Meteorological Society, Boston, pp. 193–5.

Easterling, D. R., and T. C., Peterson, 1995, "A New Method for Detecting Undocumented Discontinuities in Climatological Time Series", *International Journal of Climatology*, 15, pp. 369–77.

Easterling, D. R., T. R. Katl, J. H. Lawrimore and S. A. Del Greco, 1999, "United States Historical Climatology Network Daily Temperature, Precipitation, and Snow Data for 1871–1997", ORNL/CDIAC-118, NDP-070, Carbon Dioxide Information Analysis Center, Oak Ridge National Laboratory, Oak Ridge, Tennessee.

Gullett, D. W., W. R. Skinner, and L. Vincent, 1992, "Development of a Historical Canadian Climate Database for Temperature and Other Climate Elements", *Climatological Bulletin*, 26, pp. 125–31.

Karl, T. R., and C. N. Williams, 1987, "An Approach to Adjusting Climatological Time Series for Discontinuous Inhomogeneities", *Journal of Climate Applications*, Meteorology, 26, pp. 1744–63.

Krige, D. G., 1966, "Two-Dimensional Weighted Moving Average Trend Surfaces for Ore Evaluation", *Journal of the South African Institute of Mining and Metallurgy*, 67, pp. 13–79.

Kohler, M. A., 1949, "Double Mass Analysis for Testing the Consistency of Records and for Making Adjustments", *Bulletin of the American Meteorological Society*, 30, pp. 188–9.

Matheron G., 1967, "Kriging, or Polynomial Interpolation Procedures?" *Canadian Mining and Metallurgy Bulletin*, 70, pp. 240–4.

McKee, T. B., N. J. Doesken, C. A. Davey and R. P. Pielke Snr., 2000, "Climate Data Continuity with ASOS – Report for Period April 1996–June 2000", Climatology Report No.00:3, Colorado Climate Center, Colorado State University.

Mekis, E., and W. D. Hogg, 1999, "Rehabilitation and Analysis of Canadian Daily Precipitation Time Series", *Atmosphere–Ocean*, 37(1), pp. 53–85.

Peterson, T. C., D. R. Easterling, T. R. Karl, P. Groisman, N. Nicholls, N. Plummer, S. Torok, I. Auer, R. Boehm, D. Gullett, L. Vincent, R. Heino, H. Tuomenvirta, O. Mestre, T. Szentimrey, J. Salinger, E. J. Forland, I. Hanssen-Bauer, H. Alexandersson, P. Jones and D. Parker, 1998, "Homogeneity Adjustments of in situ Atmospheric Climate Data: A Review", *International Journal of Climatology*, 18, pp. 1493–517.

Pettitt, A. N., 1979, "A Non-Parametric Approach to the Change-Point Problem", *Journal of Applicational Statistics*, 28, pp. 126–35.

Potter, K. W, 1981, "Illustration of a New Test for Detecting a Shift in Mean in Precipitation Series" *Mondial Weather Review*, 109, pp. 2040–5.

Rhoades, D. A., and M. J. Salinger, 1993, "Adjustment of Temperature and Rainfall Records for Site Changes", *International Journal of Climatology*, 13, pp. 899–913.

Schrumpf, A. D., and T. B. McKee, 1996, "Temperature Data Continuity with the Automated Surface Observing System", *Atmospheric Science Paper*, 616, Department of Atmospheric Science, Colorado Sate University.

Snell, S. E, S, Gopal and R. K. Kaukman, 2000, "Spatial Interpolation of Surface Air Temperatures Using Artificial Neural Networks: Evaluating their Use in Downscaling GCM's", *Journal of Climate*, 5, pp. 886–95.

Sneyers, R., 1990, "On the Statistical Analysis of Series of Observations", Technical Note, 143, World Meteorological Organization, Geneva, Switzerland.

Solow, A., 1987, "Testing for Climatic Change: An Application of the Two-Phase Regression Model" *Journal of Climate Application and Meteorology*, 26, pp. 1401–05.

Szentimrey, T., 2001, *Multiple Analysis of Series for Homogenization (MASH v 2.0)*, (Budapest: Hungarian Meteorological Service), p.38.

Tuomenvirta, H, 1998, "The Influence of Adjustments on Climatological Time Series", Proceedings of the Second Seminar for Homogenization of Surface Climatological Data, Budapest, pp.73–86.

Vincent, L. A., 1998, "A Technique for the Identification of Inhomogeneities in Canadian Temperature Series", *Journal of Climate*, 11, pp. 1094–104.

Vincent, L. A., and D. W. Gullett, 1999, "Canadian Historical and Homogeneous Temperature Datasets for Climate Change Analyses", *International Journal of Climatology*, 19, pp. 1375–88.

Zurbenko, I., Porter, P. S., Rao, S. T., Ku, J. Y., Gui, R. and R. E. Eskridge, 1996, "Detecting Discontinuities in Time Series of Upper Air Data: Development and Demonstration of an Adaptive Filter Technique", *Journal of Climate*, 9, pp. 3548–3560.

CLIMATE FORECASTS, MANAGING VARIATIONS, AND DERIVATIVE PRICES

6

The Nature of Climate Uncertainty and Considerations for Weather Risk Managers

Mark Gibbas[1]

Applied Insurance Research, Inc.

Throughout the ages, weather and climate have exerted an influence, both helpful and harmful, on nearly everything mankind has ever done. Weather's influence is felt on a range of scales from infrequent catastrophes to fairly frequent episodes of adverse conditions that affect profits. Furthermore, specific weather regimes can benefit some, while causing harm to others; cold snowy winters are a boon for utilities and ski areas, while at the same time a financial burden to municipalities and airlines; the wrath of a hurricane can bring massive destruction, unsettling the lives of many, while at the same time improving the profits of construction companies and building supply stores. Weather is dynamic and seemingly in a constant state of change and as it evolves around us; it is a force that is simultaneously both friend and foe.

Indeed it is this very nature that has lead to the development of trading weather risk and the emergence of the weather risk market. Through better understanding of climate and its variability, many organisations can manage the impacts of adverse weather on their business and plan to maximise the benefits of favourable weather with lead-times on the scale of months and seasons.

Introduction

From droughts to floods, normal to extreme and back again, why is it that weather is so dynamic and so varied? This chapter explores this question for the benefit of individuals from various disciplines who now find themselves in the midst of the weather risk market. Starting with a panel section on the basic mechanics of weather, this chapter expands on basic weather fundamentals to discuss climate variability. Important drivers of season climate variation (such as El Niño) are discussed in detail, including a description of the basic mechanics of each phenomenon, how they are observed, and how climate can be predicted with respect to the phenomenon. Finally a section on managing climate variability is provided which discusses important issues for using historical weather data to estimate future climate and the use of climate forecasts.

Climate variability

Over the course of the last two decades, extensive scientific research has led to the discovery of a number of large-scale oceanic and atmospheric phenomena that

THE NATURE OF CLIMATE UNCERTAINTY AND CONSIDERATIONS FOR WEATHER RISK MANAGERS

PANEL 1

AIR DENSITY, PRESSURE AND WINDS

Weather vs climate

Weather has been defined as the instantaneous state of the atmosphere. Examples of weather events are a sunny day, a snowstorm and a hurricane. Climate on the other hand has been defined as the statistics of weather. Examples of climate are monthly mean temperature and seasonally averaged precipitation. In the not so distant past there existed a fairly solid boundary between meteorologists who studied weather and climatologists who studied climate. Furthermore, while weather was considered a dynamic subject, climate on the other hand was considered to be generally static with changes occurring so slowly as to not be noticed. In the last two decades this perspective has changed. Climate is no longer viewed as virtually static; the scientific community currently acknowledges that climate does change and it does so on timescales ranging from seasons to millennia. Furthermore, with recent advances in the understanding of how the Earth's climate system works and identification of phenomenon such as El Niño and La Niña (see explanation later), which are now known to influence short-term climate (timescales of months and seasons), climatology is now seen as a dynamic science similar to that of meteorology.

Weather and climate: background

Avoiding the more advanced physics and mathematics, the basic mechanics of weather can be understood by grasping two key concepts:

1. air expands as it is heated; and
2. locations around the world heat air differently.

The atmosphere is essentially an ocean of air, an invisible gaseous fluid. The density of air is governed by gravity and by heat energy. Panel *a* of Figure *a*, depicts the relationship of air density under the influence of gravity. In this case, gravity is pulling the air down such that the air closest to the ground is denser bearing the weight of all

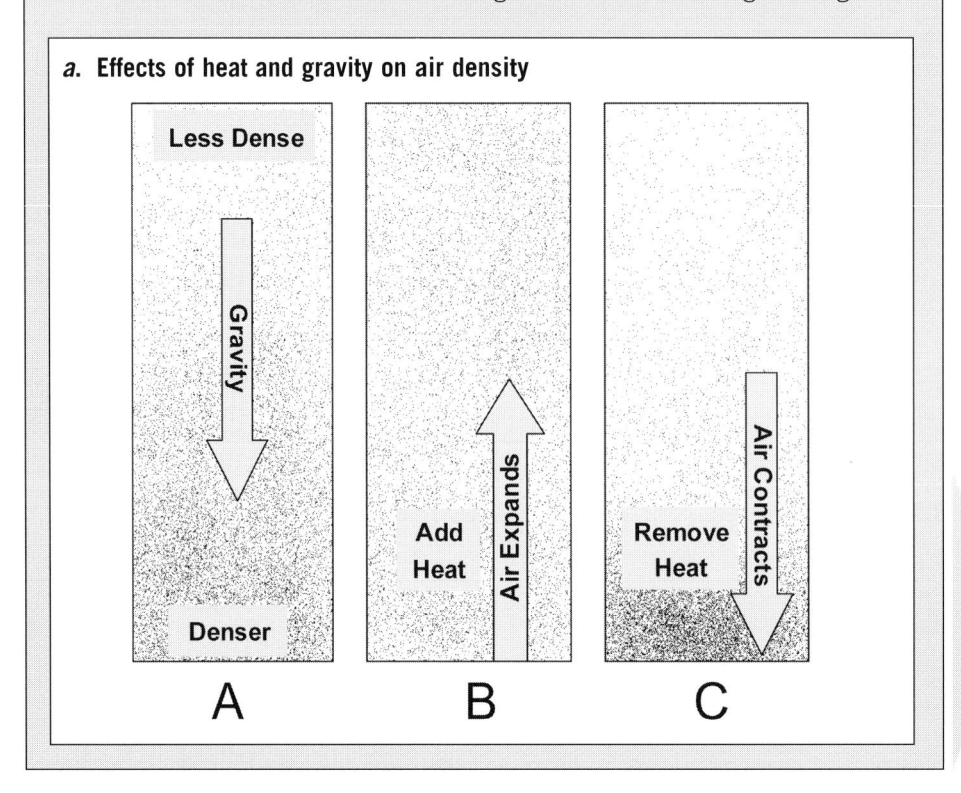

a. **Effects of heat and gravity on air density**

A: Less Dense / Gravity / Denser
B: Add Heat / Air Expands
C: Remove Heat / Air Contracts

RISK BOOKS

b. The relationship between air density and air pressure

Light (Low) Pressure | Heavy (High) Pressure

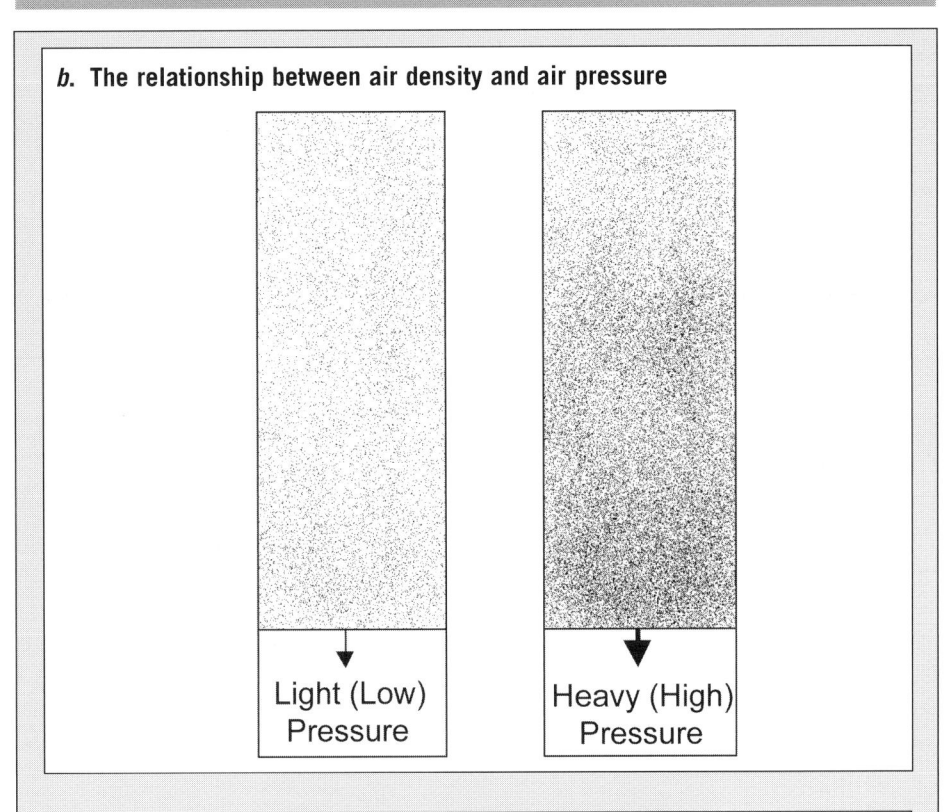

c. Flow of air from high-pressure to low-pressure areas

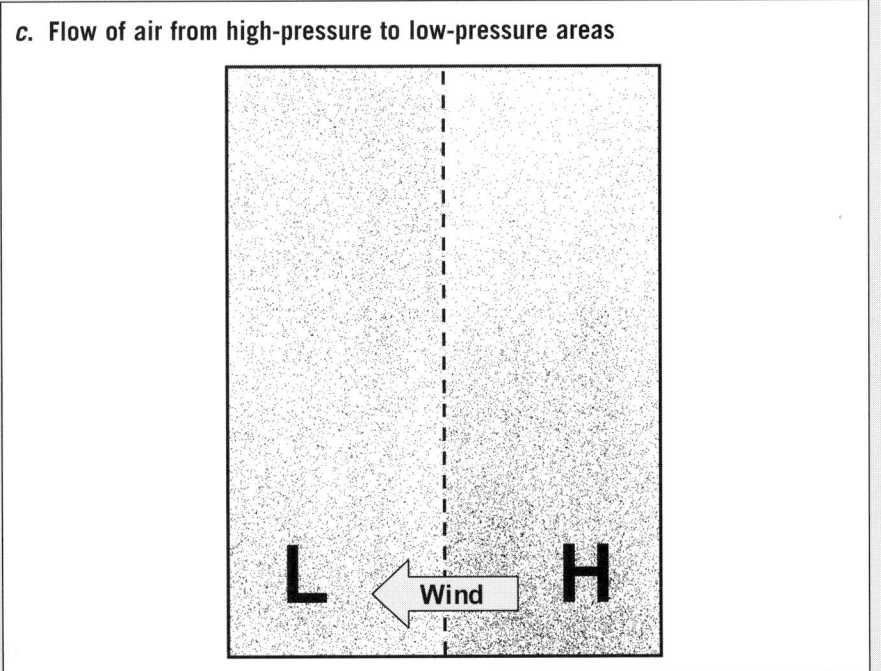

L ← Wind H

the air above, and air higher up is less dense. Add energy to air and it expands, becoming less dense (Figure *a*, Panel B). Remove energy and it contracts becoming denser (Figure *a*, Panel C).

Within the field of meteorology, typically air pressure is measured rather than air density. Figure *b* shows the relationship between air density and air pressure. In the left hand panel the air is less dense and does not have much mass under the influence of gravity to create heavy or high pressure. In contrast, the panel on the right has denser air, which means there is more mass under the influence of gravity to produce high pressure. Dense air and mass from high-pressure areas flows into less dense low-pressure areas (Figure *c*), causing wind. The interface between these regions of different air pressures is called a "pressure gradient".

reoccur periodically to affect climate at different timescales over various parts of the world. Such phenomena are referred to as "modes of climate variability". Further research has shown that for some modes, forecast methodologies can be developed to produce skilful forecasts of short-term climate variation on the scale of months and seasons.

In this section we will review some of the most significant modes of climate variability. We will assume a north-hemispheric perspective and thus, when the seasonal terms such as summer or winter are used, we will be referring to the *boreal* or northern hemisphere seasons. The qualifier "austral" will be used when referring to southern hemisphere seasons.[2]

EL NIÑO–SOUTHERN OSCILLATION
The El Niño–Southern Oscillation (ENSO) is a term used to describe an oscillation in the state of the ocean–atmosphere system in the tropical Pacific Ocean. The El Niño part of the name refers to the warming of the sea surface temperature (SST) along the equatorial Pacific. The Southern Oscillation part of the name refers to an oscillation in the difference between the sea-level air pressure (SLP) at Darwin, Australia and the Island of Tahiti in the South Pacific.[3] El Niño and the Southern Oscillation are closely related as the air pressure oscillation occurs as a result of the air reacting to the change in SST. The positive (or warm) ENSO phase is referred to as El Niño and the negative (or cold) phase is referred to as La Niña.[4] In recent years the more scientific term, ENSO has become more popular in referring to the oscillation in SST in the equatorial Pacific.

ENSO observations
There are a number of oceanic and atmospheric indexes, based on observations of SST and SLP in the tropical Pacific region, for monitoring ENSO. The Southern Oscillation Index (SOI) is one index that has been used since the earliest ENSO studies. El Niño corresponds to a prolonged negative SOI and La Niña corresponds to a prolonged positive SOI. One undesirable feature of the SOI is that it is very sensitive to small-scale atmospheric phenomena. Because of this, the scientific community has developed a set of SST-based indexes for five key regions in the Pacific Ocean that in total better represent the ENSO phenomenon. The Niño indexes include Niño-1, Niño-2, Niño-3, Niño-4 and Niño-3.4 regions. Each of these indexes refers to the average SST anomaly in a defined region of the equatorial Pacific. The index values for Niño regions 1 and 2 are often combined into a single number and referred to as Niño-1+2. Niño-3.4 is a defined region that includes parts of regions 3 and 4. A map of the regions is provided in Figure 1.

Another useful index for studying or monitoring ENSO activity is the Multivariate ENSO Index (MEI), which represents a composite view of ENSO based on sea-level pressure, surface winds, SST, surface air temperature and per cent cloudiness.

ENSO dynamics
Of all of the identified modes of climate variability, the ENSO is the most well understood to date.

The primary mechanism for development of El Niño or La Niña conditions in the tropical Pacific is related to the strength of the Trade Winds that blow east to west across the Pacific Ocean along the equator.

La Niña conditions are created when the Trade Winds in the tropical Pacific are stronger than normal (see Figure 2). In this case, the force of the wind pushes the surface water across the Pacific to the West to effectively create a pile of warm water in the western Pacific.[5] As the surface water is pushed westward away from the coast of South America, cold water rises up in a process called "up-welling" to fill the space left by the departing sun-warmed water. With the sea surface waters now colder in the eastern tropical Pacific and warmer in the western tropical Pacific, the Trade

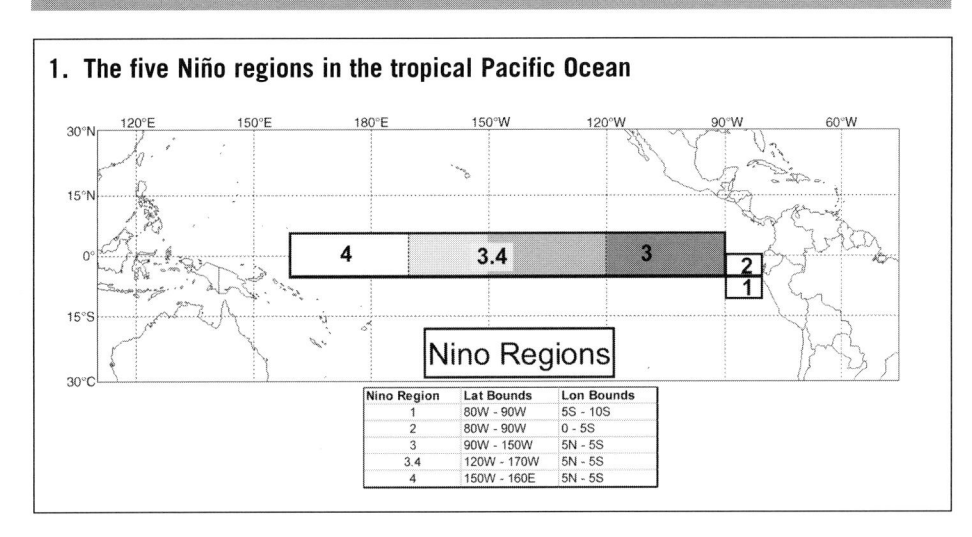

THE NATURE OF CLIMATE UNCERTAINTY AND CONSIDERATIONS FOR WEATHER RISK MANAGERS

1. The five Niño regions in the tropical Pacific Ocean

Nino Region	Lat Bounds	Lon Bounds
1	80W - 90W	5S - 10S
2	80W - 90W	0 - 5S
3	90W - 150W	5N - 5S
3.4	120W - 170W	5N - 5S
4	150W - 160E	5N - 5S

Winds respond to this temperature difference by strengthening further. In this way a positive ocean–atmosphere feedback is established, which serves to maintain the La Niña conditions. Analysis of the historical records of SST indicates that La Niña events generally last between one and two years, however there have been a few events that have lasted for three and four years.

El Niño conditions on the other hand are created when the tropical Pacific Trade Winds are weak. In this case the constant push from the Trade Winds relaxes and piled-up warm water in the western Pacific is allowed to flow back across the ocean along the equator. In response, the up-welling of cold water off the coasts of Peru and Ecuador is shut off, which allows the surface waters in the Eastern Pacific to warm further in the tropical heat. This creates a SST pattern that serves to further weaken the Trade Winds and thus sustain the El Niño conditions. Typically El Niño events start in the spring when the sun is directly over the equator. Once established, the El Niño conditions build over the next nine months and typically peak in December. After the December peak, the El Niño conditions usually start to wane over the next few months. There are of course some historical exceptions to the norm. The El Niño of 1997–8 became quite strong early and, according to some

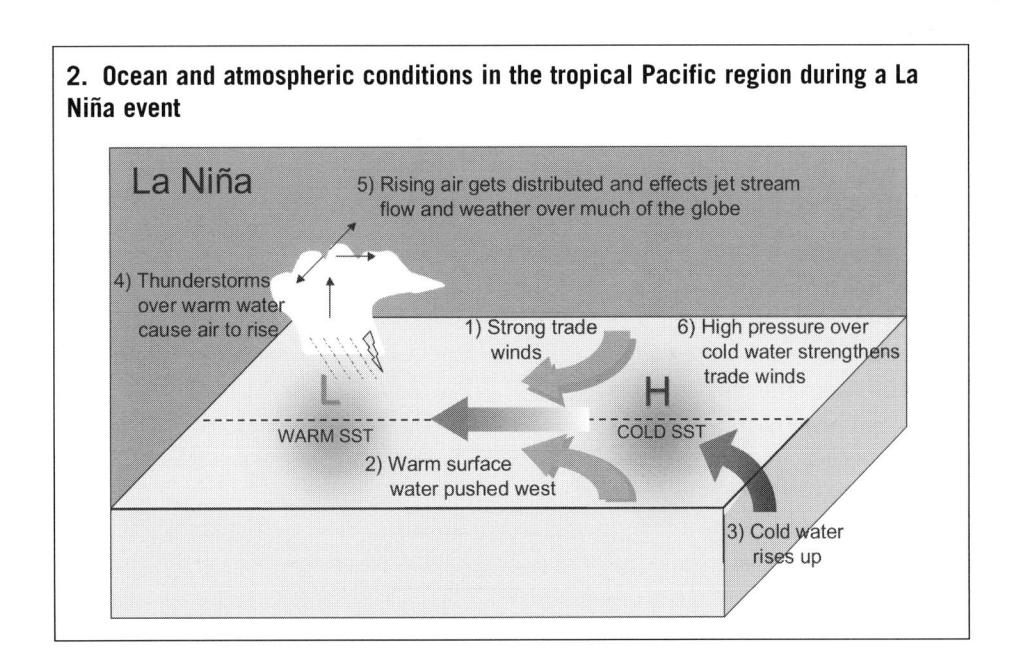

2. Ocean and atmospheric conditions in the tropical Pacific region during a La Niña event

indexes, peaked once in August and again in December.[6] In other cases, El Niño conditions have persisted for two years back to back. The El Niño that started in the spring of 1986, peaked in December of 1986, started subside, but then resurged to peak again even stronger in December of 1987.

El Niño and La Niña are variations in tropical Pacific SST resulting from variations in the base state dynamics of the tropical Pacific oceanic–atmospheric system. But how do these events affect weather in North America and Europe for instance? The answer to this has to do with three key processes:

1. the change in the position of warmest SST regions along the tropical Pacific;
2. subsequent impacts on the tropical atmospheric conditions; and
3. communication of these atmospheric impacts to higher latitudes and across the globe through atmospheric teleconnections.[7]

Regions of very warm SST (often close to 30°C or 86°F in the tropics) are quite vast (almost the size of a continent like Australasia). These regions are breeding grounds for vigorous tropical thunderstorms, convective systems that contain strong vertical winds. A massive area of convection is formed over a warm SST region, pushing a tremendous volume of air up to the top of the atmosphere like a big atmospheric fountain. The air flow at the top of the fountain is distributed to the north and south, which in turn affects the Jet Streams (see Panel 2) that play a key role on mid-latitude weather like that of North America and Europe.[8] When the warm SST moves across the Pacific in response to the El Niño/La Niña activity, the convective atmospheric fountain moves with it. When this happens, the Jet Stream pattern is affected as mid-latitude weather can be. This concept is shown schematically in Figure 3. During the El Niño winter, the warm SST anomaly in the eastern tropical Pacific generates convective outflow, which strengthens the flow of the subtropical Jet Stream. During the La Niña winter, the convective centre is further west in the Pacific, in this case the subtropical Jet Stream is diminished and the polar Jet Stream is strengthened.

ENSO effects
When trying to clarify the effects of ENSO, it is important to understand that real weather and climate are much too complex to follow simple "if this, then..." rules. Instead, one must adopt a more complex probabilistic view of the phenomenon, such as: "If there is a strong El Niño, then there is in increased probability (but not a guarantee) that the winter in Mississippi will be wetter than normal". The other thing to keep in mind is that the (potential) effects vary over time and space, and therefore must be qualified as such (as in the last example of winter in Mississippi). Furthermore, it is important to understand that not every El Niño (La Niña) is identical. Additionally, there may be influences from other climate modes that can alter the effects of El Niño (La Niña) making them atypical.

ENSO can affect a sizeable fraction of the earth's climate; the regions that are affected most by ENSO are those regions immediately near the Pacific Ocean including North and South America, Southeast Asia, Indonesian and Australia. A discussion on the all-global effects of ENSO is beyond the scope of this chapter, however given that ENSO effects in North America have been thoroughly researched and the significant size of the North American weather risk market, we will limit the review of ENSO effects to North America.

North America is most strongly affected during the winter months when the weather tends to be dominated by the polar and subtropical Jet Stream flows. During the summer, the sharp temperature contrast from equator to North Pole is reduced, as a result, Jet Stream flows tend to be much weaker and therefore do not have as much capacity to distribute ENSO influence as compared to that of the winter Jet Stream flows. Common characteristics of ENSO are:

North American El Niño winter:

3. Effects of El Niño and La Niña on the atmospheric general circulation

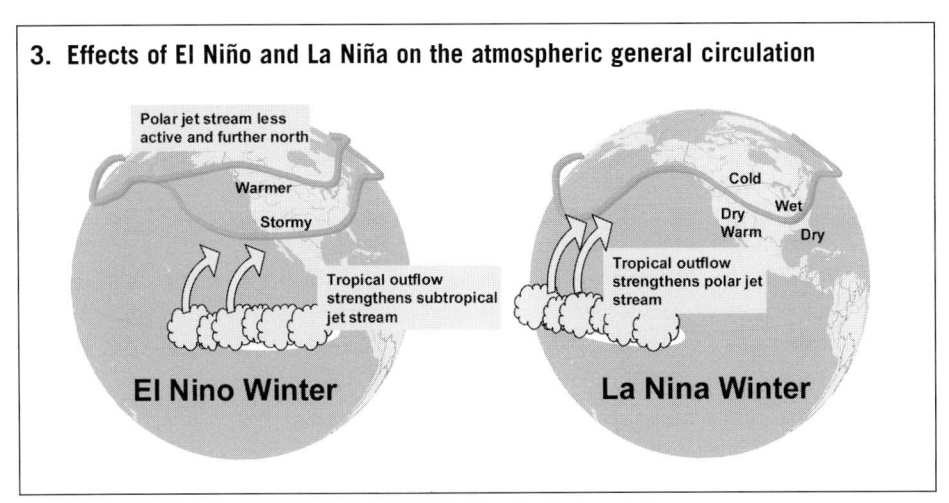

[Figure: Two globe diagrams comparing El Nino Winter and La Nina Winter patterns. El Nino Winter shows polar jet stream less active and further north, warmer and stormy conditions, with tropical outflow strengthening subtropical jet stream. La Nina Winter shows cold, dry, warm, wet and dry regions, with tropical outflow strengthening polar jet stream.]

❏ Increased probability of above average precipitation in the Southeast US and California.
❏ Slightly increased probability of below average precipitation in the Pacific Northwest including areas of British Columbia, Canada and the US states of Washington, Oregon, Idaho, Montana and southern Alaska.
❏ Increased probability of warmer than normal temperatures across the northern tier of the US and southern Canada from coast to coast.
❏ Increased probability of cooler than normal temperatures in the southwestern US and Mexico.

North American La Niña winter:

❏ Increased probability of above average precipitation in Pacific Northwest, along the US–Canadian border and in the Mississippi and Ohio River valleys.
❏ Increased probability of below average precipitation in the US Southwest, along the US Gulf Coast and up the US East Coast up to Virginia.
❏ Increased probability of warmer than normal temperatures across the southeastern US from Arizona to Massachusetts.
❏ Increased probability of cooler than normal temperatures along the US west coast.

ENSO predictability
The biggest challenge in ENSO prediction is forecasting the onset of an ENSO (El Niño or La Niña) event. Fortunately, the climate science community's in-depth understanding of ENSO dynamics, together with dedicated observing systems such as the TAO/TRITON buoy array in the tropical Pacific Ocean, has lead to some success in predicting the onset of ENSO events with several months lead-time.[9] In the cases where the prediction of onset is limited, the 6–9 month period leading up to an ENSO event peak is often enough lead-time to prepare for anticipating ENSO's effects.

PACIFIC DECADAL OSCILLATION
The Pacific Decadal Oscillation (PDO) is another mode of climate variability found within the Pacific Ocean. Similar to ENSO, the PDO is also marked by variation in the SST conditions. Compared to the ENSO cycle, however, the PDO is a much longer term oscillation (20–30 years). Figure 4 depicts a time-series of five-month averaged PDO index, which represents the pattern of SST in the north Pacific.[10] While there is some fine-scale structure on short timescales, the positive–negative shading highlights the long-term component of the signal. The period from the early 1960s

to late 1970s was dominated by the cool phase (negative PDO index). In the early 1980s the PDO switched to the warm dominant phase (positive PDO index). Recently (since the late 1990s) it appears that the PDO has again switched back to the cool phase.

PDO effects
Figure 5 illustrates the effects of the two phases of the PDO on the North American climate. In the warm phase of the PDO, the north Pacific Ocean exhibits a cool sea surface temperature anomaly (SSTA) while the central tropical Pacific exhibits a warm SSTA. The cool phase exhibits the reverse pattern with warm anomalies in the north Pacific and cool anomalies in the central tropical Pacific. When the PDO pattern is either positively or negatively strong, North American weather can be affected. During a strong warm phase period, conditions in northern and western North America tend to be warmer and dryer while conditions in southern and eastern North America tend to be cooler and wetter. The reverse is true for the strong cool-phase periods.

PDO prediction
Currently there does not exist a significant prediction system (forecast models, observing systems, etc) dedicated to PDO prediction. However, there are two key attributes about the PDO that do allow for fairly successful forecasts based on

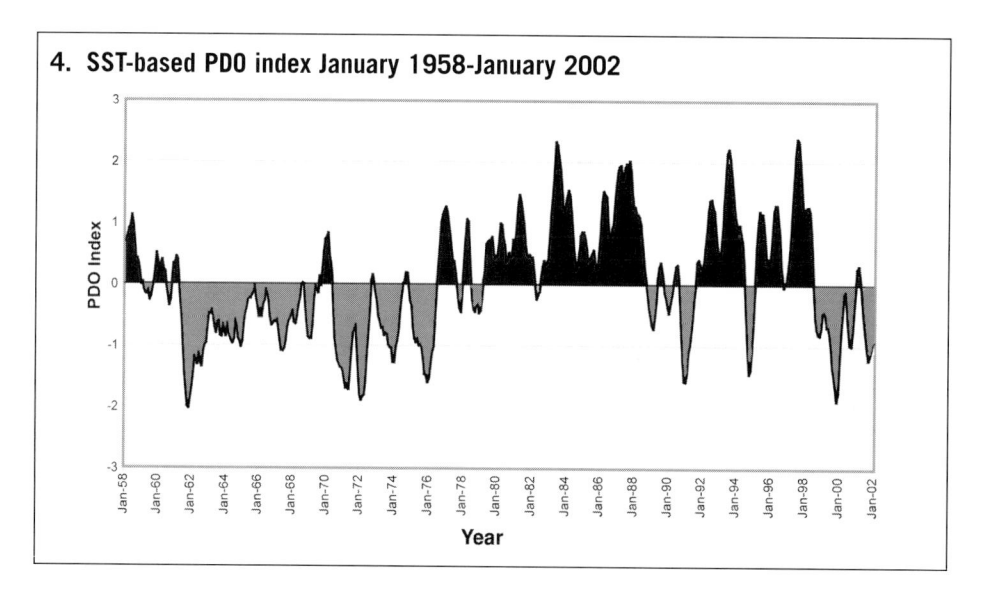

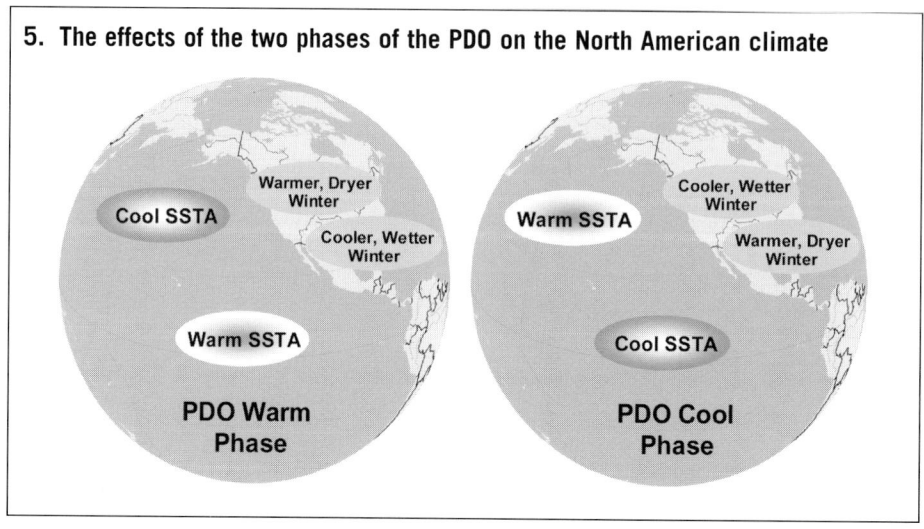

4. SST-based PDO index January 1958-January 2002

5. The effects of the two phases of the PDO on the North American climate

persistence. First, since the PDO is an ocean-based phenomenon (as are La Niña and El Niño), and as SST anomalies generally vary slowly over periods on the order of months, persistence forecasting methods can be used to make seasonal forecasts of the PDO and its effects. Additionally, the 20- to 30-year oscillation component can be leveraged to make longer-range (multi-year) PDO persistence forecasts.

NORTH ATLANTIC OSCILLATION
While the ENSO and PDO are prominent examples of ocean-based modes of climate variability, there are other modes that are not related to the oceans. The Arctic Oscillation (AO) and its closely linked sibling the North Atlantic Oscillation (NAO) are examples of modes of variability that are largely seen as atmospheric anomalies. Here we will first discuss the NAO and then the AO.

The North Atlantic Oscillation (NAO) is an oscillation in the pressure gradient between the North Atlantic and the mid Atlantic Ocean. Because of its strong effect on temperatures and storm tracks on both sides on the North Atlantic, the NAO has been a subject of study since the late 1800s. The NAO index is defined as the normalised pressure difference between Iceland and either Portugal or the Azores. A positive NAO index indicates the presence of a deep Icelandic surface low-atmospheric pressure system and a strong subtropical surface high atmospheric pressure system. A negative NAO index indicates a weak Icelandic low and weak subtropical high. Figure 6 shows the time-series of the NAO index from January 1950 to May 2002. The fine line represents the monthly averaged NAO index; the heavy line represents the one-year running mean. Because the NAO index is computed via the surface pressure difference between two locations and because surface pressure varies significantly over time and space, the monthly signal tends to be noisy, and as such the monthly time-series appears to offer very little predictability. The running mean however does exhibit some persistence, which can be leveraged for some predictability.

NAO effects
During the positive phase of the NAO, the westerly flow across the Atlantic Ocean is intensified. This has implications for both the Eastern North American and the European climates. While northern and western Europe tend to experience more storms and above-normal temperatures, eastern North America tends to be less stormy, with above-normal temperatures. During the negative phase of the NAO, Europe tends to be less stormy, but colder than normal, whereas, eastern North America tends to be more stormy and colder than normal.

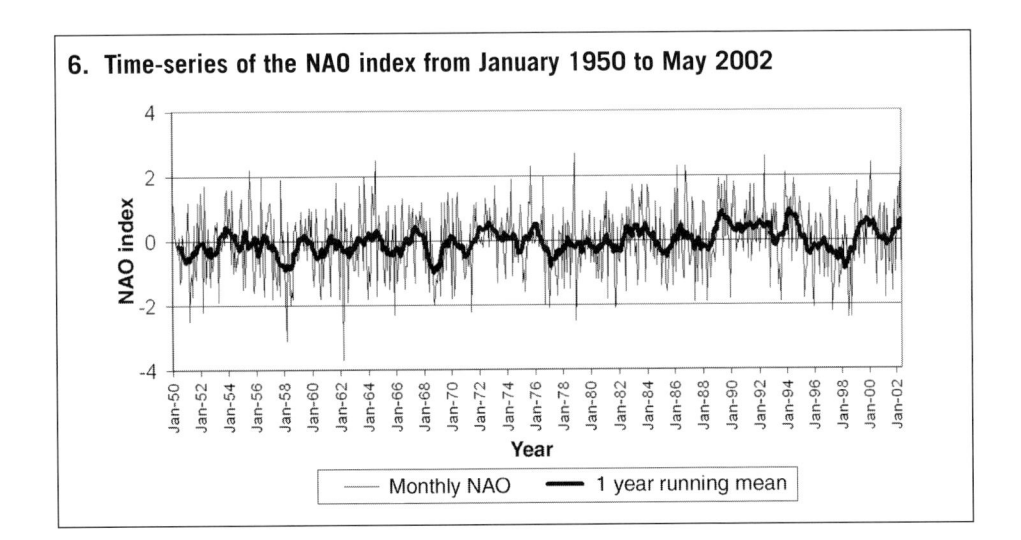

6. Time-series of the NAO index from January 1950 to May 2002

NAO prediction

While there have been a number of proposed NAO prediction methodologies, skilful NAO prediction beyond a few weeks remains elusive as analysis of the NAO data has not revealed any dependable predictors of the NAO changes. Currently the best hope for improved NAO prediction lies in improving our scientific understanding of the larger-scale Arctic Oscillation discussed below.

ARCTIC OSCILLATION

The Arctic Oscillation is the term used to describe a variation in the atmospheric circulation around the North Pole. In the growing list of identified modes of variability, the AO is a relatively new one, as Thompson and Wallace only recently identified it in 1998.[11]

The cool phase of the AO is characterised by higher than normal pressure at the North Pole and weaker westerly winds circling the North Pole. In the warm phase these conditions are reversed with lower than normal surface pressure at the North Pole and stronger westerly wind circling the North Pole.

AO effects

During the cold phase, the combination of a strong arctic high, and weak westerlies allows for cold arctic air to migrate south into North America, Europe and Asia producing frequent cold air outbreaks during winter. During the warm phase, the combination of strong westerlies and weaker high pressure at the pole lead to a condition where the cold air remains confined to the Arctic. As a result, winter temperature in North America, Europe and Asia tend to be warmer.

AO research

Since the AO was only recently identified in 1998, research into this phenomenon is still in the early stages; still much has been learned about the characteristics of the AO and how it influences climate. There are two noteworthy items, which have been uncovered through AO research thus far. First, there is strong evidence that the AO is the dominant mode of variability in the northern hemisphere and that the NAO is actually a localised North Atlantic manifestation of the larger AO phenomenon. Secondly, there is evidence that AO activity is caused by variations in stratospheric warming over the Arctic, which could lead to improved AO–NAO prediction.

The four modes of climate variability discussed above represent some of the more important drivers of variations in the North American and European climates, from year to year. However, there exist other modes of climate variability that are beyond the scope of this chapter that can also exert influence on the climate. Two other most influential modes of variability are,

1. the Pacific North American pattern (PNA), which has some correlation to the PDO and the ENSO phenomena;[12] and
2. the Madden Julian Oscillation (MJO), which is a wave of convection that travels along the equator and can affect weather patterns with its strong convective outflow similar to that of El Niño and La Niña.[13]

Managing climate variability

In order to mitigate the effects of climate variability, we need to estimate the future state of climate with some degree of certainty. Two approaches can be employed. The traditional approach is based on statistical analysis of historical weather data. However, because of data quality issues, there could be a significant error associated with the estimates derived from this method. To reduce uncertainty associated with the estimated climate, it is critical to use weather data that has been cleaned,[14] deshifted[15] and detrended[16] to produce a continuous and temporally homogenous

PANEL 2

GENERAL CIRCULATION OF THE ATMOSPHERE

Because of the variability of the Earth's surface (ie, ocean, land, ice, etc) and variation of solar warming from equator to pole, air is heated unevenly across the globe. With the sun nearly directly overhead, the Tropics receive tremendous amounts of energy while icy polar regions receive significantly less. This uneven heating from equator to pole and land to ocean to ice is responsible for setting up semi-permanent pressure patterns across the globe, which in turn generate semi-permanent or persistent global wind patterns. The figure below is a simplified representation of the global atmospheric circulation pattern (also referred to as the "atmospheric general circulation pattern").

In the above figure the cells on the left illustrate the vertical flow of the lower part of the atmosphere (known as the "troposphere").[17] The arrows on the surface of the earth show the persistent wind patterns near the ground. Starting at the equator, the persistent influence of the sun causes the air to warm and rise. This action all around the equator facilitates the generation of the Trade Winds and the emergence of an area of low pressure along the equator known as the "Intertropical Convergence Zone" (ITCZ), which can often be seen on satellite images as a ribbon of clouds along the equator. As air rises along the equator it begins to follow a path known as the "Hadley Cell".[18] Moving north away from the equator, the next cell is known as the "Ferrel Cell", which generates surface winds known as the "westerlies".[19] Finally, near the North Pole is the Polar Cell, which generates surface winds known as the "Polar easterlies". Where the Hadley and Ferrel Cells meet, the "subtropical Jet Stream" is formed. Furthermore, at this juncture, air mass flows down toward the ground to

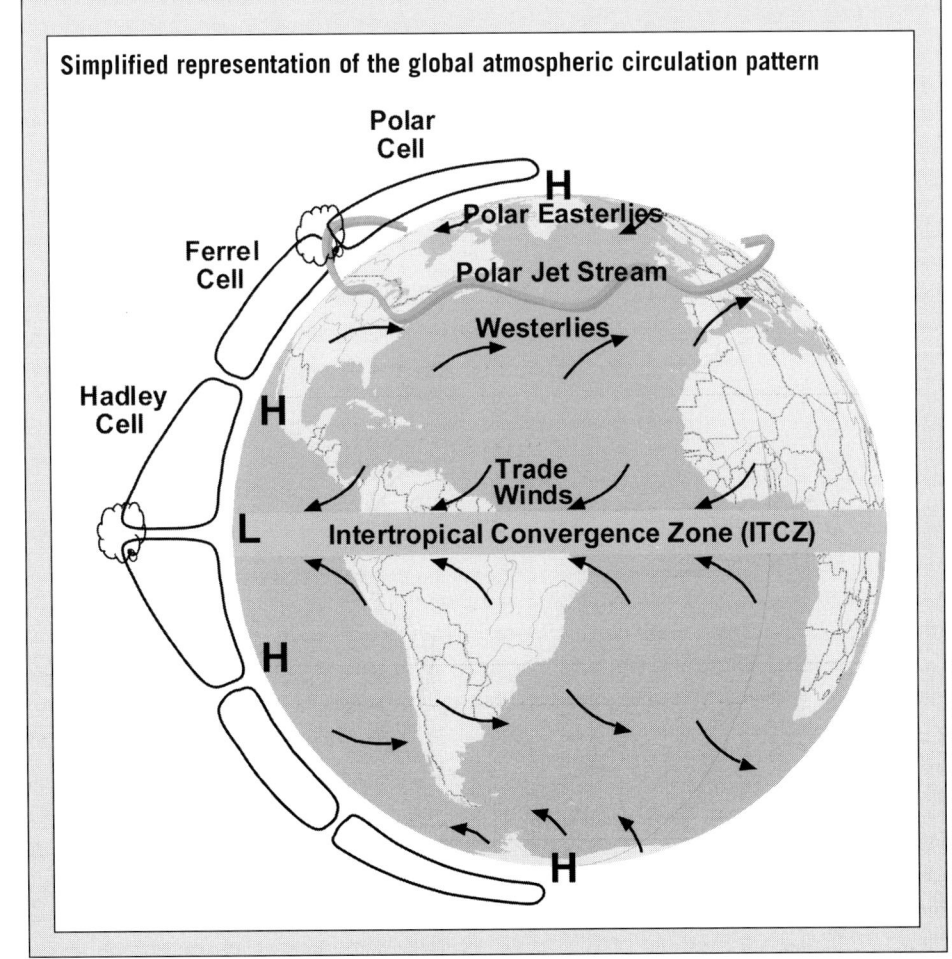

Simplified representation of the global atmospheric circulation pattern

create areas of high pressure and dry clear conditions. Where the Ferrel and Polar Cells meet, cold arctic air collides with warm air to create the Polar Front and the Polar Jet Stream, which is a very fast flow of air high in the atmosphere. The position and configuration of the Polar Jet Stream (referred to as simply the Jet Steam in the context of this chapter) has a significant influence on weather in the US and Europe.

The persistent wind patterns of a region dictate the weather for that region; change the wind patterns and the weather changes in response. Occasionally the persistent wind patterns of a region get interrupted, when this happens the weather deviates away from normal conditions into some anomalous regime.

But what causes the persistent weather patterns to vary into atypical flows? As alluded to above, the static nature of the Earth's large-scale surface features together with the seasonal – yet persistent – energy of the Sun creates seasonal, yet otherwise regular and persistent weather patterns. While the Earth's large-scale surface features and the Sun explain much of the Earth's climate, there are identifiable climate system components that relate to why climate can vary so greatly. Formally the Earth's climate system is composed of five internal components and one external component, the Sun. The internal components include the hydrosphere (oceans, lakes, rivers, etc), the lithosphere (solid Earth), the cryosphere (frozen water, ice caps, glaciers, etc), the biosphere (all living things) and the atmosphere (the air we breath).

Of the five components of the Earth's climate system, which has an influence on climate on the scale of months and seasons? For the most part the lithosphere is thankfully quite stable on these timescales. The biosphere, while it can be quite dynamic, largely reacts to the influence of the other components, and does not have a significant capacity to alter the short-term climate on its own. A similar argument can be made of the cryosphere. The hydrosphere (primarily the oceans) on the other hand can impose a significant influence on short-term climate. Approximately two thirds of the Earth's surface is ocean water and therefore the oceans are a significant component simply by proportion. In addition, large ocean currents can transport tremendous amounts of heat energy on timescales of months and seasons, and thus can influence short-term climate. Lastly, like the oceans, the atmosphere can be thought of a fast flowing ocean of air. Because of its ability to transport energy quickly, the atmosphere can influence daily weather as well as short-term climate. Thus, the oceans and the atmosphere have the capability to move significant amounts of energy, and can therefore alter the energy equilibrium established by the Earth–Sun interaction to create short-term climate variations.

time-series. However, this approach can be augmented with the modern techniques of estimating future climate and its variability, using climate-forecasting technologies.

ESTIMATING FUTURE STATE OF THE CLIMATE BASED ON HISTORICAL WEATHER DATA

Data collected over a number of years at a weather station can serve as a first-order means of estimating probability distributions that describe the nature and frequency of weather observed at a station. If the future reflected similar patterns to the past, these distributions could serve as probabilistic estimates of the future climate. The caveat here is that historical data is often inconsistent with current observations and as such, it must be corrected before it is used to generate probability estimates.

Station time-series inconsistencies generally fall into two categories; shifts and trends. Shifts occur as a result of a sudden change to a station such that the observations are somehow interrupted. Changes such as station relocation and instrument replacement are two common causes for sudden shifts in a time-series. Figure 7 illustrates a shift in observational readings resulting from the relocation of the Orlando, Florida station. The top chart shows the time-series of averaged January minimum temperature for the period before and after the station move. The difference in the mean between the before and after portions of the time-series is approximately 6°F. The lower panel shows distributions derived from different portions of the time-series. The darker distribution line was derived from the period before the move and the lighter distribution line was derived from the period after the move. The distribution represented by the dashed line was derived from the entire 1976–2000 time-series. With respect to estimating the probability of future observations, it can clearly be seen from this example that data inconsistencies such as this can lead to inappropriate distributions, thus emphasising that such inconsistencies must be corrected before the data is used to generate forecast distributions.

Trends occur as a result of slow changes within the station environment such that older observations are no longer consistent with more recent observations. Global warming is one well-known cause of this phenomenon, however, there are others

7. An example of the effects of a station move on the station time-series and associated probability distribution functions derived from the data

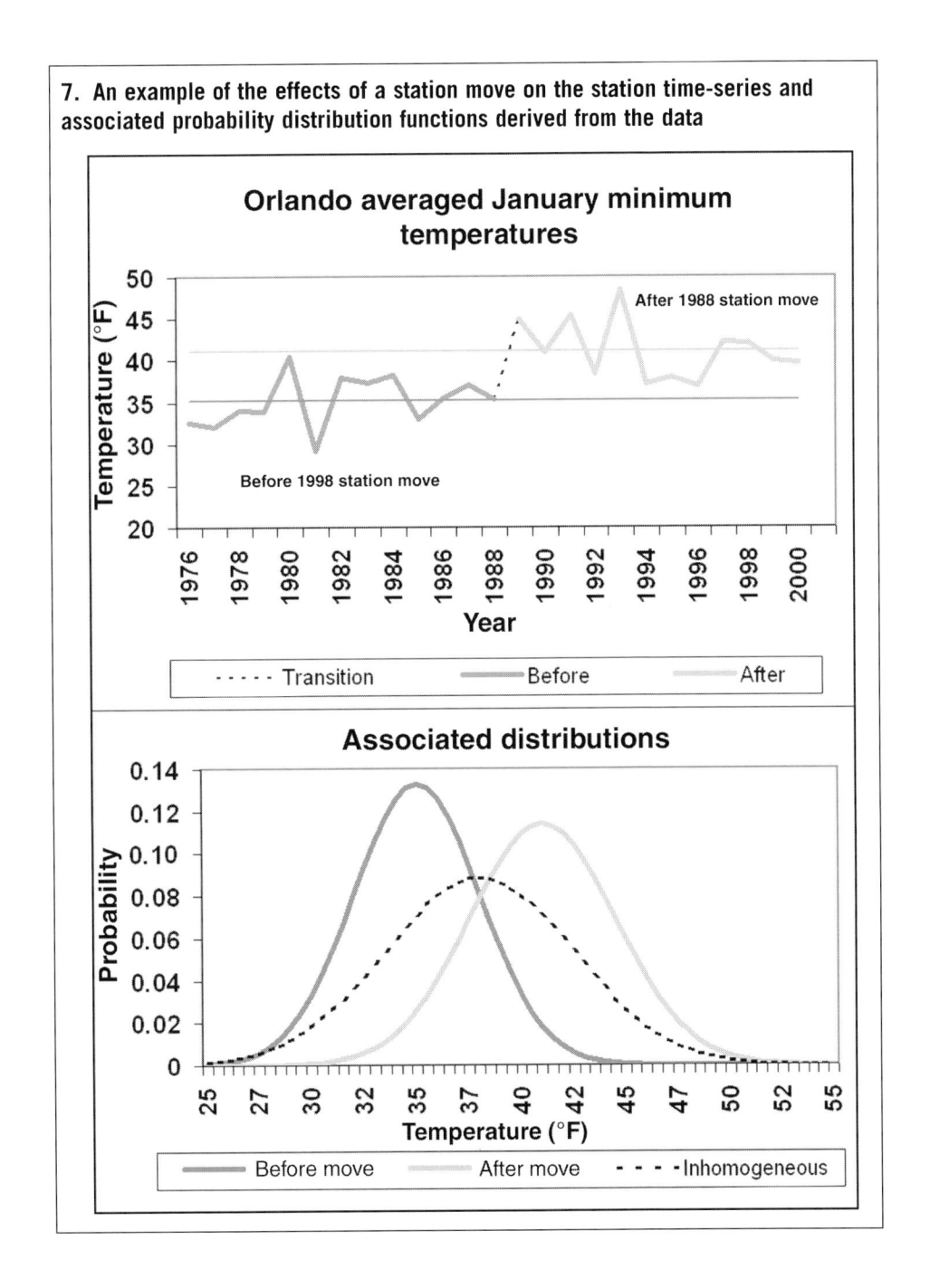

causes that can be more significant such as urbanisation and other local environmental changes. In response to demand for high quality station time-series, companies such as AIR, Inc. have developed technologies to clean, de-shift and de-trend station time-series to make them more suitable for probability estimation.

SHORT-TERM CLIMATE FORECASTING

While historical weather data can serve as a first-order means for estimating probability of future weather at a station, the resulting distributions are often quite broad due to the inclusion of a diverse range of weather including that caused by El Niño, La Niña and other weather regimes. To get a more focussed view of the expected climate variability over future months and seasons, many weather traders and weather risk managers utilise short-term climate forecasts, probabilistic monthly or seasonal forecasts with lead-times of 1–12 months.

Climate forecasts differ from traditional weather forecasts in that they forecast the probability of what the average weather should be over the course of months and seasons as opposed to attempting to predict the details of specific weather events over hours and days. For example, a short-term climate forecast (a forecast for the season ahead) may say that there is a significant probability that the next winter will be unusually warm or that the frequency of heat waves will be higher than normal next summer.

Specific weather events can be well predicted on the scale of days due the fact that the evolution of the atmosphere on the scale of days is fairly linear (ie, multiple diverse solutions are rare). However, when trying to make predictions beyond a couple of weeks, the chaotic nature of the atmosphere leads to increasing divergence in the number of possible solutions. This makes it almost impossible to forecast specific weather events beyond a couple of weeks. But even though specific weather events cannot be predicted accurately beyond a couple of weeks, predictions about the expected distribution of possible weather outcomes can be made months and seasons ahead. Based on the distribution of possible outcomes, climate forecasts provide probabilistic information about what the average weather should be over a period of time. A probabilistic climate forecast could therefore offer another probability distribution that can be used to help structure and price weather contracts.

In order to effectively use a climate forecast one needs access to:

1. the probability distribution derived from the forecast output; and
2. a quantitative assessment of the skill, or accuracy, of the underlying forecasting system.

Based on this information, the user would be able to determine the appropriate weight that should be assigned to the forecast.

Quantification of the skill of a forecasting system is usually conducted by the forecast provider. This involves a comprehensive evaluation of the accuracy of the system based on a set of previously made forecasts or "hindcasts" (forecasts that were made using historical data as opposed to real-time data). In climate prediction, skill studies are often conducted with hindcast data because it would take several years to generate and evaluate a sufficient number of forecasts that are based on real-time data. Typically forecast skill is evaluated by comparing how well the forecasts did compared to climatological distributions in explaining what actually occurred. (At AIR, forecasts are compared to climatological distributions based on reconstructed data since this has been shown to be more skilful than those from non-reconstructed data.) In other words, does the forecast provide better guidance than climatological distributions derived from historical data?

One objective means of answering this question is by computing the Linear Error in Probability Space (LEPS) score. Essentially LEPS scores are normalised between

–100% and 100%. Positive scores indicate the climate forecasts have performed better than climatological distributions. A score of 0% means the climate forecasts have had the same skill as climatological distributions and negative scores indicate the climate forecasts performed worse than climatological distributions. To be useful at all, climate forecast systems must demonstrate positive LEPS skill, otherwise it is best to use climatological distributions derived from reconstructed historical data.

Depending on the specific application, climate forecast distributions and distributions derived from reconstructed historical data could be merged to create a suitable distribution for the contract. Obviously, the higher the LEPS score of the forecast system, the more weight should be given to the forecast distribution.

Over the past decade numerous climate-forecasting methodologies have been developed.[20] The various methodologies can essentially be divided into either statistical or dynamical modelling systems. Statistical models leverage lagged correlations in historical time-series between predictors (SST, SLP, etc) and predictands (atmospheric temperature, precipitation, etc) to produce forecasts. Dynamical models on the other hand are based on mathematical representations of the physics of the Earth's climate system and produce forecasts by initialising the system with observed climate data and evolving the climate model forward in time. Regardless of the type of model used, the climate model is typically run in "ensemble mode" to produce a distribution, or ensemble, of possible outcomes. To do this the model is run numerous times with slightly different sets of initial conditions that are consistent with the current climate, yet cover the range of uncertainty inherent in the initial conditions.

While both statistical and dynamical models have their own strengths and weaknesses, both are capable of producing skilful forecasts. It has been shown in various studies, however, that the most accurate forecasts are made by incorporating guidance from a number of skilful dynamical and statistical climate models.[21] Climate forecasting systems that utilise multiple models and numerous ensemble members are called "multi-model super-ensemble prediction systems" (see Figure 8).

As shown in Figure 8, in this forecast system, initial conditions are used to initialise a number of dynamical and statistical models. The initial conditions are also used in a post-process model weighting application where outputs from the various models are weighted based on their historical skill in forecasts from climate regimes similar to the current regime. For instance, if the current climate is dominated by El

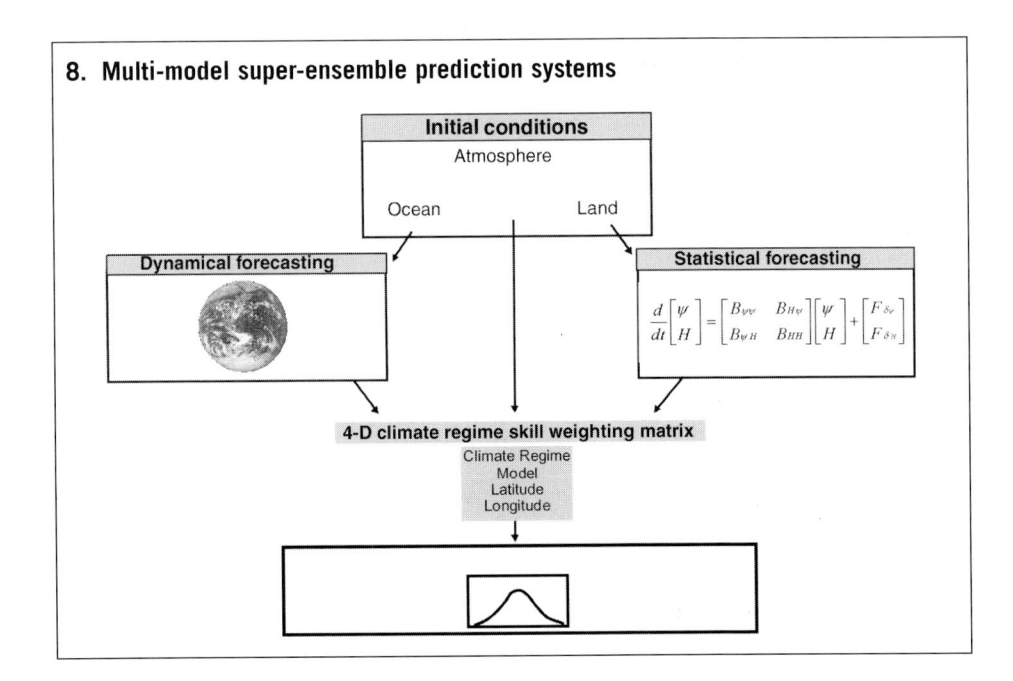

8. Multi-model super-ensemble prediction systems

Initial conditions
Atmosphere
Ocean   Land

Dynamical forecasting

Statistical forecasting

$$\frac{d}{dt}\begin{bmatrix}\psi\\H\end{bmatrix}=\begin{bmatrix}B_{\psi\psi} & B_{H\psi}\\B_{\psi H} & B_{HH}\end{bmatrix}\begin{bmatrix}\psi\\H\end{bmatrix}+\begin{bmatrix}F_{\delta_\psi}\\F_{\delta_H}\end{bmatrix}$$

4-D climate regime skill weighting matrix
Climate Regime
Model
Latitude
Longitude

Niño conditions, and model A has demonstrated that it has performed 50% better than model B under these conditions, than model A would have 50% more weight than model B in contributing to the final forecast distribution. By taking advantage of the demonstrated strengths from each model, the forecast is optimised.

Conclusion

The complexity of the Earth's climate system is formidable. However despite this complexity, weather and climate variability can be quantified. Traditionally, climate and its variations have been quantified via analysis of historical weather data. While this methodology is helpful, the result is often more uncertain than the natural uncertainty due to inconsistencies in the historical data used in the analysis. Today, improved historical data is available, which has been corrected for shifts and trends, to support a more accurate assessment of climate variability with more certainty.

In the last two decades, the climate science community has significantly advanced its understanding of the Earth's climate system and has identified important modes of climate variability that play a critical role in variations in climate from year to year in different regions around the world. Based on these advances it is now possible to reduce uncertainty in estimating the future state of the climate through climate forecasting. Weather-sensitive businesses and businesses involved in the weather market should secure sound meteorological and climatological consulting and make use of the latest data and forecast technology to support their weather and climate information needs.

1 *The author would like to acknowledge Dr Maryam Golnaraghi for her immensely valuable support and contributions in the preparation of this chapter – thank you Maryam.*

2 *In this paper, there will be frequent references to seasons at various locations around the globe. To avoid the ambiguity between similarly named seasons in northern and southern hemispheres that occur a half-year out of phase, it is important to establish a season–hemisphere reference frame. An example of the potential ambiguity can be seen in the statement: "In the winter, the influence of El Niño is more pronounced". In this case there would likely be some question as to whether the statement was made in reference to the northern hemisphere winter months of December, January, February and March, or the southern hemisphere winter months of June, July, August and September. An established convention in the geosciences is to use the terms "boreal" when referring to the northern hemisphere and "austral" when referring to the southern hemisphere. Thus the term "boreal summer" would refer to summer in the northern hemisphere, which includes the months of June, July, August and September.*

3 *This difference between the normalised air pressures at these two locations is known as the Southern Oscillation Index (SOI).*

4 *The local Peruvians who have fished this area of the Pacific for centuries named the warm phase El Niño as in "the boy". Since the cold phase, in loose terms, is the opposite of the warm phase, the natural choice of name for the cold phase was La Niña or "the girl".*

5 *The westward force of the wind causes the ocean water to pile up such that the water level in the western tropical Pacific is higher than that of the eastern tropical Pacific. This piling of warm water is a normal result of the tropical Pacific atmospheric–oceanic system. During La Niña, the western waters are piled higher than normal.*

6 *See Niño1+2, Niño 4 and MEI indexes for 1997–8.*

7 *The term teleconnection is often used to describe atmospheric and/or oceanic action and reaction pairs that are spatially far removed from each other.*

8 *The term mid-latitude is used to describe the region bounded by 30° and 60° latitudes. In the northern hemisphere this region covers the majority of the populated areas in North America, Europe and Asia.*

9 *The Total Atmosphere Ocean / Triangle Trans Ocean Buoy Network (TOA/TRITON) buoy array is an array of buoyed atmospheric and oceanic sensors covering the equatorial Pacific from Indonesia to South America.*

10 *The Pacific Decadal Oscillation (PDO) Index is defined as the leading principal component of North Pacific monthly sea surface temperature variability.*

11 *See Thompson and Wallace (1998).*

12 *Information on the PNA pattern can be found at http://www.cpc.ncep.noaa.gov/data/teledoc/pna.html.*

13 *Information on the MJO can be found at http://nws.met.psu.edu/research/TRAINING/MJO_lecture.ppt.*

14 *Data cleaning involves filling in missing observations and replacing erroneous observations with high quality estimates derived from highly correlated surrounding stations (see Chapter 8).*

15 *De-shifting is the process of identifying and removing shifts in the data time-series, such as those caused by instrument changes and station moves.*

16 *De-trending is the process of removing trends from a time-series, such as those resulting from global warming and urbanisation.*

17 *Cells of circulating gas are common to most planets with gaseous atmospheres. On Jupiter, which has many cells, the cells make a colourful pattern of stripes across the Jupiter surface.*

18 *The northern and southern hemisphere each have a set of circulation cells that are essentially mirror images of each other, however, for simplification we will limit our discussion to the northern hemisphere.*

19 *Westerlies refer to winds blowing from the west and easterlies refer to winds blowing from the east.*

20 *See Barnston (1994), Barnston et al. (1994), Cane and Zebiak (1987), Graham et al. (2000), Ji, et al. (1993), Ji, Kumar and Leetmaa, (1994), Kirtman et al. (1995) and Shukla (1981).*

21 *See Krishnamurti et al. (2000).*

BIBLIOGRAPHY

Barnston, A. G., H. M. van Dool, S. E. Zebiak, T. P. Barnett, M. Ji, D.R. Rodenhuis, M. A. Cane, A. Leetmaa, N. E. Graham, C. R. Ropelewski, V. E. Kousky, E. O. O'Lenic and R. E. Livesey, 1994, "Long-lead Seasonal Forecasts – Where do we Stand?", *Bulletin of the American Meteorological Society*, 75, pp. 2097–14.

Barnston, A. G., 1994, "Linear Statistical Short-term Climate Predictive Skill in the Northern Hemisphere", *Journal of Climate*, 7, pp. 1513–64.

Cane, M., and S. E. Zebiak, 1987, "Prediction of El Niño Events using a Physical Model" in: *Atmospheric and Oceanic Variability*, H. Cattle, (ed), (Royal Meteorological Society Press), pp. 153–82.

Graham, R. J., A. D. L. Evans, K. R. Mylne, M. S. J. Harrison and K. B. Robertson, 2000, "An Assessment of Seasonal Predictability using Atmospheric General Circulation Models", *Quantative Journal of the Royal Meterological Society*, 126, pp. 2211–40.

Ji, M., A. Kumar and A. Leetmaa, 1994, "An Experimental Coupled Forecast System at the National Meteorological Center: Some Early Results", *Tellus*, 46A, pp. 398–419.

Ji, M., A. Kumar, A. Leetmaa and M.P. Hoerling, 1993, "Coupled Ocean-Atmosphere Climate Forecast System for ENSO Predictions", Proceedings of the Workshop on Numerical Extended Range Weather Prediction, Airlie, Virginia, pp. 141–4.

Kirtman, B. P., J. Shukla, B. Huang, Z. Zhu, and E. K. Schneider, 1995, "Multiseasonal Predictions with a Coupled Tropical Ocean/global Atmosphere System", *Monthly Weather Review*, 125, pp. 789–808.

Krishnamurti, T. N., C. M. Kishtawal, T. LaRow, D. Bachiochi, Z. Zhang, C. E. Williford, S. Gadgil and S. Surendran, 2000, "Multi-Model Super-Ensemble Forecasts for Weather and Seasonal Climate", *Journal of Climate*, November.

Shukla, J., 1981, "Dynamical Predictability of Monthly Means", *Journal of Atmospheric Science*, 38, pp. 2547–72.

Thompson, D. W. J, and J. M. Wallace, 1998, "The Arctic Oscillation Signature in the Wintertime Geopotential Height and Temperature Fields", Geophysical Research Letters, 25, pp. 1297–300.

7

Weather and Seasonal Forecasting

Mark S. Roulston; Leonard A. Smith

University of Oxford; University of Oxford and London School of Economics

History of forecasting: statistics vs dynamics

"When it is evening ye say, it will be fair weather for the sky is red. And in the morning, it will be foul weather today for the sky is red and lowering." (Matthew: Chapter 16, Verse 2)

The Biblical quotation above is an ancient example of attempting to make meteorological predictions based on empirical observations. The idea it contains tends to work as a forecasting scheme and it has been expressed in many different ways in the subsequent 2,000 years. It is one of many sayings and rhymes of folklore that provide a qualitative description of the weather. In the 17th century, however, the introduction of instruments to measure atmospheric variables meant that meteorology became a quantitative science.

By the 19th century meteorological data was being collected all over Europe. It was during the 1800s that many scientists began lamenting the fact that the collection of meteorological data had far outpaced attempts to analyse and understand this data (Lempfert, 1932). Attempts had been made to find patterns in tables of meteorological data, but these had usually ended in failure. One example is the efforts to link weather to celestial motions made by the Palatine Mteorological Society of Mannheim (the world's first such society) in the late 18th century.

The breakthrough came when scientists began plotting the growing meteorological database on maps; the tendency of mid-latitude low-pressure systems to advance eastward was discovered: clouds in the west provide a red sky at dawn before bringing bad weather, clouds in the east redden the sky at sunset before making way for clear skies. The discovery of mid-latitude eastward flow not only explained the success of ancient folk wisdom, combined with the telegraph it also provided a means to extend predictability beyond the horizon. The late 19th and early 20th century saw many efforts to identify cycles in time-series of weather data. One the most prolific cycle seekers, and ultimately one of the few successful ones, was Sir Gilbert Walker. Walker collated global weather data and spent years seeking correlations (Walker, 1923, 1924, 1928, 1937 and Walker and Bliss, 1932). During this work he discovered that the pressure in Tahiti and in Darwin were anti-correlated. This discovery withstood subsequent tests and is called the "Southern Oscillation", which is now known to be the atmospheric component of El Niño-Southern Oscillation climate phenomenon.

Many other claims for the existence of weather cycles were made, but few withstood the scrutiny of statistical tests of robustness. By the 1930s, mainstream meteorology had largely given up attempting to forecast the state of the atmosphere using statistical approaches based solely on data (Nebeker, 1995). Since then, estimates of the *recurrence* time of the atmosphere have implied that globally "similar" atmospheric states can only be expected to repeat on timescales far greater than the age of the Earth (van den Dool, 1994). This result suggests that there will *never* be enough data available to construct pure data-based forecast models (except for very short lead-times for which only the local state of the atmosphere is important).

Towards the end of the 19th century, a growing number of scientists took the view that the behaviour of the atmosphere could be modelled from first principles, that is, using the laws of physics. The leading proponent of this view was a Norwegian physicist called Vilhelm Bjerknes. Bjerknes believed that the problem of predicting the future evolution of the atmosphere could be formulated mathematically in terms of seven variables: three components of air velocity, pressure, temperature, density and humidity – each of these variables being a function of space and time. Furthermore, using the established laws of dynamics and thermodynamics, a differential equation could be formulated for each of these seven quantities (Richardson, 1922). The equations describe the flows of mass, momentum, energy and water vapour. These equations, however, form a set of non-linear partial differential equations (PDEs) and so an analytic solution was out of the question. Meanwhile, the First World War raged and Briton Lewis Fry Richardson was developing a scheme for solving Bjerknes' equations of atmospheric motion.[1] In 1911, Richardson developed a method to obtain approximate solutions to PDEs. The method involved approximating infinitesimal differences as finite differences, that is dividing space into a finite number of grid boxes and assuming the variables are uniform within each grid box. The solution obtained is not exact, but becomes more accurate as the number of grid boxes increases. Richardson applied his approach to the atmospheric equations, producing a set of finite difference equations that could be solved by straightforward arithmetic calculations.

The difficulty of Richardson's achievement cannot be understated. In creating his scheme for numerical weather prediction (NWP), he had to ensure the problem was formulated in terms of quantities that could be measured, sometimes developing new methods of measurement when they were required. He also had to develop a way of dealing with turbulence. He related vertical transfer of heat and moisture to the vertical stability of the atmosphere, as measured by a dimensionless quantity now called the Richardson number. Richardson was truly years ahead of his time – and therein lay the problem. Although his recipe was straightforward, in that era, all forecasts had to be done by hand and it was therefore incredibly tedious. It took him six weeks to produce a single six-hour forecast for just two European grid points! Richardson envisaged an army of clerks doing the calculations that would be necessary to generate an operational forecast, but this did not happen.

The Second World War saw the invention of the digital computer. After the war, the computer pioneer, John von Neumann, was trying to persuade the US government of the usefulness of this new device. Though not a meteorologist, von Neumann identified weather forecasting as an ideal application to demonstrate the power of the computer. Thanks to Richardson, the problem had been formulated in an algorithm that could be executed by a computer, but was impractical to do without one. Furthermore, the potential benefits of successful weather forecasting could be appreciated by laymen, generals and politicians. In 1950, von Neumann's team, led by meteorologist Jules Charney, ran the first numerical weather prediction programme on the ENIAC computer. Thus began an intimate relationship between meteorology and the leading edge of computer science – a relationship that continues to this day. See Nebeker (1995) for a comprehensive account of the history of weather forecasting.

Since we never know the precise state of the atmosphere, it would be foolhardy to expect to produce a precise forecast of its future state. This has motivated operational forecast centres to develop probability forecasts, a set of possible outcomes based on slightly different views of the current state of the atmosphere.

These probability forecasts come closer to the type of information required for effective risk management and the pricing of weather derivatives.

Modern numerical forecasting

Weather forecasting can be divided, somewhat arbitrarily, into three categories: short-, medium- and long-range. In the context of this chapter, short-range forecasting refers to forecasting the weather over the next day or two, medium-range forecasting covers lead-times of three days to about two weeks, while long-range, or seasonal, forecasting aims to predict the weather at lead-times of a month or more. While short and medium-range forecasts are valuable to many users, including energy companies in planning their operations, it is seasonal forecasts that are of most interest to the weather derivatives markets.

Nowadays it is increasingly common for seasonal forecasts to be made using computer models which are essentially the same as those used to produce tomorrow's forecasts. In this section we shall outline the key features of these models so that the reader will have some understanding of the origin of forecast products. We also hope to familiarise the reader with some of the technical language that meteorologists use to describe their models.

Short-range forecasts can be made using limited area models. These models use grid boxes to cover restricted parts of the globe. In a period of two weeks, however, weather systems can travel halfway round the globe. Therefore forecasting in the medium range or beyond requires a global model of the atmosphere. The most advanced global model used for operational forecasting belongs to the European Centre for Medium Range Weather Forecasting (ECMWF) funded by 22 countries and based in Reading, UK.[2] Today, ECMWF makes daily forecasts out to 10 days. These forecasts are distributed through the national meteorological offices of the member countries.

The current ECMWF global model is a T511 spectral model (see Panel 1), equivalent to a horizontal resolution of 40 km, with 60 vertical levels. The complete model state, at a given time, is described by approximately 10 million individual variables. The state is evolved forward in time by taking time steps of about 10

PANEL 1

COMPUTER MODELS: GRID-POINT VS SPECTRAL

Computer models of the atmosphere come in two types: grid points and spectral. Grid-point models represent the atmosphere as finite boxes centred on a grid point and calculate the changes in mass, energy and momentum at each grid point as a function of time. The size of the boxes determines the *resolution* of the model.

Spectral models represent the distribution of atmospheric properties as sums of *spherical harmonics*. These harmonics are similar to sine and cosine functions, except they are two dimensional and exist on the surface of a sphere. The resolution of a spectral model is determined by the wavelength of the highest spherical harmonic used in the model. Spectral models are denoted by labels such as "T511"; this means that the highest harmonic used has 511 waves around each line of latitude. A T511 model has a horizontal resolution of $1/(2 \times 511)$ times the radius of the Earth (about 40 km). The state of the atmosphere, as represented in a spectral model, can be converted to a grid-point representation using a mathematical transformation. Both grid-point and spectral models divide the atmosphere vertically into layers.

minutes. The evolution of the atmosphere is thus represented as a trajectory in the ultra-high dimensional state space of the model.[3]

The approximations introduced by representing continuous fields on a finite grid are, in a sense, well-defined or mathematical approximations, sometimes referred to as "errors of representation". There are, however, another set of physical approximations that all numerical weather models contain. These approximations are called "parameterisations". Numerical models have a finite spatial resolution. As mentioned above, the ECMWF global model cannot represent the details of weather, or of topography, at scales less than 40 km. It would be nice to be able to know the distribution of rainfall on a much smaller scale, but even having information on weather averaged over tens of kilometres can be very useful. Even if one is content to accept weather forecasts averaged over relatively large regions, weather on smaller "sub-grid" scales can have a profound impact on the weather at larger scales. For example, thunderstorms are too small for global models to resolve, but the convection and rainfall associated with them has a major impact on the energy balance of the atmosphere, and consequently on the weather over an area much larger than the storm itself. Therefore, the impact of sub-grid processes must be parameterised and included in the equations that describe the evolution of the atmosphere at the larger scales. Essentially, a parameterisation scheme for cumulus convection must predict the amount of convection in a grid box purely as a function of the meteorological variables averaged over the grid box (and possibly surrounding grid boxes), and then predict the effect that this amount of convection will have on the time evolution of those meteorological variables. Parameterisation schemes are usually designed based partly on a physical understanding of the processes involved and partly on the study of empirical observations. There are many processes that must be parameterised in numerical models of the atmosphere, eg, surface evaporation, drag due to topography and sub-grid turbulence. Designing better parameterisation schemes is one of the most active and important areas of modern meteorological research.

DATA ASSIMILATION

Before a forecast can be made with a numerical model of the atmosphere, the current state of the atmosphere, as represented within the model, must be estimated. The process of using observations to make this estimate of the model's initial condition

---

PANEL 2

OBSERVING THE ATMOSPHERE

The World Meteorological Organization's World Weather Watch (WMO WWW) manages the data from 10,000 land stations, 7,000 ship stations and 300 buoys fitted with automatic weather stations. All these stations are maintained by National Meteorological Centres.

The most important development in meteorological observation in the last 40 years has been the advent of weather satellites. The WMO WWW incorporates data from a constellation of nine weather satellites that provide global coverage. It is a testimony to the importance of these satellites that weather forecasts now tend to be more skilful in the southern hemisphere, where surface stations are quite sparse, than in the more densely observed northern hemisphere. The fact that southern hemisphere skill is actually slightly *higher* is probably because there is less land south of the equator. Topography and other continental effects make the job of forecasting northern hemisphere weather harder, even with the greater data coverage in the North.

is called "data assimilation". The estimate of the state of the atmosphere derived from data assimilation is called the analysis.

To initialise the model, one must effectively know the value of all the relevant meteorological variables as represented on the grid points of the model. Even with the vast amounts of weather data that are collected every day there are still massive gaps in the observational data set. The simplest approach to overcoming these gaps is interpolation, as Richardson did in his early in numerical forecasting experiments (Richardson, 1922). More sophisticated approaches actually combine the observations with the knowledge of atmospheric dynamics that is contained in the numerical model itself. Any state of the model can be converted into an estimate of the observations that would result if the atmosphere were in that particular state, by using the "observation function".

At ECMWF a data assimilation technique called "variational assimilation" is used. This method involves trying to find a model trajectory that leads to the closest match of the model to the actual observations that were made. The trajectory of the model is the path the model state traces out in time within the state space of the model. This space is a high dimensional space, defined by the millions of variables that describe the model state. Optimisation by variational assimilation is performed by trying to find the state of the model that leads to the best match with observations over the subsequent assimilation period. The state of the model somewhere in the middle of the assimilation period is then used as the analysis with which to initialise the forecast. A lower resolution, T159 model, is used for the data assimilation.

Data assimilation is also used by forecasting centres to produce reanalysis products. These products are historical reconstructions of the state of the atmosphere, projected into the model grid-point representation. They are constructed using similar techniques that are used to generate the analysis used to initialise forecasts. When producing a reanalysis, however, it is possible to use observations made *after* the time for which the state estimate is required, in addition to those from before, to estimate the model state. Because reanalysis products are reconstructions that are complete in time and space, they can be used to estimate weather at locations for which direct historical observations are not available. Reanalysis projects have been undertaken by ECMWF and NCEP.[4]

CHAOS AND WEATHER FORECASTING
In the late 1950s the meteorologist, Edward Lorenz, was experimenting with a numerical model of the atmosphere at the Massachusetts Institute of Technology (MIT). During a set of experiments, he started a new run by resetting the state of the model to the state obtained halfway through a previous run. To his surprise, the behaviour of the atmosphere in the second half of the new run was substantially different from the second half of the initial run. Eventually Lorenz realised that, while the computer was evaluating the model to six decimal-place accuracy, it was printing out the model state to only three decimal-place accuracy, thus resetting the model with the printed output had introduced a tiny discrepancy (less than one part in 1,000) between the two runs. This error was enough to cause a big difference in their evolutions (Lorenz, 1993). While such behaviour had been known for centuries, the existence of digital computing made such behaviour more amenable to study. In 1975 the word "chaos" was coined to describe such sensitivity of these models to initial conditions, a property popularly known as the "butterfly effect".

To study the model sensitivity to initial conditions, experiments in which artificial errors, consistent with known observational uncertainty, are introduced into computer models of the atmosphere. The model results using these alternative initial conditions can then be compared with the original runs; the comparisons suggest that it is unlikely that we will ever predict the *precise evolution* of the atmosphere for longer than a few weeks (Lorenz, 1982).

The impact of chaos on meteorology has not been entirely negative. Instead, it has led to a shift in emphasis – a shift likely to be useful to weather risk management professionals. The atmosphere is not uniformly sensitive to initial conditions; its sensitivity depends on its current state. On some days the atmosphere can be approximated by a relatively predictable linear system over a short enough period of time. On other days, the inevitable uncertainty that exists in the analysis can lead to rapid error growth in the forecast. The key point is that *the predictability of the atmosphere depends upon the state of the atmosphere*. Predicting predictability has now become a major component of operational weather forecasting.

ENSEMBLE FORECASTING
Both the ECMWF and the US National Centre for Environmental Prediction (NCEP) have been running daily ensemble forecasts since 1992. The idea behind ensemble forecasting is simple: run several forecasts using slightly different initial conditions generated around the analysis. The relative divergence of the forecasts indicates the predictability of the atmospheric model in its current state – the greater the divergence the lower the confidence in the forecast. Thus, ensemble forecasts should provide *a priori* information on the reliability of that day's forecast. The difficulty arises in deciding what perturbations (errors introduced into the initial condition) to make to the analysis. Limited computing power means that only a few dozen forecasts can be produced (even when the ensemble forecasts are produced using a lower resolution model than the main forecasts).

The ensemble formation methods discussed above attempt to account for errors in the initial condition. Other ensemble forecasts use models with different combinations of parameterisation schemes for sub-grid processes, in an attempt to estimate how the uncertainty in the choice of scheme affects the final forecast (Stensrud, 2000). Another way in which ensemble members may differ is by introducing random terms into the dynamical equations of the model. This approach attempts to model the uncertainty in the future evolution of the atmosphere at each time step of the model. In such a stochastic parameterisation, the impact of sub-grid processes on the resolved flow is not assumed to be a deterministic function of the model state. Instead, it is a random variable. The parameters of the distribution from which this variable is selected (such as its mean and variance) can be determined by the model state. Due to the non-linearity of NWP models, introduction of these stochastic terms can actually improve the mean state of the model in addition to helping to assess the uncertainty in its evolution (Palmer, 2001). Stochastic parameterisation is a new feature of numerical weather prediction, and is partly a manifestation of meteorology's willingness to accept uncertainty and its attempt to quantify it.

For the purpose of pricing weather derivatives, ensemble forecasts are much more useful than traditional single forecasts. For example, each member of an ensemble forecast can be used to calculate the number of heating degree-days (HDDs) accumulated in a period: this provides a rudimentary distribution of future HDDs. At present, however, the relatively small size of the ensembles and the fact that they represent quantities averaged over tens of kilometres rather than at individual weather stations, means that using ensembles in this manner would be ill-advised. We shall discuss how the predictability information contained in current ensemble forecasts might be extracted in the next section.

Once uncertainty in the atmospheric state has been accepted as a fact of life that will not go away, the extension of medium-range forecasting techniques to longer-range seasonal forecasting is not a major leap. As noted above, the sensitivity of the evolution of the atmosphere's state to its initial conditions prohibits the precise prediction of the trajectory of this state for longer than, at best, a few weeks into the future. This means that there is little hope of forecasting whether it will rain on a specific day in a few months time. This does not mean, however, that useful forecasts at lead-times of several months are not possible. It is possible to predict whether a season will be wetter or colder than average at lead-times of over three months and to estimate of the probability of magnitudes of change – the probability it will be at least 1°C warmer than average, for example.

The crucial extra ingredient required for seasonal forecasting is a computer model of the ocean. The timescales on which the state of the ocean changes are quite long compared to the lead-time of a medium-range forecast. Because of this, when making such a forecast, the state of the ocean can be held fixed. Beyond a couple of weeks, however, the changing state of the ocean is an important influence on the behaviour of the atmosphere. To make progress in seasonal forecasting, a model of the ocean must be coupled to the atmosphere model. The atmosphere forces the ocean through wind stress at its surface, while the ocean forces the atmosphere by exchanging heat with it, especially through the evaporation of water – which forms clouds – and radiation.

The sea surface temperature (SST) is an important influence on the behaviour of the atmosphere, particularly in the tropics. It is the coupling of the ocean and atmosphere that lies behind much inter-annual climate variation such as El Niño–Southern Oscillation (ENSO).[5] ENSO is characterised by a large-scale cycle of warming and cooling in the Eastern tropical Pacific that repeats on a timescale of 2–7 years. The warm SST of the El Niño phase of the cycle influences the atmospheric circulation over large parts of the globe. In particular, El Niño events are associated with heavy rainfall in Peru and southern California, mild winters in the Eastern US and drought in Indonesia and northern Australia (Glantz, 1996). These are all probabilistic associations – ENSO is just one influence of the atmosphere's behaviour, although an important one at lower latitudes (see Philander, 1990 for further reading).

In many ways, the existence of ENSO is a blessing, it imposes some degree of regularity on tropical climate that helps seasonal prediction. The numerical ocean models used in seasonal forecasting are not fundamentally different from their atmospheric cousins; they rely on the division of the oceans into finite elements, horizontally and vertically; the equations of mass, momentum and energy conservation are integrated numerically, and sub-grid processes are parameterised. Seasonal forecasts at mid-latitudes are not as skilful as in the tropics. Although there are thought to be mid-latitude climate cycles, such as the North Atlantic Oscillation (NAO), they are not as regular and well defined as ENSO. The existence of these cycles has allowed the development of statistical seasonal forecasting models which have skill at lead-times of up to six months (eg, Penland and Margorian, 1993). Like all statistical models, however, the availability of historical data is a major constraint on their refinement. Improvements in seasonal forecasting will require better information about the state of the ocean. Observations of the ocean are not as dense as atmospheric observations. Better ocean data, such as that obtained from the TOPEX-POSEIDON satellite, and its successor JASON-1, which measure sea surface height, should enable better estimation of the state of the ocean, and thus improved seasonal forecasts.[6]

Beyond seasonal timescales, forecasts that have more skill than climatology are elusive. It is possible, however, that coupled numerical models of the ocean–atmosphere system can help to improve the climatological distributions that are used to assess weather risk. The instrumental records for most locations are quite short, often only extending back a few decades. Extended runs of oceanic–atmospheric general circulation models (GCMs) could enable better estimates of the risk of extreme events that may only have occurred on a handful of occasions in recent history. This is particularly true if secular changes to forcings of the ocean–atmosphere system – such as enhanced radiative forcing due to increased carbon dioxide and other greenhouse gases (Harries *et al.*, 2001 and IPCC, 2001) – reduce the relevance of the historical record. Before GCMs can be used for this type of risk assessment, it must be demonstrated that they can reproduce observed historical climate variability on regional scales, not just average global temperatures.

Interpretation and post-processing model output

The output generated by numerical models, should not be considered as "weather forecasts". The output merely represents the state of the model, which contains information about the weather but is not immediately relevant to quantities that are actually observed. There are many ways in which extra processing of the raw forecast products produced by forecast centres can substantially increase the value of these forecasts. For example, even the highest resolution global models cannot resolve the details of mountain topography or small islands, yet these physical features can have a substantial impact on the local weather conditions. One of the ways in which human forecasters can add value to a forecast is by using their experience of the weather in a particular locale to predict the likely conditions there, given the larger-scale weather picture that the numerical forecast provides.

The finite resolution of numerical models, that is, the grid size, limits its application to areas no smaller than the grid. The user of the model forecast, however, is likely to need to know the values of forecast variables on a scale smaller than the model grid. Pricing a weather derivative may require the temperature at the London Heathrow weather station, for example, but the ECMWF model predicts a temperature that is averaged over a grid box of 40 km by 40 km. "Downscaling" is the term used to describe a variety of quantitative methods that use the values of forecasted model variables for estimating the values of specific variables on scales smaller than the model grid.

One common method is the use of model output statistics (MOS) (see Glahn and Lowry, 1972). A small number of model variables is chosen as the set of predictors of the desired variable. These predictors are extracted from past numerical forecasts, and correlated with the corresponding observational record of the desired variable. A statistical model can then be constructed that predicts the desired variable, using the forecast variables as predictors. This statistical model should remove any systematic biases in the numerical model. How MOS should be extended to ensemble forecasts is not obvious. One could use each member of the ensemble as the predictor in a traditional statistical MOS model. However, the forecast uncertainty represented by the ensemble must also be downscaled. This is because the forecasts are averaged over grid boxes that are tens of kilometres in size. The uncertainty of a forecast averaged over a grid box should be lower than the uncertainty of a forecast at a single weather station (that averaging reduces variability is a much exploited fact in both science and finance). So, while traditional MOS can be used to downscale the mean of the ensemble, obtaining an appropriate estimate of uncertainty is trickier. Methods for extracting the information about predictability that exists in the ensemble forecasts are now being developed (eg, Roulston *et al.*, 2002).

Another method for downscaling a forecast is "nested modelling". This uses a local area numerical model, covering a much smaller region than the global model, but with a much higher resolution. The local model is "nested" inside the global model. This means that it is integrated forward like the global model, but that on the edges of the region it covers the values of its variables are obtained by using the corresponding values from the global model (interpolated onto the local model's higher resolution grid). An example of using a nested model in a forecast application can be found in Kuligowksi and Barros (1999).

FORECAST UNCERTAINTY

Many applications of weather forecasts require an estimate of the uncertainty associated with the forecast. A single, best-guess forecast can be converted into a probabilistic forecast by adding a distribution of historical forecast errors. The historical forecast errors are the differences between previous forecasts and the

PANEL 3

EVALUATION OF FORECASTS

When choosing forecast products to help manage their weather risk, users should obviously be concerned with the skill level of the forecasts. Unfortunately, it is difficult to assess skill when forecasting multiple variables with numerical models.

A common measure of skill used by meteorologists is the mean square error (MS error). This is just the mean of the square of the difference between the forecasted value and the observed value.

The mean can be an average over time for a univariate forecast (eg, temperature at Heathrow airport), or a time-space average for a multivariate forecast (eg, the 500hPa height field over the northern hemisphere). The MS error, however, is not a good way to evaluate an ensemble or probabilistic forecast. The ensemble average can be calculated and assessed this way but this approach completely fails to take into account the information about uncertainty inherent in the ensemble forecast, information of great interest to risk managers.

Verifying probabilistic forecasting systems requires different measures of skill. The class of measures most relevant to risk manages is probabilistic "scoring rules". To use these rules the number of possible outcomes is usually a finite number of classes. Examples of such classes might be rain/no rain or a set of temperature ranges. A probabilistic forecast will assign a probability to each of the possible outcomes. Scoring rules are functions of the probability that was assigned to the outcome that actually occurred. If this probability is p then the quadratic score of the forecast is p^2. The logarithmic score is log p.7 The quadratic scoring rule forms the basis for the "Brier score" and the "ranked probability score" (Brier, 1950). The logarithmic scoring rule is connected to the information content of the forecast, and also to the returns a gambler would expect if they bet on the forecast (Roulston and Smith, 2002).

Another method for evaluating probabilistic forecasts is the cost-loss score (Richardson, 2000). The cost-loss score is based on the losses that would be incurred by someone using a probabilistic forecast to make a simple binary decision. Consider the situation where the user must decide whether to grit the roads tonight. The cost of gritting is C. If it does not freeze, no loss is suffered if the roads are not gritted. If it does freeze, ungritted roads will lead to a loss, L. If p is the forecast probability of it freezing tonight, then the expected loss if the user does not grit is pL. If p>C/L then this expected loss exceeds the cost of gritting and a rational user would send out the gritting trucks. The cost-loss score illustrates how important it is to have a probabilistic forecast. A user with a C/L ratio far from 0.5 would be ill-advised to take the course of action suggested by a best-guess forecast. Best-guess forecasts can be easily converted into probabilistic forecasts by estimating the historical forecast error distribution (ie, the errors of previous forecasts). Such a conversion should always be performed before evaluating best-guess forecasts using any skill score designed for probabilistic forecasts. Not doing so will artificially inflate the advantage of using an ensemble system. Users that can formalise their decision-making can directly estimate the value of forecasts by determining their impact on decisions. Decision-making processes for real users will be more complex than the cost-loss scenario described above but, the principle of utility maximisation should still apply. Katz and Murphy (1997) contains detailed articles on common evaluation methods for weather forecasts.

corresponding value of the variable that was later observed. Historical error distributions can be refined by allowing them to vary seasonally. For example, in the winter, only forecast errors from previous winters should be used to estimate the distribution. In principle, one can go further and choose only the historical errors that occurred when the atmospheric state was similar to the current state. One must then confront, however, similar problems faced when trying to build purely statistical forecasting models, namely too little data and too few similar days. The data problem for forecast error prediction is even more acute than for statistical weather prediction. Frequent upgrades of operational numerical models mean that the time-series of past forecast errors for current models are seldom longer than a couple of years.

Ensemble forecasts contain information about the state-dependent forecast uncertainty. Ideally, this information should be propagated through any downscaling procedure in an "end-to-end" forecast. In an end-to-end forecast the output of a weather forecasting model is processed to obtain a final forecast for the (weather-dependent) variable in which the user is actually interested. This could be electricity demand or wind energy supply, for example (Smith *et al.*, 2001). A statistical downscaling model based on MOS can be applied to each member of an ensemble forecast. This, however, will probably lead to an underestimate of forecast uncertainty. The uncertainty associated with a weather variable measured at a particular location will be higher than the uncertainty in that variable averaged over the resolution of the numerical model that generated the ensemble. Therefore, the forecast uncertainty contained in the ensemble must also be downscaled. One possible approach is to include statistics describing the entire ensemble (eg, ensemble spread) in the MOS predictors that are used to construct the statistical downscaling model.

Conclusion

Over the last 2,000 years, the prediction horizon of state-of-the-art weather forecasts has advanced significantly from seeing one day ahead based on the colour of the sky to almost a two-week outlook. The risk management community moved more quickly, as in only a few years, the very concept of a weather forecast has changed from a single best-guess of the future to a distribution of likely future weather scenarios.

This chapter has outlined the methods used to produce modern numerical weather forecasts. It was also claimed that the modelling techniques used for making short-range and medium-range forecasts are not fundamentally different from the approaches that must be adopted to forecast climate on seasonal scales and beyond. The main difference is that, for longer-range forecasts, a model of ocean dynamics must be coupled to the atmospheric model. This chapter has also stressed the importance of ensemble forecasting for quantifying forecast uncertainty. Ensemble forecasting is on its way to becoming the standard approach to forecasting the evolution of the atmosphere and the ocean. This is a welcome development for the risk management community because ensemble forecasts contain information about the uncertainty of forecasts, an uncertainty that is synonymous with risk.

The ultimate aim of accurate probability forecasts of commercially relevant variables is being pursued on timescales from a few hours to several months (Palmer, 2002). For those who understand how to interpret it, this information will give a competitive edge at a range of lead-times, from pricing next week's electricity futures to pricing next year's weather derivatives. Skilled and timely forecasts can be financially rewarding.

1 *Richardson's work led to his book "Weather Prediction by Numerical Processes", which is considered to be the foundation of modern meteorology (Richardson, 1922).*

2 *All operational forecast centres frequently upgrade their forecast models. Details of these upgrades and of the present configuration of the ECMWF model can be found on their website http://www.ecmwf.int.*

3 *A description of the ECMWF and its output products can be found in Persson (2000).*

4 *Details of reanalysis projects and data can be found at http://ecmwf.int/research/era/ and http://www.cdc.noaa.gov/ncep_reanalysis.*

5 *Gilbert Walker discovered the atmospheric Southern Oscillation in the 1920s. The relationship between this variation and the oceanic El Niño phenomenon was discovered in the 1960s by Jacob Bjerknes, son of Vilhelm.*

6 *Information on these satellites is available at http://topex-www.jpl.nasa.gov/mission/jason-1.html.*

7 *Readers may be curious as to why the simplest linear skill score =p has been omitted. It turns out that this score is "improper" in the sense that a forecaster increases their expected score by reporting probabilities that differ from what they believe to be true.*

BIBLIOGRAPHY

Brier, G. W., 1950, "Verification of Forecasts Expressed in Terms of Probabilities", *Monthly Weather Review*, 78, pp. 1–3.

van den Dool, H. M., 1994, "Searching for Analogues, How Long must we Wait?" *Tellus A*, 46, pp. 314–24.

Glahn, H. R., and D. A. Lowry, 1972, "The Use of Model Output Statistics (MOS) in Objective Weather Forecasting", *Journal of Applied Meteorology*, 11, pp. 1203–11.

Glantz, M. H., 1996, *Currents of Change: El Niño's Impact on Climate and Society*, (Cambridge University Press).

Harries, J. E., H. E. Brindley, P. J. Sagoo and R. J. Bantges, 2001, "Increases in Greenhouse Forcing Inferred from the Outgoing Longwave Radiation Spectra of the Earth in 1970 and 1997", *Nature*, 410, pp. 355–7.

IPCC, Intergovernmental Panel on Climate Change, 2001, *Climate Change 2001: Synthesis Report*, (ed) R. T. Watson, (Cambridge University Press).

Katz, R. W., and A. H. Murphy, 1997, *Economic Value of Weather and Climate Forecasts*, (Cambridge University Press).

Kuligowski, R. J., and A. P. Barros, 1999, "High-Resolution Short-Term Quantitative Precipitation Forecasting in Mountainous Regions Using a Nested Model", *Journal of Geophysical Research (Atmospheres)*, 104, pp. 31, 553–64.

Lempfert, R. G. K, 1932, "The Presentation of Meteorological Data", *Quarterly Journal of the Royal Meteorological Society*, 58, pp. 91–102.

Lorenz, E. N., 1982, "Atmospheric Predictability Experiments with a Large Numerical Model", *Tellus*, 34, pp. 505–13.

Lorenz, E. N., 1993, *The Essence of Chaos*, (London: UCL Press).

Murphy, A. H., 1997, "Forecast Verification", in: *Economic Value of Weather and Climate Forecasts*, (eds) R. W. Katz and A. H. Murphy, (Cambridge University Press), pp. 19–70.

Nebeker, F., 1995, *Calculating the Weather*, (San Diego: Academic Press).

Palmer, T. N., 2000, "Predicting Uncertainty in Forecasts of Weather and Climate", *Reports on Progress in Physics*, 63, pp. 71–116.

Palmer, T. N., 2001, "A Nonlinear Dynamical Perspective on Model Error: A Proposal for Non-Local Stochastic-Dynamic Parameterization in Weather and Climate Prediction Models", *Quarterly Journal of the Royal Meteorological Society*, 127, pp. 279–304.

Palmer, T. N., 2002, "The Economic Value of Ensemble Forecasts as a Tool for Risk Assessment: From Days to Decades", *Quarterly Journal of the Royal Meteorological Society*, 128, pp. 747–74.

Penland, C., and T. Magorian, 1993, "Prediction of Niño 3 Sea Surface Temperature Using Linear Inverse Modeling", *Journal of Climate*, 6, pp. 1067–76.

Persson, A., 2000, "User Guide to ECMWF Forecast Products", European Centre for Medium Range Weather Forecasting, Reading.

Philander, S. G., 1990, *El Niño, La Niña, and the Southern Oscillation*, (San Diego: Academic Press).

Richardson, L. F., 1922, *Weather Prediction by Numerical Processes*, (Cambridge University Press).

Richardson, D. S., 2000, "Skill and Relative Economic Value of the ECMWF Ensemble Prediction System", *Quarterly Journal of the. Royal Meteorological. Society*, 126, pp. 649–67.

Roulston, M. S., and L. A. Smith, 2002, "Evaluating Probabilistic Forecasts using Information Theory", *Monthly Weather Review*, 130, pp. 1653–60.

Roulston, M. S., D. T Kaplan, J. Hardenberg and L. A. Smith, 2002, "Using Medium Range Weather Forecasts to Improve the Value of Wind Energy Production", *Renewable Energy*, Forthcoming.

Smith, L. A., M. S. Roulston, and J. Hardenberg, 2001, "End to End Ensemble Forecasting: Towards Evaluating the Economic Value of the Ensemble Prediction System", ECMWF Technical Memorandum 336, ECMWF, Reading.

Stensrud, D. J., J. W Bao and T. T. Warner, 2000, "Using Initial Condition and Model Physics Perturbations in Short-Range Ensemble Simulations of Mesoscale Convective Systems", *Monthly Weather Review*, 128, pp. 2077–107.

Walker, G. T., 1923, "Correlation in Seasonal Variations of Weather VIII", Memorandum of the Indian Meteorological Deptartment, 24, pp. 75–131.

Walker, G. T., 1924, "World Weather IX", Memorandum of the Indian Meteorological Deptartment, 24, pp. 275–332.

Walker, G. T., 1928, "World Weather III", Memorandum of the Royal Meteorological Deptartment, 17, pp. 97–106.

Walker, G. T., 1937, "World Weather VI", Memorandum of the Royal Meteorological Society, 4, pp. 119–39.

Walker, G. T., and E. W. Bliss, 1932, "World Weather", *Memorandum of the Royal Meteorological Society*, 4, pp. 53–84.

8

Weather Derivative Modelling and Valuation: A Statistical Perspective

Anders Brix, Stephen Jewson and Christine Ziehmann

Risk Management Solutions

Weather derivatives are usually priced by analysing the historical outcomes of the underlying weather index. In this chapter we review statistical and actuarial methods for such analyses and discuss the relevance of arbitrage pricing. We look at reasons for trends in historical data and describe how to estimate and remove them. Statistical methods for modelling and validating models for weather indexes and daily temperatures are discussed, and in particular we show that traditional ARMA time-series models are not adequate for modelling daily temperatures. We show how dependencies between different indexes and different locations can be modelled and we review some of the methods for risk loading of actuarial prices.

Introduction
Weather derivatives are different from most other derivatives in that the underlying weather cannot be traded. Furthermore, the weather derivatives market is relatively illiquid. This means that weather derivatives cannot be cost-efficiently replicated with other weather derivatives, ie, for most locations the bid–ask spread is too large to make it economical to hedge a position. One of the consequences of this is that valuation of weather derivatives is closer to insurance pricing than to derivatives pricing (arbitrage pricing). For this reason it is important to base valuation on reliable historical data, and to be able to model the underlying indexes accurately.

In the future, the weather derivatives market may become more liquid, and at that point it may be possible to use other weather derivatives for hedging and thereby derive prices from the market for at least some contracts. However, the main purpose of this chapter is to review how weather derivatives are priced using historical data, and to highlight some of the challenges that arise when doing so. The presentation is statistical in its focus on the choice of models and model validation.

Before discussing the topics of detrending (removing trends from a time-series), index and daily temperature modelling, portfolio modelling and risk loading (the risk premuim added to the expected payoff in order to compensate the risk bearer for taking on risk), this chapter begins with with a discussion of the relationship between index and payoff distributions, which will be useful in the later sections.

Throughout the chapter the methods described will be illustrated with a relatively commonly traded contract: a New York LaGuardia, May 1–September 30 cooling degree-day (CDD) call option. Because more than 90% of the weather derivatives currently traded are based on temperature (WRMA, 2002), the main focus of this chapter will be on models for such indexes. Most of the index-based methods

WEATHER DERIVATIVE MODELLING AND VALUATION: A STATISTICAL PERSPECTIVE

described would, however, apply to indexes based on other weather variables. The models for daily temperatures, on the other hand, would probably not apply to any other index types. The formulae are illustrated with some commonly used distributions in Panel 1.

Deriving the payoff distribution

The simplest way to understand the distribution of possible payoffs related to a weather index is to model the distribution of the index rather than the payoff. The reason for this is that limits on the payoff result in payoff distributions that are mixtures of discrete and continuous distributions in the sense that they will have discontinuities at zero and at the limit (see the following sub-section "A call option example"). Index distributions on the other hand are usually either discrete or continuous. This means that it is often possible to find a standard statistical distribution which models the index in a satisfactory way. The question of finding such a distribution is discussed in more detail in later. Alternatively, we could model daily temperatures, derive the corresponding index distribution. This will be discussed further in in the "Modelling daily temperatures" section.

Once we have an estimate of the index distribution we can derive the payoff distribution of the weather derivative. Exact and approximate expressions can be derived in many cases, but it is often simpler and faster to simulate realisations from the estimated index distribution, and convert each simulated index into a payoff. Nevertheless, knowing how the payoff distribution is derived theoretically from the index distribution can be useful for model validation and understanding which aspects of the index distribution are the most important.

In the following, I denotes the underlying index and we denote that the distribution of I has cumulative distribution function (CDF) F. As an example we will consider a call option on a degree-day index, but other contract types are treated in similar ways.

A CALL OPTION EXAMPLE

We consider a call option with strike S (in degree days), limit L (in degree-days) and tick d (in US dollars per degree-day). The payoff P of the call option is given by

$$P = d(\min(I, L) - \min(I, S)) \quad (1)$$

$$= \begin{cases} 0 & I \leq S \\ d(I - S) & S < I \leq L \\ d(L - S) & I > L \end{cases}$$

Using the tick we can convert the limit, L, into a USdollar limit $L_\$ = d(L-S)$, and the CDF G of the payoff can be expressed as:

$$G(P) = \begin{cases} F(S) & P = 0 \\ F\left(S + \dfrac{P}{d}\right) & 0 < P \leq L_\$ \\ 1 & P > L_\$ \end{cases} \quad (2)$$

Because of the strike and limit, the payoff P is partly discrete with point masses at zero and the limit $L_\$$. Between these two points the distribution is discrete or continuous depending on whether the underlying index distribution is discrete or continuous.

The mean of the payoff can be conveniently calculated using the limited expected value (LEV) L_I for the index I[1]:

$$L_I(m) = E \, min(I, m)$$
$$= \int_{-\infty}^{m} I \, dF(I) + m \, (1 - F(m))$$

Here, m is the arguement of L_I and it is seen that if we let m tend to infinity, $L_I(m)$ tends to the expected value of I. Taking the expectation of Equation (1) and using the definition of $L_I(m)$, we see that the mean payoff is given by:

$$EP = d \, (L_I(L) - L_I(S))$$

This expression shows that the LEV function is one of the fundamentally relevant properties of the index distribution when pricing weather derivative options. If other moments of the payoff distribution are of interest, they can be calculated using higher-order LEV functions such as $E \, min(m, I)^k$.

The formulae for calculating payoff distributions and LEV functions are illustrated with some commonly used distributions in examples 1–3 in Panel 1.

Adjusting for warming and cooling trends

One of the common requirements for accurate modelling of weather indexes is that the series of historical indexes is stationary, which roughly means that the distribution of indexes does not change over time. Obviously, stationarity cannot be assumed: climate change and urbanisation may both be reflected in data. Climate change refers to variations in the Earth's climate on large spatial scales due to either natural variation or anthropogenic changes in the composition of the atmosphere. In contrast, urbanisation, which can also greatly alter a measurement station's temperature record, is a local and regional effect.

In global temperature trend studies, correction for local effects is often made either by using rural stations alone or by using rural stations to "correct" measurements from stations in urban areas. Several papers describe such studies, and their main findings are:

1. The global average surface temperature has increased by 0.6 (+/– 0.2) degrees Celsius since the late 19th century (IPCC, 2001).
2. Trends depend on the period chosen. For the US, for example, the temperature history since 1910 can be divided into three periods: a warming period until 1940, a cooling period from 1940 to 1970, and the recent warming period from 1970 to the present (Knappenberger *et al.*, 2001).
3. Trends depend on location. The recent period of warming has been almost global, but the largest increases of temperature have occurred over the mid- and high latitudes of continents in the northern hemisphere. Year-round cooling is only evident in the northwestern North Atlantic and the central North Pacific Oceans (although the North Atlantic cooling appears to have reversed recently, see for example Hansen *et al.*, 1996 and IPCC, 2001).
4. Trends for maximum and minimum temperatures are different. The diurnal range, the difference between daily maximum and daily minimum temperatures, is decreasing, although not everywhere. On average, minimum temperatures are increasing at about twice the rate of maximum temperatures (see for example Easterling *et al.*, 1997 and IPCC, 2001).
5. A recent study of Knappenberger *et al.* (2001) shows that the trends are not uniform; cool days are much warmer than they used to be, whereas warm days are not.

Urbanisation effects are best demonstrated by comparing temperature time-series from neighbouring stations where one is from an area with little urban change while major changes have been made to the surroundings of the other. Such a comparison

PANEL 1

DERIVING THE PAYOFF DISTRIBUTION

Example 1
Consider the case where the underlying index follows a *log normal distribution* with parameters μ and σ2, ie, log *I* follows a normal distribution with mean μ and variance σ^2. The payoff CDF is:

$$G(P) = \begin{cases} \Phi\left(\dfrac{\log S - \mu}{\sigma}\right) & P = 0 \\ \Phi\left(\dfrac{\log(S + P/d) - \mu}{\sigma}\right) & 0 < P \le L_\$ \\ 1 & P > L_\$ \end{cases}$$

where Φ is the standard normal CDF. The LEV function is:

$$L_I(m) = \exp\left(\mu + \frac{\sigma^2}{2}\right) \Phi\left(\frac{\log m - \mu}{\sigma} - \sigma\right) + m(1 - G(m))$$

Hence the expected payoff is:

$$EP = d\exp\left(\mu + \frac{\sigma^2}{2}\right)\left(\Phi\left(\frac{\log L - \mu}{\sigma} - \sigma\right) - \Phi\left(\frac{\log S - \mu}{\sigma} - \sigma\right)\right)$$
$$+ dL(1 - G(L)) - dS(1 - G(S))$$

Example 2
If the index follows a *normal distribution*, with mean μ and variance σ^2, the situation is a bit more complicated because the limited expected moments are less tractable. The payoff CDF is simple:

$$G(P) = \begin{cases} \Phi\left(\dfrac{S - \mu}{\sigma}\right) & P = 0 \\ \Phi\left(\dfrac{S + P/d - \mu}{\sigma}\right) & 0 < P \le L_\$ \\ 1 & P > L_\$ \end{cases}$$

The LEV function is given by

$$L_{\mu,\sigma}(m) = \sigma L_{0,1}\left(\frac{m - \mu}{\sigma}\right) + \mu\Phi\left(\frac{m - \mu}{\sigma}\right) + m\left(1 - \Phi\left(\frac{m - \mu}{\sigma}\right)\right)$$

Here the LEV $L_{0,1}$ for the standard normal is given by

$$L_{0,1}(m) = \int_{-\infty}^{m} \frac{u}{\sqrt{2\pi}} \exp\left(-\frac{1}{2}u^2\right) du$$

which can be calculated numerically.

Example 3
The general formulae above apply equally well when the underlying index follows a discrete distribution. Consider the case where I follows a *negative binomial distribution*. Although it is easy to calculate numerically, there is no closed-form expression for the CDF of this distribution. In the following we will denote the CDF of the negative binomial with mean $r(1-p)/p$ and variance $r(1-p)/p^2$ by $F_{r,p}$. The payoff CDF is given by Equation (2) with F replaced by $F_{r,p}$ and the LEV function is given by:

$$L_{r,p}(m) = \frac{rp}{1-p} F_{r+1,p}(m) + m(1 - F_{r,p}(m))$$

can only be made if the two stations are in the same microclimate region. As an example, this chapter will compare the CDD index for New York LaGuardia Airport with the corresponding index for Central Park, New York. These two stations are in the vicinity of each other, but whereas much has been built in the LaGuardia area over the last 30–40 years and it is at the water's edge, not much has changed structurally over the same period around Central Park, in the heart of the New York City. Figure 1 shows the historical CDD indexes for the two locations with linear trends overlaid. Visually, the difference between the two plots is striking, and t-tests of the significance of the slopes of the trendline reveal that the Central Park trend is not significant (p=34%), while the LaGuardia trend is highly significant (p=0.06%).

For the purposes of this chapter, it is not important to distinguish between local and global trends because we are interested in removing the combined trend. There are many models in the statistical literature for estimating distributions with trends. The way the trend is incorporated is often linear or multiplicative in the mean (for example, an index is typically modelled as trend plus noise for a normal distribution, and trend multiplied by noise for a log-normal distribution), and is usually chosen on the basis of mathematical convenience rather than reflecting reality (since it can be difficult to find a 'realistic' model for many applications and since the appropriateness of the model must be validated anyway). This section separates trend estimation from distribution estimation on the basis that this simplifies the calculations, and because we can then treat all distributions in the same way.

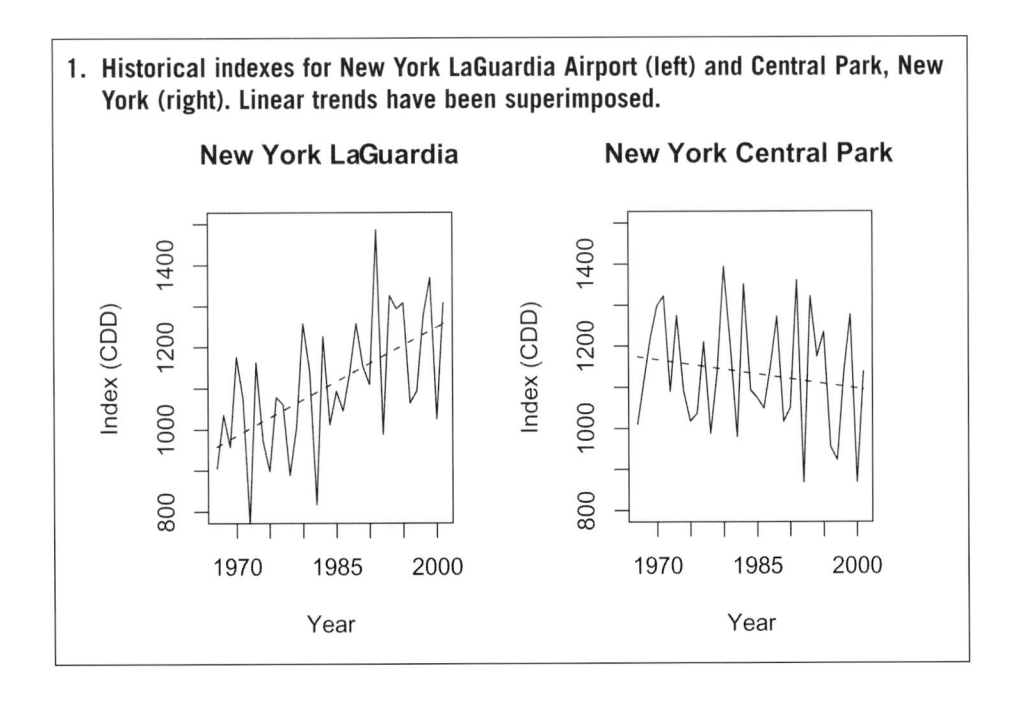

1. Historical indexes for New York LaGuardia Airport (left) and Central Park, New York (right). Linear trends have been superimposed.

Trend estimation is done in essentially the same way for daily temperatures and annually compounded indexes. Because seasonal effects can complicate the process of daily detrending, detrending of annually compounded indexes is described first, and daily detrending later. Non-stationarity due to seasonality is discussed under the heading of "Capturing seasonality of temperatures" below.

There can be other sources of non-stationarity than trends, such as, relocation of measurement stations or equipment and sudden changes to the surroundings of stations. The effects of such measurement discontinuities can be better dealt with by first enhancing the underlying data, as described in detail in Chapters 5 and 6. This chapter is concerned with the more gradual changes that remain after those that are associated with measurement discontinuities have been removed.

INDEX DETRENDING

The assumption of the trend model used here is that an index I_i can be represented as a sum of a trend R_i and a random variable e_i:

$$I_i = R_i + e_i, \qquad i=1, ..., n$$

e_i are assumed to be independent and identically distributed with mean zero. The detrended indexes, $\tilde{I}_i$, are then defined as

$$\tilde{I}_i = I_i - \hat{R} + \hat{R}_n \qquad (3)$$

where $\hat{R}$ and $\hat{R}_n$ are the estimated trends for indexes i and n.

In this way, the mean of all indexes are shifted to the estimated mean of the last index. Often the contract will commence a year (or more) after of the end of the historical indexes. If the trend is thought to continue after the last historical data point it can be extrapolated to year $n + k$, where k is the number of years to extrapolate forward. We then replace R_n by R_{n+k} in Equation (3), and get the k-year ahead forward detrended indexes. For example, suppose we are looking at data from a station which has experienced large growth in urbanisation in its surroundings in recent years and where the urbanisation is still continuing. If we are considering a contract for the winter, the most recent historical data would be from the previous winter and we would need to extrapolate the estimated trend in order to capture the trend introduced by continuing urbanisation.

Parametric trends

In this chapter, trends are assumed to be smooth and vary slowly in time since we have assumed that jumps due to for example station relocations have been removed. Therefore, it is often reasonable to approximate trends by parametric curves such as linear or polynomial functions. The standard way of estimating the parameters of the trend is by ordinary least squares (OLS), ie, minimising the sum:

$$\sum_{i=1}^{n}(I_i - R_i)^2$$

With y_i denoting the year of index i, the trend R_i could then be parameterised by, for example:

$$R_i = a + b\,y_i \qquad \text{(linear)}$$

$$R_i = a + b\,y_i + c\,y_i^2 \qquad \text{(quadratic)}$$

$$R_i = a\exp(b\,y_i) \qquad \text{(exponential)}$$

OLS estimates are known to be sensitive to extreme observations, so if outliers are present, a more robust estimation procedure may be needed – least absolute deviations for example (ie, minimise $\sum_{i=1}^{n} | I_i - R_i |$. See Huber (1981) for a detailed treatment of robust statistics).[2]

Non-parametric trends
Sometimes it may be desirable to use a non-parametric trend if there is reason to believe that parametric trends do not provide a satisfactory approximation for the period considered. In such situations various non-parametric methods are available (see, for example, Bowman and Azzalini, 1997). The simplest is called the "moving average" method, where the trend in year *i* is estimated as the average of the neighbouring years:

$$R_i = \frac{1}{2w + 1} \sum_{i=-w}^{w} I_{i+w}$$

The number of neighbouring years, $2w + 1$, is usually called the "window", and the years may be weighted such that years closer to the base year contribute more than years that are further away. The main disadvantage of moving average estimation is that it does not extrapolate the trend beyond the last historical year.

An alternative that allows extrapolation is the "loess" method (Cleveland and Devlin, 1988), which is based on local parametric regressions. Linear loess, for example, estimates the trend for year *i* by weighted linear regression, with most weight on nearby years. Loess is known to have better theoretical properties than moving average estimation, especially close to the edges of the observation window.

Figure 2 shows estimated linear trend and loess trends for the CDD example used in this chapter. The difference between the two trendlines ranges from –26 to 26 CDDs.

DAILY DETRENDING
Detrending of daily temperatures can be done using the methods described above. However, local or global warming effects may have different magnitudes in different seasons (Hansen *et al.*, 1996), thus create different trends at different times of the year. This is not a problem when modelling annual indexes since such seasonality does not appear, and similarly non-parametric trends adapt to each season. Parametric trends, on the other hand, may need to be adapted to vary by time of the year. One way this can be done is to estimate linear trends separately for each month

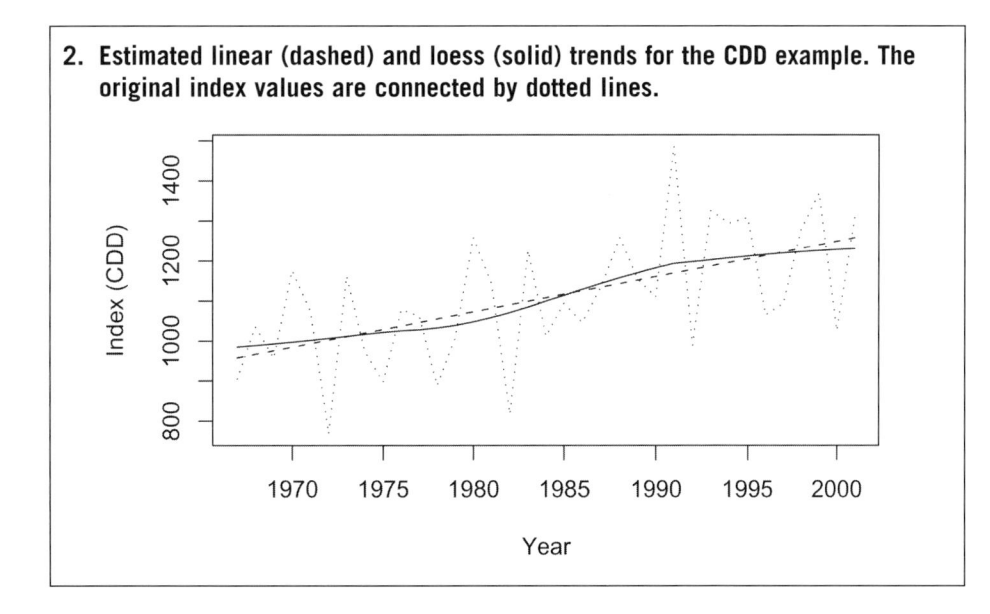

2. Estimated linear (dashed) and loess (solid) trends for the CDD example. The original index values are connected by dotted lines.

of the year and interpolate to get a trend for each day. The seasonal linear trend at day t then becomes:

$$R_t = a + b_t t$$

where b_t is the slope on day t, which will follow an annual pattern.

Even if the analysis is based on annual indexes, it may be worth using daily detrending when considering short contracts. When using index-based trends, the estimated trend for a weekly contract may be implausibly different from the estimated trend for the same contract the following week. This point is illustrated in Figure 3, which shows how estimates of linear trends of average temperature vary by week when they are estimated using only data from that week. By detrending daily values from a period longer than the week of the contract, we use information about the trend in the surrounding weeks, and thus get a more stable estimate.

HOW MANY YEARS OF DATA SHOULD BE USED?
One of the reasons for linear and higher-order polynomial trends is that they provide good approximations to smooth trends over periods of a reasonable number of years. What is a reasonable number of years? The answer varies by location, since many trends are due to local effects, but also because of spatial variation in global warming. Based on backtest studies it has been found that, on average over many stations and many years, a reasonable number of years would be between 15 and 25 if a linear index trend is used.[3] If more data is used the quality of the linear trend line deteriorates due to non-linearity, and if less is used, there is too little data to get an accurate estimate of the index trend and distribution. It must be stressed, though, that this guideline is valid only on average, and that the number of years that is appropriate for individual stations may lie outside this range.

VALIDATING THE TREND
So far this section has discussed how to estimate a given trend, but avoided giving guidelines on how to chose a trend and how to validate it. One (and arguably the best) way to do this is by graphical checks. First, a plot of the index values can give some idea about whether a trend might be present or not. Second, after estimating the trend, a residual plot should be made. The residuals are the difference between the indexes and the trend, and should be scattered around zero without any systematic variation by time, ie, the residuals must show no clear trend, no change in spread over time and there must not be a clear tendency for the points to be mainly

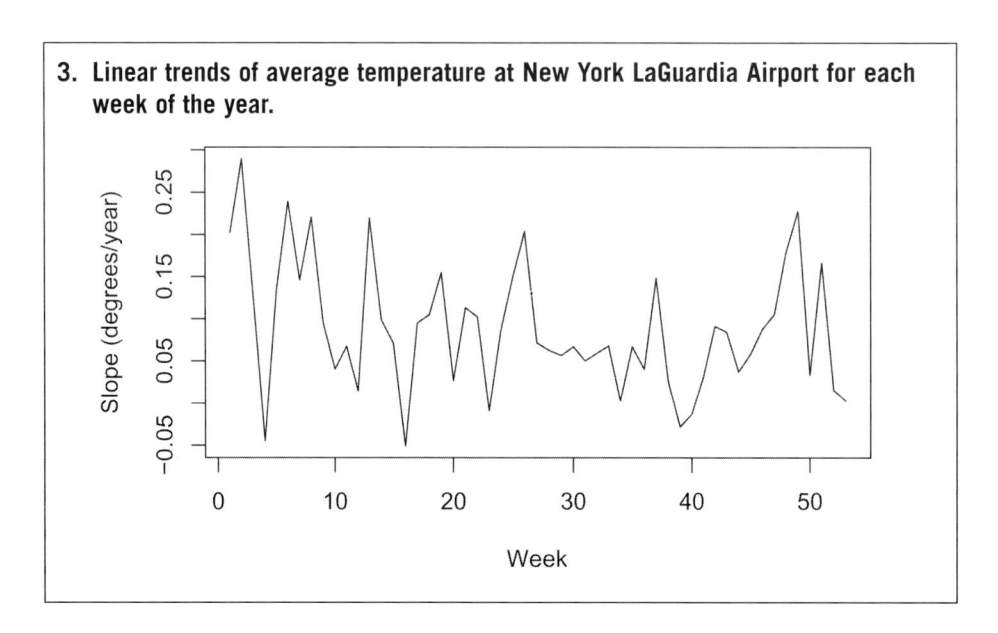

3. Linear trends of average temperature at New York LaGuardia Airport for each week of the year.

negative or positive. For all the methods outlined above, the residuals should also be approximately normally distributed. With only a few index values, validation can be a difficult task, and any external information on when changes in trends could have occurred should be used when choosing a trend estimate. For example, many temperature measurements are made at airports, or in city centres, so information about when the airport or city has grown is useful for determining when a trend could have started.

Weather index modelling

Having described detrending of historical indexes in the previous section, we will now discuss which types of statistical distributions are appropriate for modelling the detrended indexes. Both parametric and non-parametric methods will be considered and a discussion on model validation will follow.

NON-PARAMETRIC INDEX MODELLING

The simplest form of non-parametric distribution is the empirical (historical) distribution of the indexes. In the actuarial literature use of this distribution is known as "burn" analysis. However, given that we usually deal with relatively few historical indexes, the empirical distribution can become very rugged and sometimes have several modes (peaks). If the distribution of the index is thought to be smooth or unimodal, then this may not be realistic. Instead the empirical distribution can be smoothed by a process called "kernel smoothing", whereby the probability density function f of the index distribution taken at the point x is estimated by the expression

$$\hat{f}(x) = \frac{1}{n} \sum_{i=-1}^{n} \frac{1}{h} k\left(\frac{x - \tilde{I}_i}{h}\right)$$

Here, k is a probability density function (PDF), and the degree of smoothing is determined by the bandwidth, h. The effect of kernel smoothing is illustrated in Figure 4, where three Gaussian kernel density estimates have been superimposed on a histogram. Whereas the choice of smoothing function k is not very critical, the bandwidth selection is extremely important for the overall shape of the estimated distribution: the larger h, the more smoothing is obtained.

PARAMETRIC INDEX MODELLING

Even with a large degree of smoothing the kernel distribution may put too much weight on the historical data; for example, we may not believe in multimodality, the

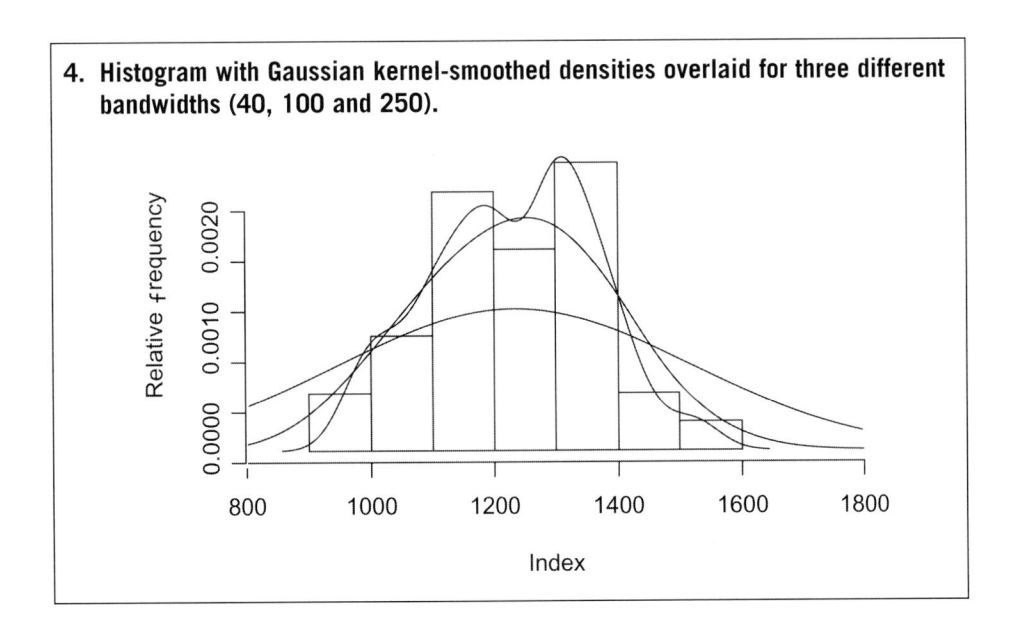

4. Histogram with Gaussian kernel-smoothed densities overlaid for three different bandwidths (40, 100 and 250).

historical distribution may not look smooth enough or we may want to extrapolate further beyond the historical observations than is possible by kernel smoothing. If so, we can use parametric distributions such as normal, gamma or Poisson, depending on what type of index is considered. A normal distribution is often an appropriate choice for degree-day and average indexes, especially for longer contract periods. The reason for this is the central limit theorem, which states that, under very general conditions, the sum of a number of outcomes (like daily degree-days or daily temperatures) approximately follows a normal distribution (see, for example, Casella and Berger, 2002). Similarly, it can be shown that extreme events, such as the number of days with temperature exceeding a high threshold, will be approximately Poisson-distributed if they are close to independent (see for example, Coles, 2001). However, extreme weather events tend to occur in clusters, and so are often better modelled by a negative binomial distribution (see for example, McCullagh and Nelder, 1989).

In general, parameters of the distribution are most efficiently estimated by the maximum likelihood method, ie, the parameter estimates are chosen to maximise the PDF (probability mass function for discrete distributions) as a function of the parameters. Alternatively, parameter estimates can be obtained by deriving expressions for the moments, and matching these with empirical estimates. For the normal distribution the two methods are equivalent and amount to calculating the empirical mean and variance.

VALIDATING INDEX DISTRIBUTIONS

The sparsity of data with which one is often faced in weather index modelling can make it difficult to distinguish between a bad fit due to sampling error and a bad fit due to wrong choice of model. This makes it particularly important to estimate how much variation can be expected due to sampling error and to apply a variety of model checks. The following paragraphs describe a number of graphical techniques, show how these can be made more quantitative through simulation and discuss goodness-of-fit tests.

PP-, QQ- and LL-plots

Two traditional ways of validating the fit of a distribution are by comparing the model PDF and CDF with the histogram and the empirical CDF, respectively. Such comparisons are, however, not without difficulties. For histograms the number of bins and the bin width have to be chosen, and for CDFs it can be difficult to distinguish between the different S-shapes that these usually take. Instead, it is customary in statistics to look at so-called PP- and QQ-plots, which contain the same information as the CDF but allow easier comparison of distributions.

A PP-plot is a plot of the empirical CDF against the model CDF; if the model is good, then the points should be close to a straight line with slope of one and intercept of zero. This way, the problem of comparing S-shapes is turned into a comparison of straight lines, which is much easier. Note that a PP-plot is also useful for getting a qualitative evaluation of lack of fit. If the points fall on a straight line with an intercept different from zero this indicates that the mean is wrong. If the points fall on a straight line with slope different from one this indicates wrong variance. The left panel of Figure 5 shows a PP-plot for the CDD example for New York.

Because all CDFs start at zero and end at one, points in the tails of the distributions on the PP-plot will tend to be close to a straight line. This makes it difficult to evaluate the quality of the fit to the tails of the distribution. A QQ-plot, where model quantiles are plotted against empirical quantiles, is a way around this problem. Again the points should lie close to the straight line with slope of one and intercept of zero if the fit is good. The advantage of QQ-plots is that a bad fit in the tails of the distribution will show more clearly. The middle panel of Figure 5 shows a QQ-plot for the CDD example.

While PP- and QQ-plots provide good tools for evaluation of the overall fit, it is

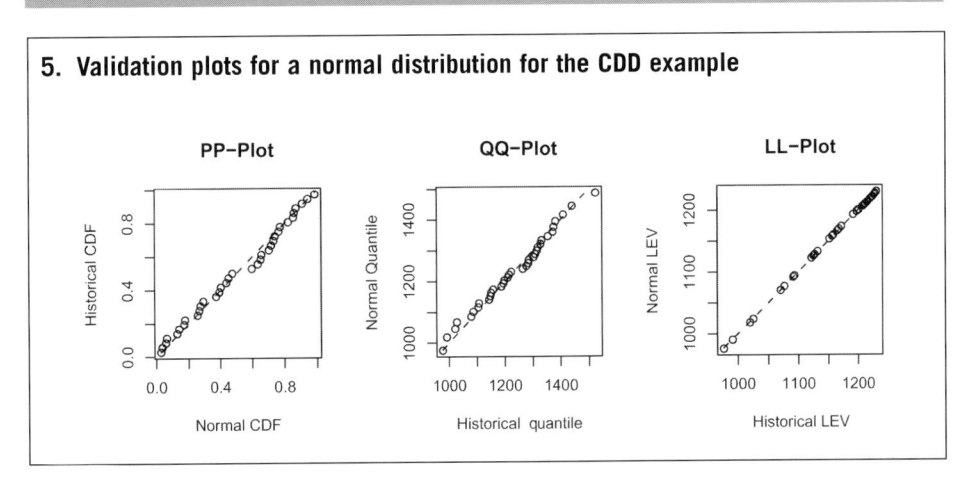

5. Validation plots for a normal distribution for the CDD example

often useful to check the fit of specific characteristics of the distribution that are more closely related to the actual purpose of the modelling. Since we are interested in modelling payoffs, it is important to capture accurately those features of the index distribution which affect the basic characteristics of the payoff distribution. Hence, if we are modelling options or swaps, we should compare the model and empirical LEV functions. As before, we prefer to look for straight lines when validating the fit, so we plot the two against each other. The plot created this way is called an LL-plot, and is shown in right panel of Figure 5 for the CDD example.

Envelopes
The validation plots described in the previous section are purely descriptive, and it can sometimes be hard to see when deviations from a straight line are due simply to sampling variation. In order to make them more quantitative we can add confidence intervals to the plots. It is usually not possible to derive exact expressions for such confidence intervals – we use simulation envelopes instead. This is done by simulating samples of the same length as the historical data from the estimated model; if the model is good, then the CDF of the samples should fall around the historical CDF.

Simulation envelopes for the PP-plot are obtained as follows:

Step 1. Simulate a sample of the same length n as the historical data from the estimated model.
Step 2. Calculate the empirical CDF from the sample at each historical index.
Step 3. Repeat Steps 1 and 2 K times and store the results.
Step 4. Sort the calculated CDF at each value of the historical indexes.

To achieve 90% confidence intervals we could simulate, say, K=100 samples and pick out the fifth lowest and the fifth highest CDF values for each historical index.[4] The envelopes created this way are pointwise confidence intervals; there is 90% probability that the historical CDF at a given point will fall within the envelopes. Because points on the simulated CDFs are highly dependent this does not mean that the probability of the full historical CDF falling within the envelopes is 90% to the power of n.

Similarly for the QQ-plot and the LL-plot, we calculate the quantiles corresponding to the quantiles of the historical indexes, and LEV at each historical index. (Figure 10 shows an example of a QQ-plot within envelopes.)

GOODNESS-OF-FIT TESTS
Apart from graphical checks several goodness-of-fit tests are available, of which the most common are chi-square (χ^2), Shapiro–Wilks, Kolmogorov–Smirnov and Anderson–Darling. Because of the small number of historical indexes that are usually

available, none of these tests are very powerful, ie, they will rarely reject a bad model. Graphical checks, such as the ones described above, often give a much better idea of how appropriate the model is.

❏ The chi-square test can be used for checking the validity of any distribution. The test is done by grouping the observations into intervals and comparing the expected number of observations in each interval with the observed number. Because the test relies on grouping the observations, it requires a substantial number of observations in order to give a powerful result (a general rule of thumb for when to use this test is that the expected number of observations in each group must be at least five).
❏ The Shapiro–Wilks test can be used only for normal distributions (and log-normal by transforming the observations with the logarithm). The test is quite powerful and gives a good indication of whether a normality assumption is reasonable.
❏ Kolmogorov–Smirnov is a classical test which compares the empirical CDF with the model CDF using the maximal vertical difference between the two. The test is not very powerful, but should it result in a low test probability then there is good reason not to rely on the model.
❏ Anderson–Darling can be considered as a modification of the Kolmogorov–Smirnov test which compares the CDFs over the whole range of the distribution. In contrast to the Kolmogorov–Smirnov test, the Anderson–Darling test probability is model-specific and hence the test is more powerful. Both this test and the Kolmogorov–Smirnov test can be used only for continuous distributions.

Modelling daily temperatures

The most common approach to analysing weather derivative contracts is to fit a distribution to the detrended annual indexes. However, often we have only a few historical values from which to estimate this distribution, resulting in significant estimation uncertainty. This uncertainty could be reduced if the data were used more efficiently, and the index approach has some inefficiencies, as shown by the following examples:

❏ For a US cooling degree-day (CDD) index, the index approach uses only information about how far above 65°F the temperature is. It does thus not distinguish between days where the temperature is far below 65°, and days where the temperature is just below 65°.
❏ Event indexes only use data from days on which events occurred. Data on all other days is discarded.
❏ One-week contracts only use data for that week of the year. Data from other weeks is discarded.
❏ For some indexes, in particular indexes relating to short periods and extreme events, it may not be possible to find a suitable model for the index distribution.

These problems could be alleviated if the underlying daily temperature distribution could be modelled. Furthermore, a temperature modelling approach would make it easier to include forecasts in the weather derivative pricing process (see Chapter 10).

To see how much more efficiently data can be used in a daily model compared with an index model, consider the May–September CDD contract for New York LaGuardia Airport. We estimate the index distribution from a daily temperature model and an index model using the same amount of historical data. Both estimates of the index distribution will have an error due to estimation uncertainty, and these errors can be quantified by looking at confidence intervals (envelopes) around the estimated CDFs. Figure 6 shows the estimated CDFs with 90%-confidence envelopes using a normal index distribution and a daily temperature model, respectively.

The daily model used for Figure 6 is the CJB model to be described later, but it must be noted that almost any model for daily temperatures would show this apparent decrease in estimation uncertainty. This is because a daily model uses the data more efficiently than an index model and because the comparison is based on the assumption that the model is correct (the simulation envelopes are created using the estimated model for both the index and the daily model). However, a bad daily model could also produce significant bias, and hence reduce the value of increased estimation accuracy. For this reason it is extremely important to validate a daily model before using it for weather derivatives pricing; validation methods are discussed in detail later.

Another reason for a thorough validation of daily validation models is that a daily model may give a different index distribution from that given by the historical index values (and from a normal distribution). Since a daily model is using data more efficiently than an index model we can place more weight on the daily model results, but only if it has been thoroughly validated.

Statistical modelling of daily temperatures has been a research subject for decades (see for example the review in Wilks and Wilby, 1999) and, more recently, a number of papers on weather derivative pricing that propose models for temperature have been published (Alaton *et al.* (2002), Brody *et al.* (2002), Caballero *et al.* (2002), Cao and Wei (2000), Davis (2001), Diebold and Campbell (2001), Dischel (1998), Moreno (2000) and Torró *et al.* (2001)). Common to all of the above referenced papers except Brody *et al.* (2002) and Cabellero *et al.* (2002) is that they use ARMA-type models which will be discussed in more detail in the following subsections.

We will start by discussing the basic steps of building a model for daily temperatures and then show how statistical validation techniques are applied. The model validation will show that ARMA-type models fail to validate well on two points:

1. The modelled auto-correlation function decays too quickly relative to reality.
2. The residuals deviate markedly from their theoretical distribution.

For this reason we also show validation results for the model discussed in Caballero *et al.* (2002) which, in general, validates much better than ARMA models for daily temperatures. We conclude the section with a brief discussion of some of the problems that still remain to be solved and which, to the authors' knowledge, apply to all published daily temperature models.

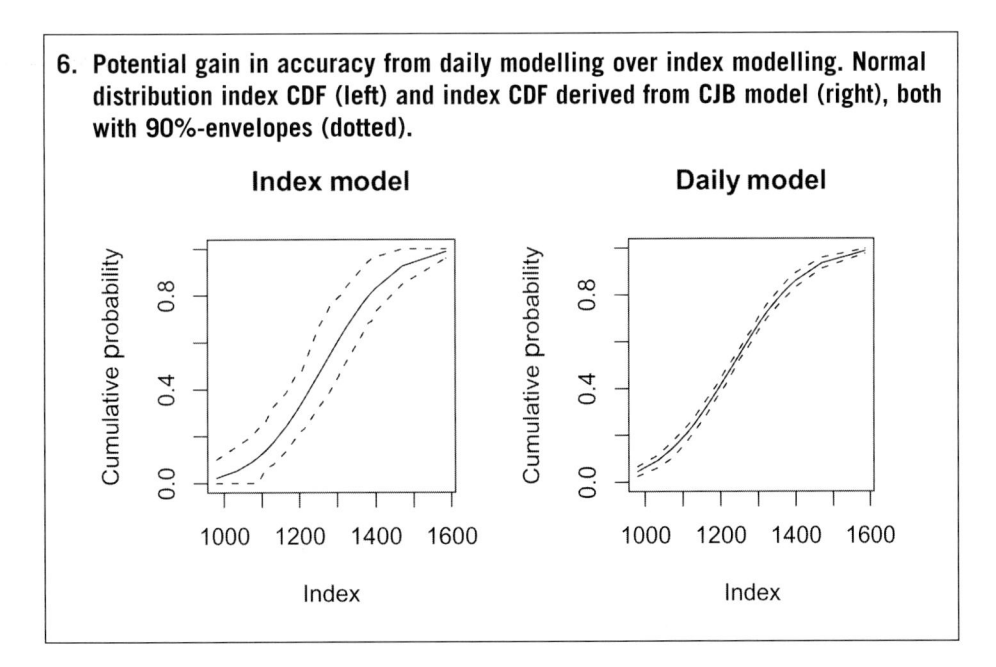

6. Potential gain in accuracy from daily modelling over index modelling. Normal distribution index CDF (left) and index CDF derived from CJB model (right), both with 90%-envelopes (dotted).

CAPTURING SEASONALITY OF TEMPERATURES

One basic principle underlying statistical models for stochastic processes is that the data should be stationary, ie, the distribution should be invariant over time. Apart from trends, daily temperatures have some obvious non-stationarities, namely those due to seasonal variation.

Seasonal variation undoubtedly affects daily temperature in many complicated ways (eg, in all the moments), and it will probably never be possible to remove all types of seasonality, let alone to check that it has been done. In the following it is assumed that most seasonality is due to seasonal variation in the mean and the standard deviation. In mathematical terms we assume that the time-series of daily temperatures can be decomposed as follows

$$T_i = m_i + s_i T'_i \qquad (4)$$

where T_i is the temperature, m_i is the mean temperature, s_i is the standard deviation and T'_i is a mean zero variance one variable that is known in meteorology as the "anomaly" for day i. The anomalies T'_i are assumed to be a stationary time-series, for which a large number of candidate models exist.

Like estimation of trends, seasonality estimation can be either parametric or non-parametric. The following section discusses a number of such methods for removing the seasonality in the mean. The seasonal variance can be estimated and removed by exactly the same method applied to the square of the difference between the temperature and the seasonal mean. For this reason we only describe the methods for estimating the seasonal mean.

Non-parametric approaches

The simplest way to estimate the seasonal mean is to average each day of the year over the historical data period, ie, the estimate for January 1 is the average of January 1 in all years and so on. When using this method, special care must be taken of leap days. If the seasonal mean estimated this way is considered too ragged it can be smoothed, by means of kernel smoothing, for example. One problem is that even with smoothing, results are often too jumpy.

Parametric approaches

Another way of estimating the seasonal mean is by parameterising the mean temperature using trigonometric functions. This has the advantage that leap days are easy to take account of. We can estimate the m_i using ordinary regression with m_i given by

$$m_i = \alpha \cos\left(\frac{2\pi}{365.25} i + \omega\right) \qquad (5)$$

where α is the amplitude and ω is the phase of the seasonal mean.

Alternatively we can estimate m_i in the frequency domain. In both cases more harmonics can be included in the parameterisation of m_i if necessary (ie, using cosines to produce

$$m_i = \sum_{k=1}^{K} \alpha_k \cos\left(\frac{2k\pi}{365.25} i + \omega\right)$$

While the parametric form in Equation (5) is very simple, it can be justified by plots of power spectra of temperature anomalies which show surprisingly clear peaks at 365.25 days (and its harmonics).

The result of applying a seasonal linear trend and parametric seasonal cycles for mean and standard deviation with two harmonics to the New York LaGuardia Airport example can be seen in Figure 7, which shows how the detrended daily temperatures are decomposed into seasonal mean, seasonal standard deviation and anomalies.

ANOMALY MODELLING

The biggest challenge in modelling daily temperatures is to find an appropriate model for the anomalies. We have seen that the index distribution's LEV function plays an important role in deriving the moments of the payoff distribution. Similarly there are functions of daily temperatures which are important for capturing moments of the index distribution from a model for daily indexes. For example, it is important to model both the seasonal cycle and the autocorrelation of temperatures accurately in order to get a good estimate of the index mean and variance of degree-day or average-temperature indexes. Consider, say, a CDD index. Using the representation in Equation (4), and assuming that the daily temperature never falls below 65°F, we have the following expressions for the mean and variance of the index:

$$E\ I\ =\ \sum_{i=1}^{n} m_i - n65$$

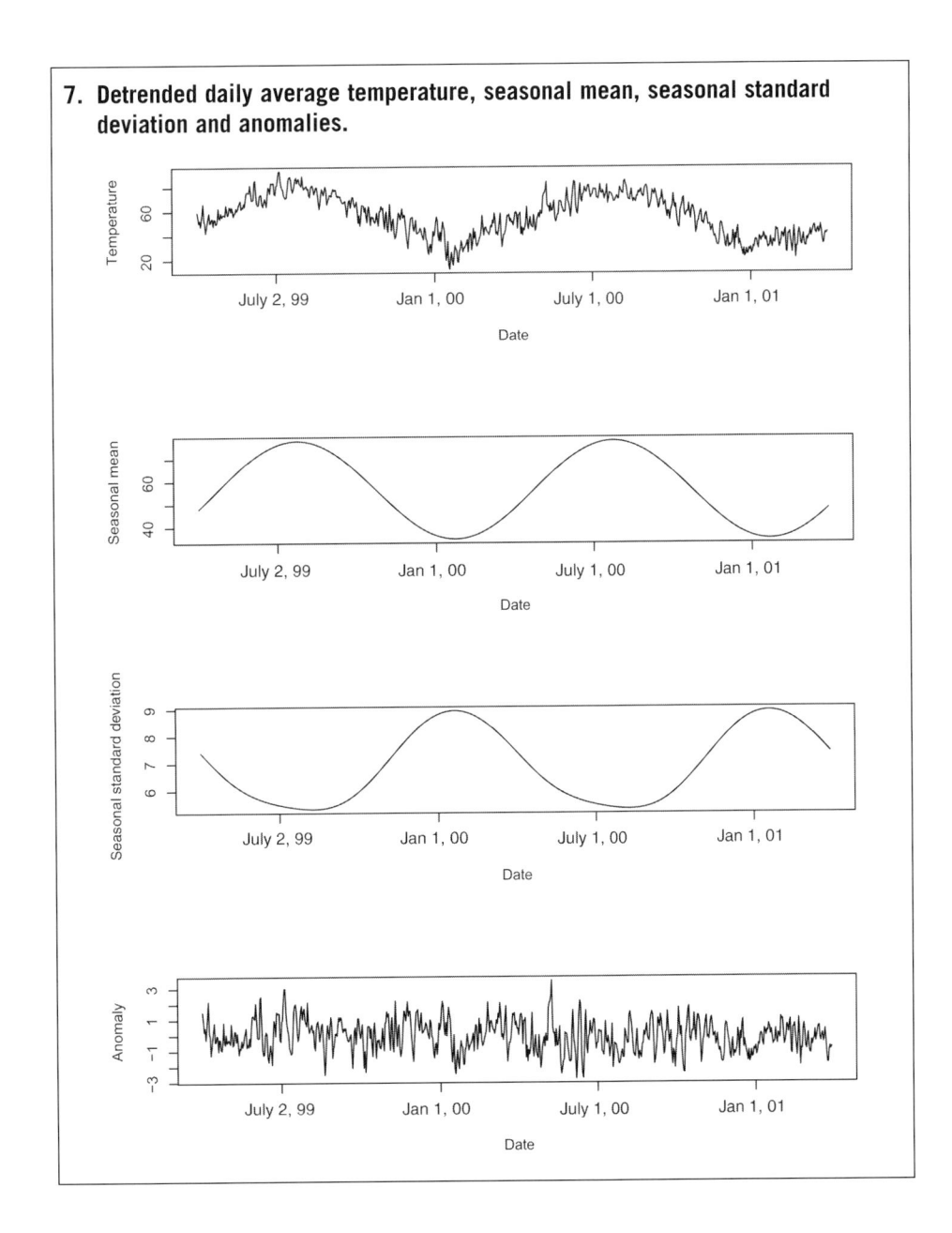

7. Detrended daily average temperature, seasonal mean, seasonal standard deviation and anomalies.

$$V_I = \sum_{i=1}^{n} \sum_{j=1}^{n} s_i s_j \gamma(i-j) \qquad (6)$$

Here, n is the number of days in the contract and γ is the autocorrelation function (ACF) of the anomalies. From Equation (6) it can be seen that we would underestimate the index variance if the model ACF, γ, decays too quickly to zero.

What kind of ACF behaviour do daily temperatures show? Dependencies in day to day temperature arise from complicated atmospheric, oceanic and land surface processes, many of which evolve slowly. In particular, ocean circulation can have cycles from years to decades and even centuries. It would thus not be surprising if temperature were to show dependencies on long timescales. As we shall see later, the ACF of temperature anomalies does indeed show relatively slow decay.

NON-PARAMETRIC TIME-SERIES MODELLING

Non-parametric time-series modelling can be done by resampling from the observed anomalies. The resampled time-series can then be used to estimate distribution statistics of interest, weather derivative indexes, for example. Resampling should not be done completely at random, since we would then lose all dependencies in the time-series. Instead we could resample blocks of anomalies and concatenate these. If the blocks are sufficiently long and if we have enough data, this should result in simulated time-series which would mimic the behaviour of the original time-series. The length of the blocks should be chosen so that the statistics of interest are not too affected by the "breaks" between blocks. For example, block lengths could be chosen to equal the contract length in the case of weather derivatives. However, it may be difficult to capture the distribution of a time-series by resampling, even if long blocks are used. See Davison and Hinkley (1997) for a survey of non-parametric resampling methods.

PARAMETRIC TIME-SERIES MODELS

Traditional time-series models are parametric models, and the simplest example is probably a first order autoregressive (AR(1)) process. For such a process the temperature at one day is given by a linear dependency on the temperature the previous day with some random noise added:

$$T'_i = \beta T'_{i-1} + \varepsilon_i$$

Here T'_i denotes the anomaly at day i and the ε_i are independent and identically distributed random variables following a mean zero normal distribution. Despite its simplicity an AR(1) process can be a useful model for a variety of problems, but is unfortunately too simple for daily temperature anomalies (one reason being that the ACF decays too quickly). A simple extension of the AR(1) process provides us with a flexible class of models, known as ARMA models, which can be used to approximate any stationary time-series model.[5] The definition of an ARMA(p, q) process is:

$$T'_i = \phi_1 T'_{i-1} + \ldots + \phi_p T'_{i-p} + \theta_1 \varepsilon_{i-1} + \ldots + \theta_{i-q} \varepsilon_{i-q} + \varepsilon_i$$

The interpretation of the model is that the temperature today depends in a linear way on the temperatures on the previous p days through the parameters $\phi_1, \ldots, \phi_p$. Just as in the AR(1) case random pertubations are added to reflect the fact that we do not expect the temperature today to be a perfect linear function of the past p days' temperatures. The $\theta_1, \ldots, \theta_q$ are parameters used to express linear dependence between the random pertubations.

Although ARMA models can, in theory, approximate any time-series model, it is actually not the best class of models for temperature anomalies. This is partly because of the slow decay of the ACF of temperature anomalies (see Figure 8 and the subsection on validation using ACFs), which by Equation (6) implies that an ARMA model would result in underestimated variances for degree-day indexes – a fact that has been recognised by many participants in the weather market for some time.[6] If we were to model a slower decay of the ACF using an ARMA model, we would need far more parameters than would be feasible to estimate in practice. A more parsimonious choice of model (in the following referred to as the CJB model) that preserves much of the ARMA model flexibility but has slower decaying ACF is described in Caballero *et al.* (2002).

In the following sections we illustrate classical statistical methods for model validation, which provide more evidence on how ARMA models fail to provide adequate modelling of daily temperatures and how the CJB model overcomes many (but not all) of the problems associated with ARMA models.

Validation of daily temperature models

Modelling daily temperatures is a difficult task and careful validation of models must be undertaken before putting them into practical use for pricing weather derivatives. In the following subsections we show some simple tools for evaluation of daily temperature models for weather derivative pricing. The methods are illustrated on the New York example, using an ARMA(3,1) model and the CJB model.

AUTOCORRELATION FUNCTION

Equation (6) and the discussion following it highlighted the importance of capturing the ACF of the temperature anomalies, making the comparison of the model ACF with the empirical ACF an appropriate step in validation of the model. However, since all ACFs start at one and usually decay quickly over the first few lags, comparison can be difficult using the usual representation of an ACF. Instead, we compare ACFs by plotting the logarithm of the ACF against logarithm of the lag. The plot in Figure 8 is a plot of this type and emphasises what we have already mentioned: that the ARMA model ACF decays too fast to capture the behaviour of the empirical ACF.

DISTRIBUTION OF RESIDUALS

Residuals are the difference between the observed anomalies and the predicted anomalies based on the past observations, ie, the historical one-step prediction errors. The parameters used for the prediction are estimated from the full time-series. For most models, including the two considered here, it is possible to derive theoretical expressions for the distribution of the residuals. We can thus check the

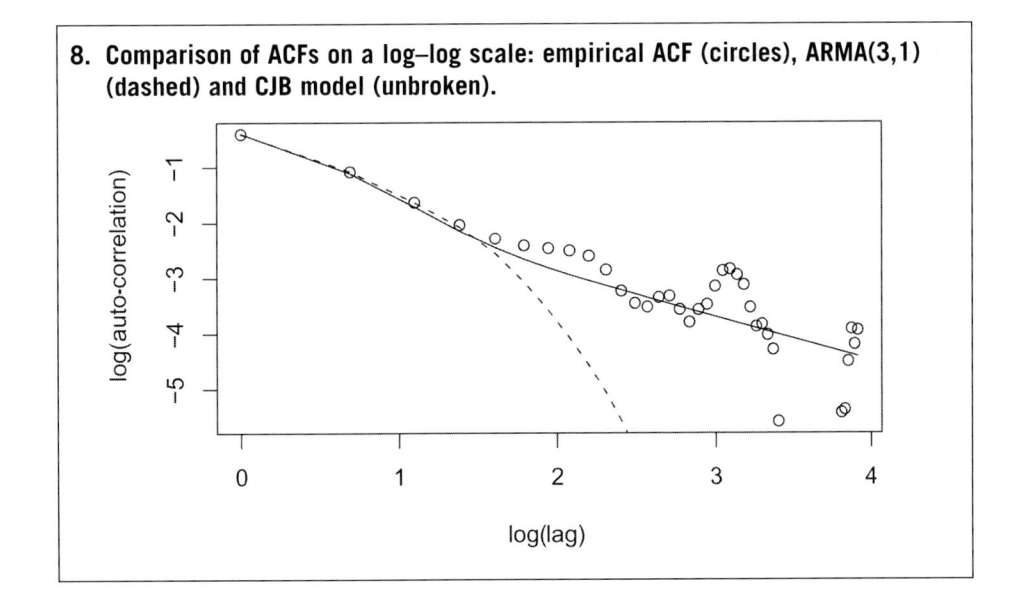

8. Comparison of ACFs on a log–log scale: empirical ACF (circles), ARMA(3,1) (dashed) and CJB model (unbroken).

time-series model by comparing the residuals from the model with their theoretical distribution. For non-parametric methods the theoretical distribution of the residuals is unknown and hence cannot be checked.

The left plot of Figure 9 shows a residual QQ-plot for the ARMA(3,1) model, and the right plot shows a QQ-plot for the CJB model. We see that whereas the CJB model shows consistency with the historical anomaly distribution, the ARMA(3,1) model results in a residual distribution with far too light tails.

DISTRIBUTION OF ANOMALIES
The next step in validation is the verification of the distribution of the anomalies. For mathematical convenience most time-series models assume a normal distribution, so a natural first step is to compare the empirical anomaly distribution with that of a normal through PP- and QQ-plots. If a normal distribution is not suitable, it is often possible to find a transformation that makes the anomaly distribution approximately normal.

Models also exist for time-series with non-normal behaviour such as heavy-tailed or skewed distributions. For such models, however, it can be difficult to verify when the model is stationary and determine exactly what the marginal distributions are.

ACCURATE MODELLING OF WEATHER DERIVATIVE INDEXES
While it is important that the time-series of temperatures is modelled accurately, the ultimate goal is to model weather derivative indexes. The transformation of temperatures to a degree-day or an event-specific index (such as a critical-day index) is non-linear, and hence it is not clear how small model deficiencies at the temperature level will propagate to errors in the index distribution. This, however, is not a problem for a period-average temperature index. We need to validate the index distribution, derived from a daily model, using the same tools as we used for validating the index model. As an example, the two plots in Figure 10 show the QQ-plots for the index distribution resulting from using an ARMA(3,1) and the CJB model.

Imposing 90%-confidence envelopes on the plots we see that the CJB model falls within the envelopes, but that the ARMA model is questionable because of the relatively large number of points that fall outside or on the boundary of the envelopes.

Some outstanding issues in daily temperature modelling
We have seen that ARMA models are not adequate for daily temperature modelling and that the CJB model is a better alternative. However, there are still some

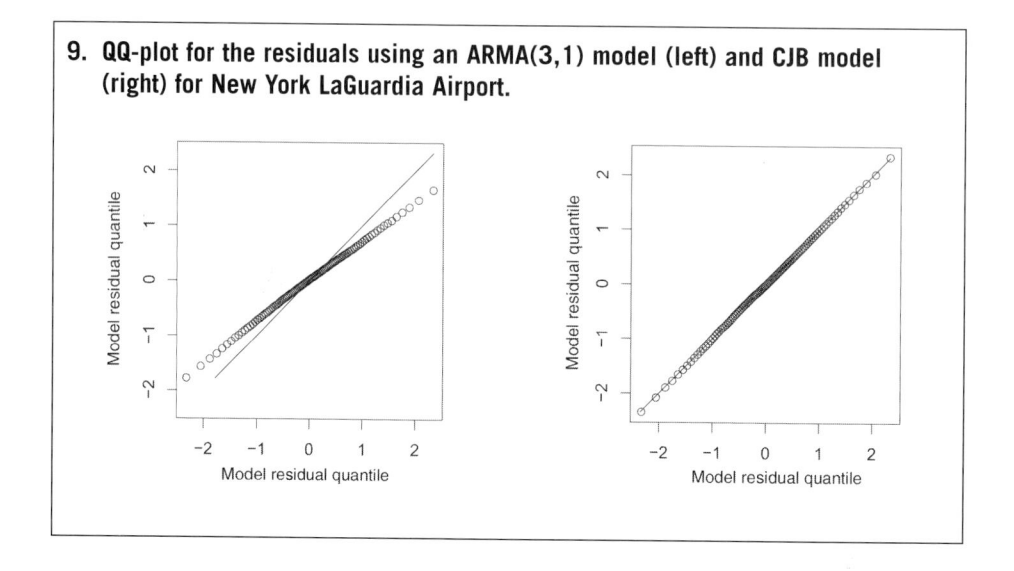

9. QQ-plot for the residuals using an ARMA(3,1) model (left) and CJB model (right) for New York LaGuardia Airport.

10. QQ-plot for the index distribution for New York LaGuardia Airport using an ARMA(3,1) model (left) and CJB model (right).

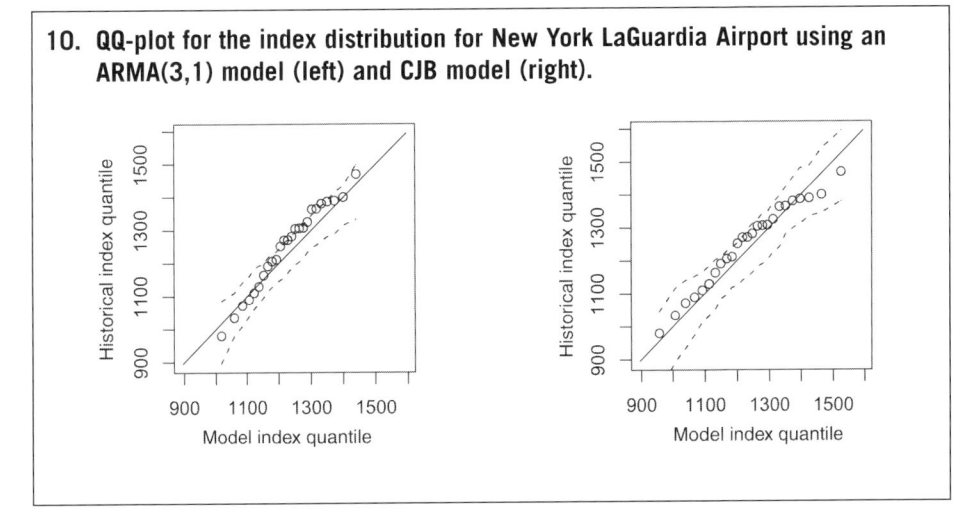

unresolved issues which means that even the CJB model may not be adequate in all situations. In particular some locations show strong seasonal variation beyond that of mean and standard deviation, which in turn means that both ARMA and CJB models validate badly for these locations.

Another issue is that modelling daily temperatures for several locations simultaneously is complicated by the fact that the spatial correlation structure is non-stationary. This makes it difficult to exploit all the information in daily temperatures and as a result dependencies between locations are usually modelled on an index level instead (see the following section).

Capturing index dependencies

Weather indexes often show strong geographical dependencies – a fact that must be captured by our modelling approach. One reason for this is that pricing of a weather contract is generally done in the context of the current portfolio: the more risky the portfolio gets by adding the contract, the less that contract is worth. Furthermore, portfolio modelling is essential to valuing a book of contracts and the principles for risk loading described in the following section rely crucially on appropriate joint modelling of the contracts in the portfolio. To model dependencies, various tools exist – mostly based on correlations. However, correlations are not just correlations!

The usual notion of correlation is linear correlation, which is what is calculated in spreadsheets and other standard software packages. Linear correlation is only an appropriate measure of dependency if the joint distribution of the indexes is close to Gaussian or elliptical (Embrechts *et al.*, 2000). If this is not the case, linear correlations may fail to accurately capture the dependency between indexes, especially in the tails of the distribution. Moreover, the correlation coefficient need not have a maximum of one and a minimum of minus one in the non-Gaussian case. Instead the minimum and maximum linear correlation will depend on the distribution and its parameters, which makes it more difficult to evaluate the degree of dependence. Fortunately other methods for capturing dependence exist, two of which are discussed here: rank correlation and copulas.

RANK CORRELATION

Rank correlation is a alternative way of measuring dependencies between distributions. All values between one and minus one are possible whatever the distribution, which makes interpretation of rank correlation easier than linear correlation in the non-Gaussian case.

Rank correlation is calculated by transforming all indexes by their empircal CDF and then computing the linear correlation between the transformed indexes. This is equivalent to computing the linear correlation between the rank of the sorted indexes (hence the name) and the method can be seen as a non-parametric approach to capturing dependence between indexes.

COPULAS

The notion of copulas is based on the same fundamental idea as the rank correlations, namely that dependency between distributions is compared after the distributions have been transformed to uniforms: a copula is simply the CDF of a multivariate distribution with uniform marginals. For example the Gaussian copula is the CDF given by transforming each variable of a multivariate Gaussian distribution by its CDF. Rank correlations can be thought of as a special type of copula, where the dependence is given by the empirical correlation of the observed indexes transformed to uniforms. In general, however, a copula is given as a parametric CDF, and the parameters of the copula must either be estimated from data or guessed from intuition. Guessing is usually not a good idea, but due to lack of data it may be necessary. Likewise the preference of one specific copula over another can be hard to justify from data, and is often made on the basis of mathematical convenience.

Risk loading principles

Every risk has its price, and the difference between this and a fair price is called "risk loading". How is an appropriate risk loading calculated? The principles outlined below are general principles for how to obtain a price of risk based on risk measures such as standard deviation and quantiles. Most of the principles can also be adapted to other risk measures.

The price of a contract does not always depend directly on how risky it is. If, for example, we know that we can buy a certain contract from somebody else in the market at a given price it would be risk-free to offer the contract at a higher price. This way of determining a price is driven by the market's perception of risk rather than one's own. This is discussed further in the paragraphs that follow.

SIMPLE ADDITIVE RISK LOADING

The simplest risk loaded price, P_r, of a contract is the expected outcome plus λ times the risk, where λ is a risk factor defining the speculator's appetite for risk. Ignoring discounting, the pricing formula is:

$$P_r = EP + \lambda R \qquad (7)$$

where R is the risk measure and EP is the expected payoff. A measure of risk is needed, and it is customary to use standard deviation or variance of the payoff for this purpose. Alternatively we could use the difference between two quantiles such as the median and the 5% quantile, whereby we get a risk measure which is more akin to Value-at-Risk.[7]

Equation (7) also allows the pricing of the contract against the full portfolio. To do this we change the risk measure R to be a measure for the additional risk that is taken on by trading the contract. Changes in any of the risk measures above could be used for this purpose. For example, we could use the difference between the standard deviation of the portfolio payoff with and without the contract as risk measure R.

INVESTMENT EQUIVALENT PRICING

Most companies will have ways of allocating capital to contracts based on their own internal measures of risk. Had the capital not been allocated to a contract it could have been invested in other assets. This gives the speculator another way of pricing a contract: the return on the allocated capital should be at least what would be expected on an investment in a similar risk elsewhere. This is the idea behind so-called Risk Adjusted Return On Capital (RAROC) (Nakada et al., 1999) and investment equivalent reinsurance pricing (Kreps, 1999).

Consider the case where an option is sold at a price Pr. The price consists of the discounted expected payoff and risk loading, L:

$$Pr = EP/(1 + r_f) + L \tag{8}$$

where r_f is the risk-free interest rate. In addition to the premium income we allocate an amount of capital A to the contract and invest $Pr + A$ in risk-free securities. The idea is now that the average return on allocated capital should equal the return from an investment in an alternative instrument with the same risk. The expected Return on Allocated Capital (RAC) is thus defined by

$$(1 + RAC) A = (1 + r_f) (Pr + A) - EP$$

Combining this with Equation (8) we get the following expression for the loading

$$L = (RAC - r_f)/(1 + r_f)A$$

Using the estimated payoff distribution we can get an idea of how risky the option is and use this to find a reasonable target RAC which, in turn, gives the risk loading. Several variations of this theme are possible, see Kreps (1999).

Other topics

We have seen in this chapter that valuation of weather derivatives is a broad subject where many factors must be taken into account. While we have tried to cover the majority of topics there are still some that we have left out and many which could have been further elaborated.

ARBITRAGE PRICING
Arbitrage pricing is the standard way of pricing financial derivatives in a liquid market. As noted in this chapter's Introduction most weather derivative contracts are not yet liquid enough to justify arbitrage pricing but some are now traded several times a day. This makes it worthwhile to consider how arbitrage pricing for weather derivatives may be done.

The basic principle of arbitrage pricing is that the cost of a derivative is the cost of creating and managing a portfolio which replicates the payoff of the derivative contract at maturity. The active management of the replicating portfolio is what is usually called dynamic hedging. However, the underlying index of a weather derivative cannot be traded and hence weather derivatives cannot be hedged this way. Instead we could replicate the payoff of a weather derivative using other weather derivatives. While this is possible in principle, the illiquidity of the current weather market makes it prohibitively expensive for most types of contracts. However, should the market become liquid enough, several arbitrage strategies would be possible.

One strategy would be to use properties of forecasts to derive a theory for how the market would price a weather swap. Since swaps can be used to hedge options, this gives us a way of obtaining an arbitrage price for an option. The ideas behind this are discussed in detail in Stephen Jewson's contribution to the End Piece.

Some authors advocate hedging of weather derivatives with other derivatives such as power or gas derivatives (Geman, 1999). However, such hedges are not likely to be complete hedges and we are thus left with a basis risk. In order to find the price of the basis risk we must model the weather derivative and the underlying of the alternative hedge jointly, for example using methods similar to those previously discussed in this chapter. Note also that whereas HDDs are highly correlated with gas

WEATHER DERIVATIVE MODELLING AND VALUATION: A STATISTICAL PERSPECTIVE

consumption, they are usually very little correlated with gas prices, which is one of the reasons for trading weather derivatives in the first place.

OTHER WEATHER VARIABLES

Another challenging topic is modelling of daily weather variables other than temperature, such as of wind speed or precipitation. These relatively uncommon underlyings (compared with temperature) are dealt with thoroughly in Chapter 3.

EXTREME VALUE THEORY

One subject that has not been mentioned is the use of extreme value theory (EVT) for evaluating risk and estimation of the distribution of extreme indexes. While this is an interesting point of discussion, the current state of the market is such that weather derivatives are mostly written on non-extreme risk, and as such EVT is less applicable.

Conclusion

This chapter has described how weather derivatives can be valued by statistical and actuarial methods based on historical data. It has given an overview of all the important topics: estimation and adjustment of trends, modelling and validation of detrended historical weather indexes and daily temperatures, accounting for index dependencies and risk loading of expected payoffs.

As trading liquidity increases for certain contracts, aspects of arbitrage pricing will become more important and in some cases replace actuarial methods. However, for most contracts, the actuarial methods described in this chapter will continue to be the main, and only reasonable, valuation approach that can be used.

1 *The LEV function is often used in insurance for calculating mean loss to an excess of loss reinsurance layer.*
2 *By "robust" we mean methods which, at the cost of accuracy, are less affected by outliers than traditional methods.*
3 *Based on the authors' own study using 50 years of data from 200 US stations.*
4 *In statistical literature the observed sample is often included in the K simulations since the hypothesis modelled is a sample from the correct model. In practice this distinction does not make a lot of difference.*
5 *It can be shown that for any autocovariance function $\gamma(\cdot)$ such that $h \to \infty \, \Upsilon(h) = 0$, and any integer $k>0$ it is possible to find an ARMA process with autocovariance function $(\cdot)$ such that $(h) = \gamma(h)$, $h=0,1,...,k$.*
6 *This is the authors' own experience based on discussions with the major market participants over the last three years.*
7 *Value-at-Risk (VAR) is a quantile (typically 5%) in the modelled profit and loss distribution over a specified time period, and is one of the most common measures of risk in finance.*

BIBLIOGRAPHY

Alaton, P., B. Djehiche and D. Stillberger, 2002, "On Modelling and Pricing Weather Derivatives", *Applied Mathematical Finance*, Forthcoming.

Bowman, A, and A. Azzalini, 1997, *Applied Smoothing Techniques for Data Analysis*, (Oxford University Press).

Brockwell, P., and R. Davis, 1991, *Time Series: Theory and Methods*, (New York: Springer-Verlag).

Brody, D. C., J. Syroka and M. Zervos, 2002, "Dynamical Pricing of Weather Derivatives", *Quantitative Finance*, Forthcoming.

Caballero, R., S. Jewson and A. Brix, 2002, "Long Memory in Surface Air Temperature: Detection, Modelling and Applications to Weather Derivative Valation", *Climate Research*, Forthcoming.

Cao, M., and J. Wei,. 2000, "Equilibrium Pricing of Weather Derivatives", Research Paper, Deptartment of Economics, Queen's University, available at: http://qed.econ.queensu.ca/pub/faculty/cao/papers.html.

Casella, G., and R. L. Berger, (2002), *Statistical Inference*, (Pacific Grove, CA: Brooks/Coles Publishing).

Cleveland, W., and S. Devlin, 1988, "Locally-Weighted Regression: An Approach to Regression Analysis by Local Fitting", *Journal of the American Statistical Association*, 83, pp. 596–610.

Coles, S., 2001, *An Introduction to Statistical Modelling of Extreme Values*, (New York: Springer Verlag).

Davis, M., 2001, "Pricing Weather Derivatives by Marginal Value", *Quantitative Finance*, 1, pp. 1–4.

Davison, A., and D. Hinkley, 1997, *Bootstrap Methods and Their Application*, (Cambridge University Press).

Diebold, F. X., and S. D. Campbell, 2001, "Weather Forecasting for Weather Derivatives", Working Paper, Department of Economics, University of Pennsylvania, available at: http://www.econ.upenn.edu/Centers/pier/Archive/01-031.pdf.

Dischel, R., 1998, "Black-Scholes Won't Do", Risk Magazine and Energy and Power Risk Management Report on Weather Risk, http://www.financewise.com/public/edit/energy/weatherrisk/wthr-options.htm.

Easterling, D., B. Horton, P. Jones, T. Peterson, T. Karl, D. Parker, M. Salinger, V. Razuvayev, N. Plummer, P. Jamason and C. Folland, 1997, "Maximum and Minimum Temperature Trends for the Globe", *Science*, 277, pp. 364–7.

Embrechts, P., A. McNeil and D. Straumann, 2000, Correlation and Dependency in Risk Management: Properties and Pitfalls, in: M. A. H. Dempster and H. K. Moffatt (eds) *Risk Management: Value at Risk and Beyond*, (Cambridge University Press).

Geman, H., 1999, "The Bermudan Triangle: Weather, Electricity and Insurance Derivatives, in: H.Geman (ed), *Insurance and Weather Derivatives*, (London: Risk Books).

Hansen, J., R. Ruedy, M. Sato and R. Reynolds, 1996, "Global Surface Air Temperature in 1995: Return to Pre-Pinatubo Level", *Geophysical Research Letters*, 23, pp. 1665–8.

Huber, P. J., 1981, *Robust Statistics*, (New York: John Wiley & Sons).

IPCC, 2001, "Climate Change 2001-The Scientific Basis", Technical Report, Technical Summary of the Working Group I Report.

Jewson, S., and A. Brix, 2001, "Sunny Outlook for Weather Investors?", *Environmental Finance*, February, pp. 28-9.

Knappenberger, P., P. Michaels and R. Davis, 2001, "Nature of Observed Temperature Changes across the United States during the 20th Century", *Climate Research*, 17, pp. 45–53.

Kreps, R., 1999, "Investment Equivalent Reinsurance Pricing", in: *Proceedings of the Casualty Actuarial Society*, Volume LXXXVI, (Arlington, VA: CAS).

McCullagh, P., and J. A. Nelder, 1989, *Generalized Linear Models*, (London: Chapman and Hall).

Moreno, M., 2000, "Riding the Temp", Research Paper, Speedwell Weather Derivatives, Available at: http://www.weatherderivs.com/Papers/Riding.pdf.

Nakada, P., H. Shah, H. Koyluoglu and O. Collignon, 1999, "P&C RAROC: A Catalyst for Improved Capital Management in the Property and Casualty Insurance Industry", *Journal of Risk*, 1, pp. 52–70.

Torró, H., V. Meneu and E. Valor, 2001, "Single Factor Stochastic Models with Seasonality Applied to Underlying Weather Derivatives Variables", Research Paper, Facultat d'Economia, Universitat de Valencia, available at: http://www.efmaefm.org/htorro.pdf.

Wilks, D.S., and R. L. Wilby, 1999, "The Weather Generation Game: A Review of Stochastic Weather Models", *Progress in Physical Geography*, 23, pp. 329–57.

WRMA, 2002, "The Weather Risk Management Industry: Survey Findings for November 1997 to March 2001", Report prepared for Weather Risk Management Association by PriceWaterhouseCoopers, Available at http://www.wrma.org.

9

The Accuracy and Value of Operational Seasonal Weather Forecasts in the Weather Risk Market

Jeffrey A. Shorter, Todd M. Crawford and Robert J. Boucher

WSI Energycast Trader

Introduction

The goal of this chapter is to educate the readers of the value of seasonal weather forecasts in weather risk management. The chapter is organised into four main sections:

1. the skilful timescales of weather forecasts;
2. methodologies used in seasonal forecasting and when they provide valuable forecasts;
3. the accuracy in operational seasonal forecasts; and
4. the usefulness of seasonal forecasts in weather derivative transactions and the hypothetical financial results.

Skilful timescales of weather forecasts

Weather phenomena occur on numerous timescales. For example, tornadoes last for approximately 15 minutes, thunderstorms for one hour, flash floods for a few hours, heat waves for days, and droughts for months. The ability to forecast these events both spatially and temporally is, in general, related to their duration. Thus it is important to understand the different spatial scales of atmospheric energy and resultant motion with which they are associated. Weather phenomena are classified by their duration and spatial extent. The classifications are, from smallest to largest: microscale, mesoscale, synoptic-scale and planetary-scale.

Weather disturbances that are fast moving and have short lifetimes are classified as microscale, tornadoes for example, which have spatial scales of <1 km. Microscale events are embedded in large mesoscale systems such as a line of thunderstorms, which have a spatial scale of 1–100 km (1–60 miles). In turn, mesoscale events are embedded in large synoptic-scale systems. Synoptic-scale weather is responsible for day to day variability and is of the order of 1,000 km (600 miles) in horizontal extent. Common examples are cold fronts, warm fronts and high- and low-pressure systems that are regularly seen during the weather segment of the evening news. Synoptic-scale systems typically last several days and sometimes as long as a week. The larger and more slowly varying long-wave patterns (eg, the location and orientation of the Jet Stream) typically have horizontal scales from 10,000–40,000 km (6,000–24,000

Table 1. Classification of atmospheric phenomenon with their characteristic spatial and temporal scales

Classification	Space scale	Timescale	Examples
Microscale	1 m–1 km	seconds–hours	Tornado
Mesoscale	1–100 km	hours–days	Thunderstorm
Synoptic-scale	100–10,000 km	days–weeks	Heat wave
Planetary-scale	10,000–40,000 km	week(s)–months	Jet stream

Source: Moran and Morgan, 1997

miles) and timescales of weeks to months and are referred to as "planetary scale". Table 1 illustrates the various scales of atmospheric circulation.

The key to skilfully forecasting weather phenomena is the utilisation of sophisticated numerical weather models. The continuing improvement in weather forecasts is due to improvements in both the weather models and the weather information to initialise the models. Two key factors in improving the models are scientific research to improve the physics in the models and the increased computational power as the semi-conductor industry continues to improve processing speed. One enhancement often made to the model is to increase its resolution, both in the horizontal and the vertical dimensions. In theory, increasing the model's resolution should better enable the simulation and resolution of smaller-scale motions, which then affect the larger-scale motions. This results in more accurate forecasts, out to timescales of 1–2 weeks (American Meteorological Society, 1998). Beyond that range, the chaotic nature of the atmosphere is the limiting factor, which was discovered in pioneering work by Lorenz (1963). Thus new techniques have been developed to extend the forecast lead-time to understand the probability of potential events instead of the deterministic view of an event happening or not (see Panel 1).

Chaos, as termed by Lorenz (1963), is a condition that prevents perfect weather forecasts. Numerical weather models are initialised with a "best guess" at the initial state of the atmosphere. Since this guess will always be imperfect due to inaccuracies in the observations and model approximations, the model result will have errors even before the simulation begins. The magnitude of the errors in the initial conditions will grow with time, doubling approximately every 2–2.5 days. Theoretical and experimental studies have determined that after 1–2 weeks these initial errors are so large that the deterministic forecasts can no longer be skilful (Lorenz, 1963). As a result of the growing errors in the forecast, the short time-frame phenomena get filtered out, then the day to day atmospheric variability begins to be overcome by the growing errors. As this higher frequency information is filtered out, the longer timescale motions should still contain information. Due to these forecasting limitations there will always be different skilful timescales associated with various types of weather phenomena.

The amount of detail desired from a forecast will dictate the effective timescale. In the first few days of a forecast the error is small enough that skilful hourly forecasts can often be made. Beyond this time, hourly forecasts lose most of their skill and the use of daily minimum and maximum temperature forecasts should be utilised. Beyond 10–14 days, the averaging periods need to become much larger to counteract the noise in the daily and hourly forecasts (American Meteorological Society, 1998). Averaging periods currently used in the energy sector are five-day periods (eg, 6–10 and 11–15 days), weekly forecasts, and balance of the month forecasts. The end-user of the forecast must appropriately determine the minimum amount of detail they will accept and then apply or choose the appropriate averaging period.

While most attention is focused on forecasts out to approximately 15 days, there is a continuing increase of interest in publicly and privately available seasonal forecasts (1–3 months). With seasonal forecasts it is important to understand both their usefulness and their limitations. Typically, seasonal forecasts are issued for (a) the next month and (b) an average for the next three months. The three-month forecasts are more skilful than the monthly forecasts due to the fact that the longer averaging period allows for the smaller synoptic-scale events to be averaged out. Understanding this fact is so important that the American Meteorological Society (AMS), an internationally recognised trade organisation for the meteorological field, felt it was important to inform the public of these limitations. The AMS stated, "Claims of skilful predictions of day to day weather changes beyond 1–2 weeks have no scientific basis and are either misinformed or calculated misrepresentations of true capabilities."[1]

With the increasing volume of forecast data now available it is important for the end-user to be able to effectively understand and interpret the results. The useful timescale of a given forecast will depend upon the amount of detail desired (eg, hourly or daily temperatures). Because of the chaotic nature of the atmosphere and the imperfection of the models, however, uncertainty will be present in every forecast. The final forecast value should not be used without a determination of the confidence placed in that value. Probabilistic forecasts provide the user with both a degree of confidence in the forecast value and the range of possibilities.

Seasonal weather forecasting techniques

In general, there are two techniques used to make seasonal forecasts. The first is the use of dynamic climate models, which integrate the fundamental equations of atmospheric motion forward in time to determine the future state of the atmosphere. The second is the use of statistical models, which employ observed correlations between certain atmospheric or land surface states and future air temperatures. Each technique is outlined below and the temporal and geographical areas where they are expected to provide skill is presented. Through the optimal blending of the techniques, there is an expectation of positive skill in all regions and all seasons. This blending should produce better forecasts than either type of model by itself.

DYNAMIC CLIMATE MODELS

Dynamic climate models are used by WSI Energycast Trader (WSI) and many research organisations in order to predict the general state of the prevailing long-wave atmospheric pattern (eg, the location and orientation of the Jet Stream) for the upcoming season. The climate model is not trying to predict, for example, if there will be a snow storm off of the East Coast of the US in 42 days, but does have the ability to predict if the general atmospheric pattern will be more or less conducive to such an event. These global models are generally of coarser spatial and temporal resolution than those used to predict short- and medium-range weather. This is due to both the relevant scales of atmospheric flow involved in seasonal forecasting and computational limitations.

A dynamic climate model is a set of mathematical equations that represent global atmospheric motions. In the model, the Earth's atmosphere is divided up into discrete volumes called "grid boxes". 10,000–30,000 of these grid boxes are needed to cover the surface of the Earth, with 18–45 layers stacked on top of the surface layer. Typically, the horizontal extent of a climate model grid box is 160–320 km (100–200 miles). The model equations are solved separately in each grid box, although each grid box is affected by the results of the solved equations in the surrounding grid boxes. The model equations are advanced, or integrated, forward at 5- to 60-minute intervals for a period of months, effectively producing a forecast of the future state of the atmosphere.

A key factor in dynamic climate models is the specification of the lower boundary conditions (ie, sea surface temperatures (SSTs), snow cover, soil moisture). This is because the long timescale motions, the planetary waves, are slowly modulated by effects from both the sea and land surfaces. The end result of a model-based seasonal forecast is an indication of the general atmospheric pattern for the entire upcoming season rather than for the daily, hourly or weekly timescales that are handled by short- and medium-range models. Since the model is driven by the boundary layer, the temporal range of skilful seasonal forecasts is currently limited by a lack of ability in the scientific community to skilfully model changes in SSTs. However, since the boundary conditions have a lagged effect on the atmosphere (on a scale of months), skilful seasonal forecasts can currently be made out to 4–5 months.

The National Center for Atmospheric Research (NCAR) Community Climate Model (CCM) is used by WSI as part of its seasonal forecasting process (as seen in Kiehl *et al.*, 1998). The model was run for 19 years (1981–2000) in order to establish a baseline skill level of the model before it was used operationally. The 19-year run was initialised with climatological values of atmospheric and land-surface parameters and with observed SSTs. The model output was compared to observations to determine the spatial and temporal variation in skill.

WSI has performed extensive research to quantify the degree of skill from the model, when used in forecast mode, by season and locations. This research involved five years of historic forecasting or "hindcasting". This process involves making "forecasts" for times that have already past using only information available at the time the forecast would have been made. For example, to make a forecast for the summer of 1999, data from May 1999 would be used to initialise the model and the forecast would be run as if it was actually May 1999; no information from after May 1999 would be used.

Specifically, the model skill in the US is greatest:

1. from the Rocky Mountains westward; and
2. the northern half of the country east of the Rocky Mountains, or approximately north of a line drawn between Kansas City and Philadelphia.

In these two regions, the NCAR CCM model produced a directionally correct forecast around 70% of the time during the five-year model testing period (1996-2000). In the remaining region, the southeast quadrant of the country, the model was directionally correct only 40% of the time.

From a seasonal perspective, the dynamic model's skill is greater in the summer and less so in the winter. Table 2 illustrates the general geographic and spatial

---

PANEL 1

ENSEMBLE FORECASTS

Since the initial state of the atmosphere can never be perfectly specified, the numerical weather prediction community has developed a method to run the models with many different sets of initial conditions to produce an ensemble of forecasts. The best guess of initial conditions is slightly perturbed to produce more sets of initial conditions, which are then used to initialise the model. The solutions produced by the various ensemble members can be then used to bound the uncertainty in the forecast. Typically the ensemble mean and standard deviation is calculated, which can then be used to construct a probability distribution function (PDF). It has been shown that the ensemble average is generally more skilful than a single particular run and that the spread of the ensemble can be used as a factor in determining how much confidence should be placed in the forecast. If the ensemble members all produce vastly different solutions, the uncertainty in the forecast is large. On the other hand, if all of the members produce similar solutions, a more confident forecast can be issued.

variation in model skill. "+" denotes an area and a time where/when the dynamic model has skill. "-" denotes an area where the dynamic model has not shown skill. The regions are as follows: SW – southwest quarter of the continental US, NW – northwest quarter, SE – southeast quarter, NE – northeast quarter. Note that the lack of winter skill of dynamic models can be overcome by incorporating statistical techniques.

STATISTICAL MODELS

Although climate models are skilful over many geographical regions and seasons, statistical models can also be used to complement the output from the climate model. Statistical models are developed by employing correlations between certain atmospheric or land surface states at one time and air temperatures at a future time. For example, one could compute the correlation between the historic sea surface temperatures in the equatorial Pacific during each March over the last 40 years to the following May average temperature in New York. Then the correlation could be utilised with the current March sea surface temperatures to predict the future May temperature in New York. Below are descriptions of many of the statistical tools utilised in seasonal forecasting. The temporal and geographic character of each tool's skill is shown in the table in Panel 1.

Climatology

Climatology represents the historical average of any meteorological parameter, such as air temperature, snow or wind at a given location during a given season. Here we focus on temperature. In the meteorological community, the average from the previous 30 years (sometimes called the "climatological normal") is used, while weather markets generally employ the average from the previous 10 years. Climatology is useful for long duration forecast since the day to day variability gets averaged out. Philadelphia in January can be used as an example: the average temperature is about 32°F and the standard deviation for the monthly average is 5°F. Thus climatology would forecast 32 ± 5°F degrees. Because of this, using climatology as a seasonal forecast usually results in skill levels just beneath those of the best seasonal forecasts. The skill of a climatology forecast is especially hard to surpass in the summer, when the weather patterns are much less variable than in the other three seasons. Figure 1 illustrates the larger variability in winter and smaller variability in summer. Because the standard deviation of the winter distribution is larger and therefore the distribution is wider, temperatures are less likely to be near the climatological mean in winter than in summer. To employ a climatological statistical model, simply calculate the mean and standard deviation over the previous 10 years (for weather derivatives) and use the result as a seasonal forecast.

Long-term trends

The last 20 years have been marked by a notable increase in temperatures across the US, especially in the winter. Whether this increase is anthropogenic or just part of a

Table 2. Geographic and spatial variation in dynamic climate model skill

	Summer	Winter
SW	+	+
SE	+	−
NW	+	−
NE	+	−

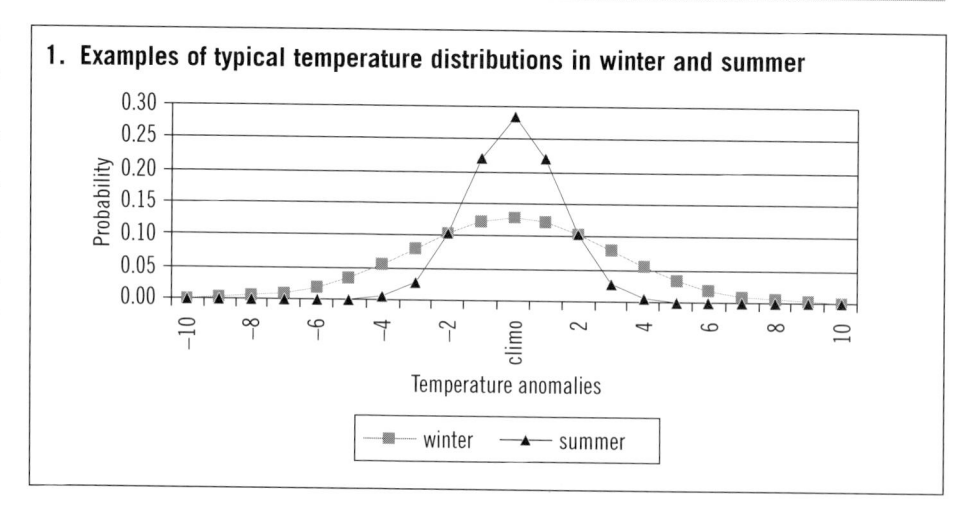

1. Examples of typical temperature distributions in winter and summer

natural variation is still the subject of vigorous debate (Spencer and Christy, 1990, Karl *et al.*, 1993, Santer *et al.*, 1995 and IPCC, 1996). Either way, these long-term trends must be accounted for when making a seasonal forecast. In order to use long-term trends as a forecast tool, the recent temperature trend is extrapolated out into the future. For example, if the temperature has risen five degrees over the last 20 years in a given location and season, this forecast tool would predict another increase of 0.25 degrees (5 divided by 20) in the following year. One must always be aware that long-term trends can change direction over time.

Persistence
Persistence means that a temperature anomaly from the current month or season will re-occur during the next month or season. An anomaly refers to the difference between the observed temperature and the normal temperature. In certain seasons and regions, persisting the previous monthly or seasonal temperature anomaly acts as a skilful seasonal forecast. Using persistence as a forecast is especially effective in the early winter, again in the late winter, and during the summer. The southeast and west US tend to exhibit unusually persistent temperature anomalies compared to the rest of the country. For example, WSI in-house research has shown that Atlanta and Phoenix exhibit 62% and 64% persistence, respectively, of monthly temperature anomalies while Dallas, Chicago, and New York are between 52 and 54%. Persistence of a monthly temperature anomaly means that the sign of the observed temperature anomaly does not change from month to month, eg, if a given month was warmer than normal then the following month will also be warmer than normal. To use this tool, the temperature anomaly from the previous month or season is used as a forecast for the next month or season. For example, if Atlanta was two degrees below normal in July, this tool would predict two degrees below normal in August.

El Niño–Southern Oscillation
The El Niño–Southern Oscillation (ENSO) pattern is a 3- to 10-year periodic but irregular oscillation in SSTs in the equatorial Pacific Ocean. When SSTs are in their warm phase (El Niño), broad changes in the northern hemispheric atmospheric circulation occur that are unusually predictable, especially in the winter and in the southern tier of the US. Other predictable phenomena occur in the cold phase (La Niña), such as a higher probability of drought in the Midwestern US. The most well-known effect of ENSO is the heavy rains in California that occur during an El Niño winter.

Pacific Decadal Oscillation
The Pacific Decadal Oscillation (PDO) is another important, stable SST pattern that has a major effect on the weather in the US. In this case the oscillation occurs in the

North Pacific Ocean and switches phase every 20–25 years. A large area of cooler than normal SSTs in the north-central Pacific surrounded by warmer than normal SSTs typifies the positive phase of the PDO. The sign of the anomalies is reversed in the negative phase. Both phases are associated with atmospheric patterns in the US. For example, the negative phase of the PDO is typically associated with cooler than normal temperatures in the major cities along the Pacific coast of the US. Thus if the PDO is in the negative phase, one would forecast cooler than normal temperatures along the Pacific coast.

Snow cover
Snow cover plays a major role in modulating air temperatures on scales of 1 day–6 months. Snow-covered areas reflect most of the incoming sunlight, leaving little solar energy left to warm the surface which, in turn, warms the lower atmosphere. Anomalous snow cover in certain geographical regions is correlated to anomalous temperatures in other regions 1–12 months later (eg, snow cover in February in the northern US is strongly correlated to March temperatures at different locations in the US. Thus by correlating the historic snow cover in February to the historic corresponding March temperatures of New York, one can use the current snow cover in February to forecast the March temperature of New York. The skill of this statistical relationship, when used as a forecasting tool, is comparable to the skill from our other forecasting tools, including dynamic model output.

Soil moisture and vegetation state
Soil moisture and vegetation state play a major role in long-term temperature patterns by influencing the partitioning of the sun's energy between evaporating water and heating the lower atmosphere. Much of the sun's energy is used to drive the important process of evapotranspiration (ET), which is the combination of evaporation from wet soil or bodies of water and transpiration of water from active,

---

PANEL 2

STATISTICAL TOOLS

The table summarises the various statistical tools available and shows the regions in which each is used. The following regions are used: SW – southwest quarter of the US; NW – northwest quarter of the US; SE – southeast quarter of the US; NE – northeast quarter of the US; MW – Midwest US, or region between the Rocky Mountains and Mississippi River; US – the entire Unites States. A dash implies that the statistical tool has little utility during a particular season.

Utility of each statistical tool for various geographic regions and seasons

Statistical tool	Summer	Winter
Climatology	US	–
Long-term trends	SW SE NW	US
Persistence	SE	US
ENSO	NW	US
PDO	US	US
Snow cover	–	SE NW NE
Soil moisture/vegetation	MW	–
Other atmospheric indexes	US	US

THE ACCURACY AND VALUE OF OPERATIONAL SEASONAL WEATHER FORECASTS IN THE WEATHER RISK MARKET

healthy vegetation. As the percentage of solar energy used for ET increases, the amount of warming in the lower atmosphere will decrease due to energy conservation. Thus areas with very wet soil are generally cooler than those with dry soil since most of the solar energy is used to evaporate the water from the soil. The state of the native vegetation is also important in modulating the temperature. Healthy, green vegetation is indicative of high ET values, similar to wet soil. During a drought, the vegetation begins to wilt and ET stops. Once this happens, most of the sun's energy is used to heat the lower atmosphere. A positive feedback process often develops, whereby extremely dry (wet) conditions are more likely to persist, along with warm (cool) temperatures (Charney, 1975). Because of this, the soil moisture anomalies and/or state of the vegetation can sometimes be used to predict the temperature pattern across the country for the following month or season.

Other atmospheric indexes
Since the atmosphere tends to organise into stable, well-documented patterns on a periodic basis (eg, North Atlantic Oscillation, Pacific North American pattern), certain atmospheric patterns can often foretell future temperature anomalies. These patterns can be used in certain situations and regions to produce skilful seasonal forecasts.

Historic performance of operational seasonal weather forecasts

The previous section outlined a number of seasonal forecasting techniques and the temporal and geographical areas where they are expected to provide skill. Through the optimal blending of the techniques, there is an expectation of positive skill in all regions and all seasons. Due to the small number of forecasts made per year, it will take a number of years to quantify the skill of operational forecasts. In this section we will provide an initial view of operational validations covering approximately a year and a half from two organisations: the Climate Prediction Center (CPC), a US government organisation, and WSI, an energy product from a private weather information company (the authors' employer). Some of the expectations in the previous section have already been borne out, but it will take more time for the rest of the expectations to come to fruition.

The operational forecasts were issued monthly, in the middle of each month, for the subsequent three months. WSI's forecasts were issued two days prior to CPC's. CPC's forecasts cover the period April 2000–November 2001 while WSI's cover the period October 2000–November 2001. For example, the average temperature forecast for Philadelphia International Airport (PHL) over the period March–May 2001 was made by WSI on February 13 and by CPC on February 15. The forecasted average station temperature over the following three months was assessed.

It is important to note that seasonal forecasting (and weather forecasting in general) is a rapidly evolving science and, as such, forecasting procedures evolve with time. Over the nearly two-year record of WSI's operational forecasts, the forecasting procedures have gone through two significant enhancements and a number of minor adjustments. The first significant enhancement occurred in October 2000 with a modification to the methodology for post-processing the dynamic model data. The second was made in October 2001 with the incorporation of statistical models into the forecasting procedures.

Before evaluating the accuracy of seasonal forecasts, it is important to take into account the expectations of seasonal forecasting. Skilful insight about the potential for a season to be above or below a climatological average is feasible, while deterministic forecasts for day to day weather three months out is not possible. Remember, the forecasts are probabilistic in nature and cover multiple months.

With this expectation of skill in mind, the validation metric chosen is binary in nature and based on the directional correctness of the forecast. In other words, if a forecast includes a probability greater than 50% to be above the previous 10-year climatological average and the temperature for that period turns out to be warmer than the 10-year average, that forecast is considered "correct". However, if the average temperature during the period is less than the 10-year average, then the forecast is "incorrect". The pivot point is based on the 10-year average since that is generally utilised in pricing weather derivatives. The 10-year average, spanning the years 1990–9, was selected to validate the operational forecasts.

RESULTS

A general examination of the results shown in Table 3 reveals a couple of interesting points. First is that the overall performance of WSI and CPC are comparable; on average both WSI and CPC have done better in the summer seasons (April–June to September–November) than in the winter seasons (October–December to March–May). Second is that CPC's performance for summers 2000 and 2001 are consistent. Finally, the validations for the two individual winters 2000–01 and 2001–02 are significantly different. (See Appendix for full data and Table 3 for a summary.)

A geographic dependence on the observed skill is also seen in the data (see Table 3). US Regions with demonstrated notable skill include the west coast, the north central, and the southeast, especially Florida. Lower skill levels were demonstrated in the mountain region and the northeast.

Winter

The winter operational skill is an example were the research communities' expectation of skill has not been borne out in the data. As such, it is hard to draw any sweeping conclusions about winter forecasting skill from the limited data set. However, insight can be gained by looking at the two winters individually.

The winter of 2000–01 began with the coldest November–December period on record and continued in many areas to be below the 10-year average all winter (as reported by the US National Weather Service). It should be noted that while this 10-year average directional correctness metric shows both CPC and WSI were near "perfect", both organisations under-predicted the magnitude of the severe cold, and this was a problem for some weather market speculators.

In the winter of 2001–02, the opposite occurred. The November–December–January period was the warmest on record according the to the US National Weather Service and the 10-year average directional correctness metric shows both CPC and WSI had very little skill, as seen in Table 3. CPC were forecasting a little below the 10-year average in the southwest US and substantially below in the north-eastern tier of the US. WSI were forecasting a little below the 10-year average nationally except for the north-central where they forecasted substantially below. When evaluating such individual forecasts and their actual forecasted temperatures, these forecasts are proving to be much better (ie, closer to the truth) than many of the weather forecasts reported by service providers other than WSI in press releases who were predicting a bitter cold 2001–02 winter.

Summer

The year to year skill in the summer has been consistent and as such provides greater confidence for weather traders to take advantage of seasonal forecasting. Table 3 shows the average skill for six geographic regions in the US where forecast skill has been highest in the north-central, southeast and west regions. On the other hand, forecast skill and year to year consistency has been the lowest in the northeast. This northeast result is surprising since research and hindcasting has shown that the northeast should be one of the more predictable regions. This difference in expectation and realisation is likely a result of the small operational data set.

Table 3 is a consolidation of the results shown in Table A1. The first number in each column is the WSI forecast validation and the second number is the CPC forecast. Note that WSI initiated city-specific seasonal forecasts in October of 2000.

Each column of the table represents a different geographic region (NE – northeast: Boston, New York, Philadelphia and Washington DC; NC – north-central: Minneapolis and Chicago; SE – southeast: Covington, Atlanta and Miami; TX – Texas: Houston and Dallas; MTN – mountain: Phoenix, Las Vegas and Denver; and WEST – west coast: Los Angeles, Sacramento, San Francisco, Portland and Seattle).

Incorporation of seasonal weather forecasts in weather risk management

In this section, we will outline the traditional approach for pricing weather derivatives, an approach for blending seasonal forecasts into the existing methodology, and the use of seasonal forecasts in making speculative derivative buying decisions.

TRADITIONAL PRICING MODELS
Pricing models for weather derivatives are typically derived actuarially and based on observed distributions of temperature data, generally from the last 10 years. Some pricing models take into account previous trades that are still active via intercity correlations and some even account for cross-commodity relationships, eg, between weather and the price of electricity.

The derivation of actuarial-based pricing requires a set of temperature data that spans a chosen period. Figure 2 shows the January–February–March (JFM) seasonal average temperature for Philadelphia International Airport (PHL) and San Francisco International Airport (SFO) for the time period of 1960–2001. Overlaid on the charts are running 10-, 20- and 30-year averages. It is clear from this chart that there has been a warming trend in all three averages and furthermore that the trailing 10 years has provided the warmest average for the last decade. Figure 3 shows the same plot for SFO for the June–July–August (JJA) season. It is seen that the difference between the trailing 10-year trend and the trailing 30-year trend in 2001 is ~1°F in the winter (JFM) and ~0.5°F in summer (JJA).

The recent observed warming shown above has driven the weather risk market to use the most recent 10 years of observations for pricing. The current approach is to use the 10-year "burn" which is basically the extension of the 10-year trend forward to the period of interest. Figure 4 shows the last 10 years of JFM observations for PHL. The 10-year trend line leads to a value for 2002 of 39.3°F. This is compared to the straight 10-year average of 37.8°F (the dashed line) and the 30-year average of 36.7°F (not shown). Note that differences of 1.5°F over a three-month winter period equate to approximately 135 degree-days and are very significant in the context of weather derivative contracts.

The second component of the temperature data that is required for pricing is the standard deviation, so that a distribution can be derived. Typically the standard deviation is derived from observed data over the last 10–30 years. Fortunately, the standard deviation has been more stable with respect to the averaging period than

Table 3. Fraction of directionally correct operational forecasts, with respect to the 1990–9 average, for each season (WSI/CPC)

Skill	NE	NC	SE	TX	MTN	WEST
Average	0.51/0.57	0.59/0.63	0.65/0.54	0.62/0.50	0.63/0.67	0.81/0.75
Summer 00	–/0.85	–/0.60	–/0.60	–/0.40	–/0.73	–/0.72
Winter 00–01	0.95/1.00	0.90/0.83	0.87/0.83	0.70/0.58	1.00/0.72	0.92/0.93
Summer 01	0.38/0.46	0.75/0.75	0.72/0.56	0.75/0.50	0.56/0.56	0.90/0.83
Winter 01–02	0.29/0.00	0.17/0.33	0.17/0.33	0.42/0.50	0.39/0.67	0.63/0.50

THE ACCURACY AND VALUE OF OPERATIONAL SEASONAL WEATHER FORECASTS IN THE WEATHER RISK MARKET

2. The trailing 10-, 20- and 30-year temperature averages for winter (JFM), showing continued warming during the last 40 years

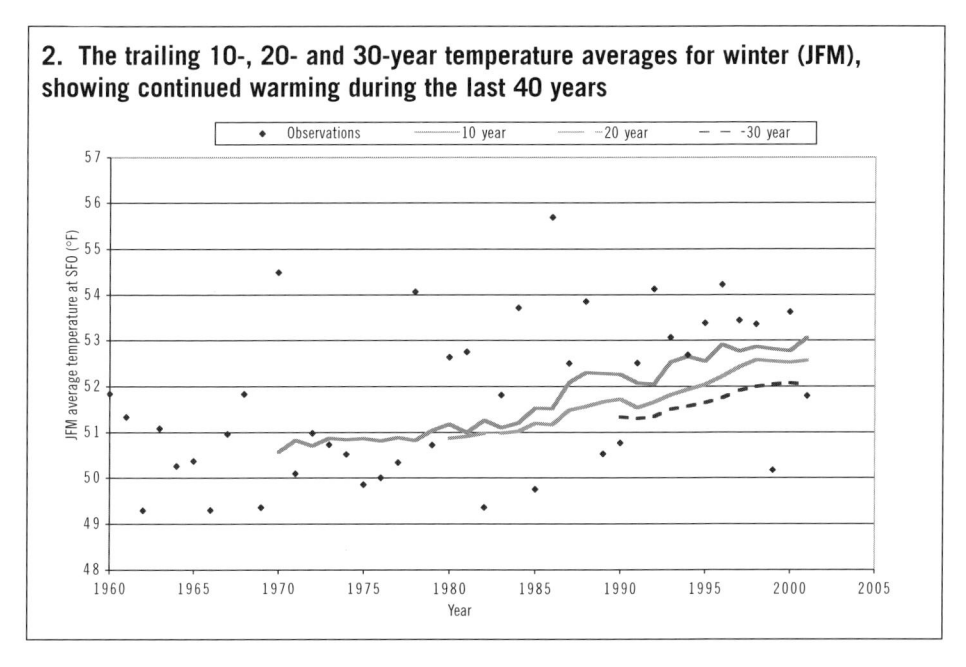

PANEL 3

THE CONVOLUTION PROCESS

The convolution process is illustrated in Figure *a* for a number of different weights. The two thick lines represent the forecast (tall and narrow) and climatological (short and broad) probability distribution functions (PDFs). The number specified in the legend is the weight given to the climatological PDF. The lines represent the blended PDF's created using weights of 0.8, 0.6, 0.4, and 0.2 for the climatological PDF and 0.2, 0.4, 0.6, and 0.8 for the forecast PDF, from left to right. To compute the strike point for a swap, the 50% point of the CDF must be computed. Figure *b* shows the integration of the PDF's shown in Figure *a*. The bold curve to the left is the pure climatology line, the narrow line to the right of it is a blend of 80% climatology and 20% forecast, and the subsequent three narrow lines are 60%, 40% and 20% climatology, respectively. The bold line to the right is the 100% forecast. The shifting of the 50% probability point is

a. Series of probability distributions derived by weighting the climatological and forecast distributions

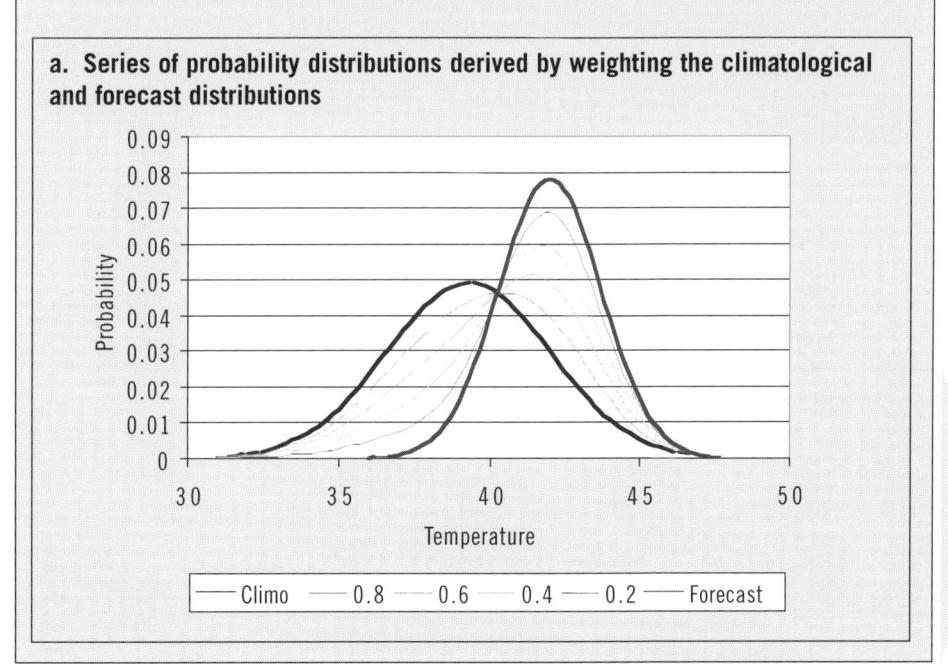

b. Blending climatology and seasonal forecasts in the form of cumulative probability distributions

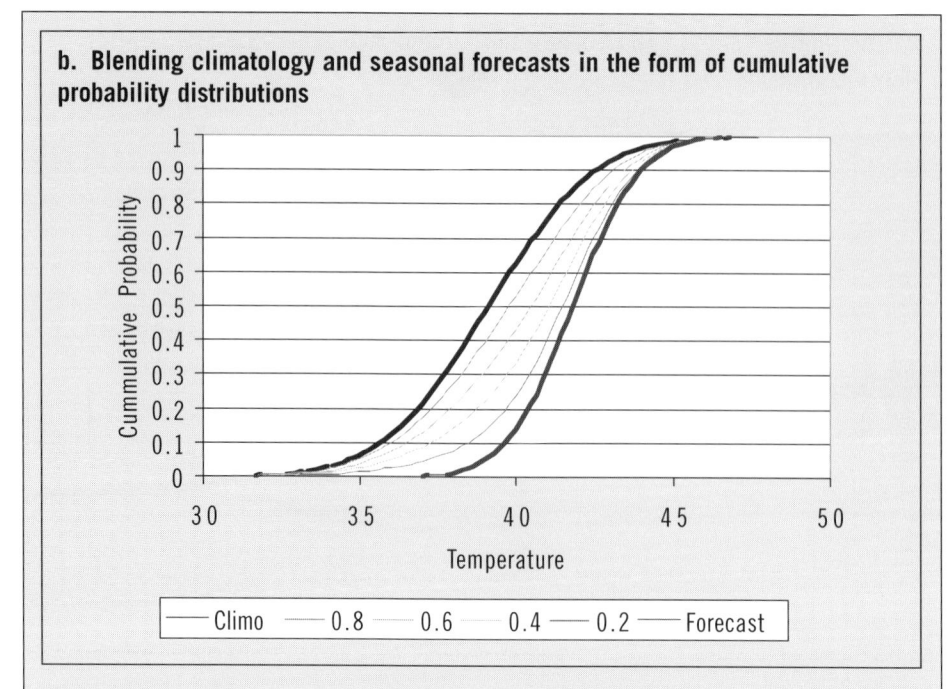

clearly seen in Figure a and in Table 4. It is apparent that blending in only 20% of the forecast PDF creates a new PDF that is substantially shifted from the climatology PDF. Furthermore, it is seen that as more of the forecast PDF is blended with the climatology PDF, the relative change to the 50% probability point decreases (eg, a 0.7 degree change for the first 20% and a 0.4 degree change for the last 20%). This strong initial impact is primarily due to the narrower PDF (higher confidence) of the forecast.

the mean temperature. It is important to note that the variability of temperatures is much greater in the winter than in the summer and as a result the standard deviations are larger in the winter than in the summer. For example, in PHL, the 20-year standard deviation in JFM is 2.7°F while in JJA it is only 1.8°F. The degree of the winter to summer variability varies with geographic location.

For the classic "swap" contracts, the mean of the distribution is used with the

3. The trailing 10-, 20- and 30-year temperature averages for summertime (JJA) showing continued warming during the last 40 years

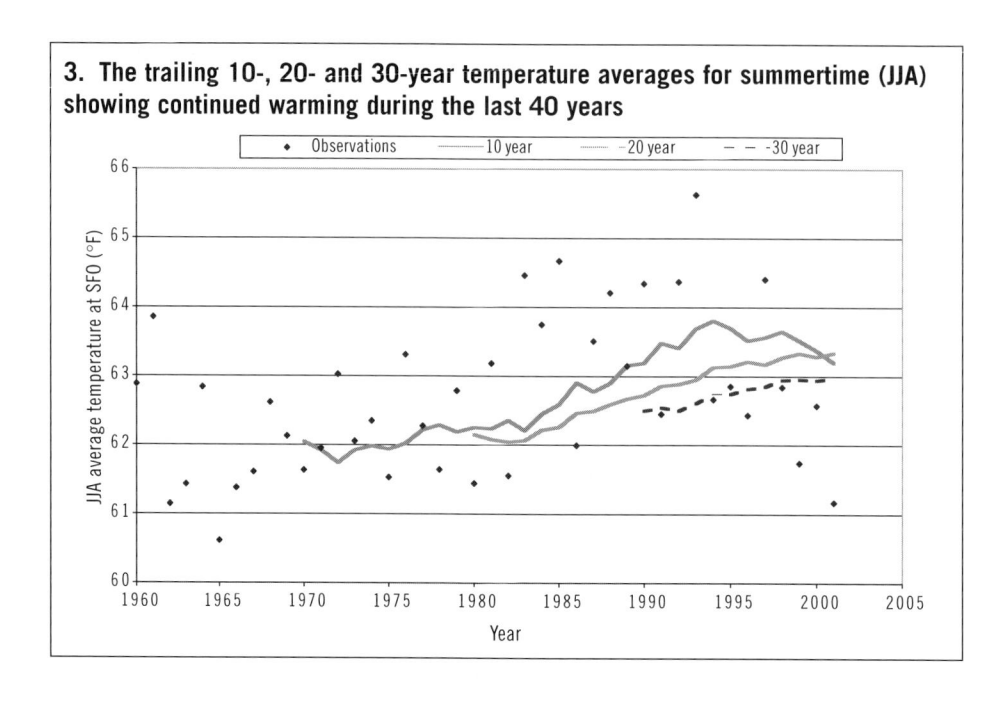

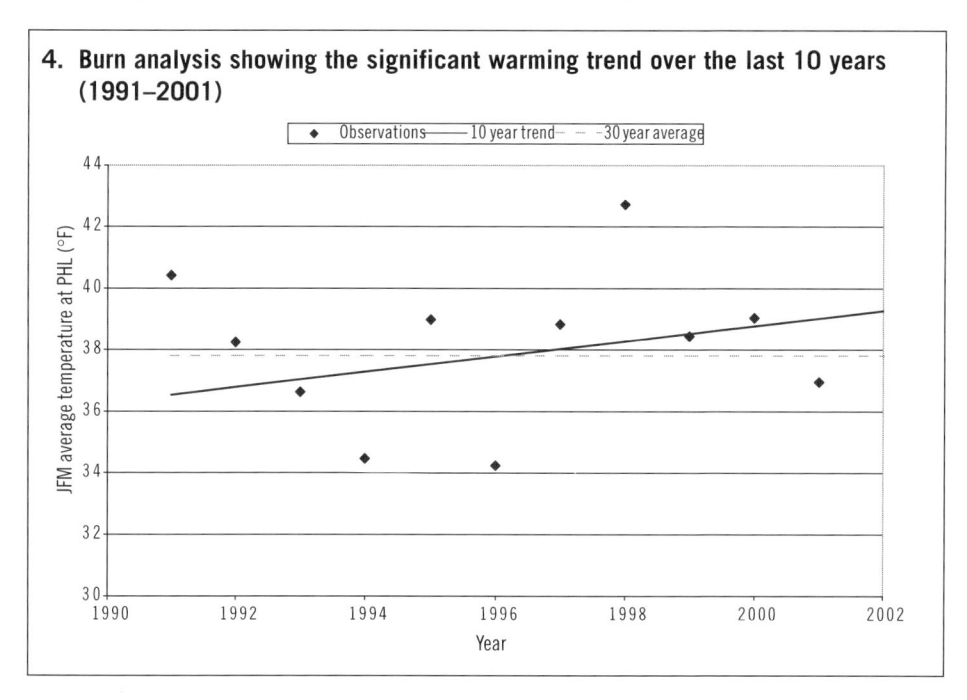

4. Burn analysis showing the significant warming trend over the last 10 years (1991–2001)

addition of a bid–ask spread. For example, if the 10-year burn temperature was 39.3°F and the bid–ask spread was 0.4°F, then the short strike would be 39.3 − 0.2 = 39.1°F and the long strike would be 39.3 + 0.2 = 39.5°F. For "puts", "calls", and "extreme weather" contracts, the trader uses the distribution to calculate the probability of return and sets the strike accordingly. It should be noted that each weather desk has their own proprietary pricing models that are variations of the presented approach.

INCORPORATING SEASONAL FORECASTS INTO THE TRADITIONAL PRICING MODELS
Seasonal forecasts are delivered as probability distributions and as such could be directly substituted for the climate data in the pricing models. This may eventually be the case, but until forecasts are more skilful and traders embrace seasonal forecasting, the more likely scenario is a blending of the forecast with a chosen climatology, such as the 10-year burn with a 15-year standard deviation. The forecast and climatological PDFs can be blended by applying a weight to each PDF and averaging them. The relative weights for the forecast PDF and the climatological PDF can vary depending on the trader's confidence in the seasonal forecast. Once the convolved distribution is computed, the 50% point of the cumulative PDF (CDF, the integral of the PDF) would be computed and used for pricing standard "swap" strike points (see Panel 3).

Incorporating seasonal forecasts into the speculative buying decisions

In this section we describe a simulated trading exercise in which seasonal forecasts are used for making buying decisions of weather derivatives. We will describe the surrogate weather derivative, the decision process for either going long or short, and present the results.

The simulated trading exercise used a standard swap for the derivative. The strike point was centred at the 10-year average (1990–9) with a bid–ask spread of 0.4°F. The payoff was US$10,000 per degree away from the strike and there were no caps. This financial swap structure was adapted from the current offering on the Intercontinental Exchange. Thus if the 10-year average was 37.2°F, the bid strike would be 37.0°F and the ask strike would be 37.4°F.

The two trading rules for the simulation are simply:

Table 4. The movement of the mid-point of the temperature distribution as a seasonal forecast is blended into the climatology with weights as indicated

Climatology/forecast blend	50% point
Pure climatology	39.3
80% climatology + 20% forecast	40.0
60% climatology + 40% forecast	40.6
40% climatology + 60% forecast	41.1
20% climatology + 80% forecast	41.6
Pure forecast	41.6

1. if the mean of the forecast distribution is outside the bid–ask spread, then a trade will be executed at either the bid or ask strike depending on which side forecast mean lies; and
2. once a trade is executed, it is carried to maturity.

Continuing with the example in the previous paragraph, where the bid–ask was 37.0°F and 37.4°F respectively, and assuming the mean of the forecast was 37.8°F for illustration purposes, a trade would be logged as going long at 37.4°F. If the average temperature for the period and location was observed at 39.0°F, then the profit from the simulated trade would be (39.0–37.4) × US$10,000 = US$16,000. This evaluation was performed for all 19 cities shown in Table 3 for each seasonal forecast period using forecast from both CPC and WSI. Then all the payoffs were summed up and analysed.

The results clearly demonstrate that there is significant financial value in these seasonal forecasts. Furthermore, it is also evident that the financial results could have been increased by selectively buying weather derivatives for cities and seasons that have greater demonstrated skill. Overall, the simulated trading exercise produced an average payoff per city, per forecast of US$4,300 and a net income of US$1.4 million, using the WSI forecasts, and US$2,200 and US$1.7 million, using the CPC forecasts. Note that WSI did not perform city-specific forecasts until September 2000. In the remaining three seasons evaluated, WSI either produced more profit or fewer losses. Though this sounds impressive, it is important to understand both the temporal and geographic volatility in the payoffs. For example, the largest gain in one season was US$1.8 million and US$1.7 million using the WSI and CPC forecasts, respectively, and the largest loss for one season was US$788,000 and US$1,113,000 using the WSI and CPC forecasts, respectively.

As expected, temporal and geographical volatility mimic the skill of the forecasts discussed in the "Historic Performance of Operational Seasonal Weather Forecasts" section. WSI and CPC forecasts both produced the largest profit in the winter season of 2000-1, with gains of US$1.8 million and US$1.2 million, respectively. They both produced the largest losses in the 2001–2 winter US$788,000 and US$1,113,000 losses, respectively.

The summer results from CPC are more consistent year to year and we expect the same will be true with WSI's summer forecasts in the future. CPC's forecasts generated an income in both summers (US$715,000 and US$374,000 for 2000 and 2001, respectively) while WSI's forecasts produced an income of US$433,000 in 2001. (WSI started city specific forecasts in October 2001 and does not present validation information for summer 2000.)

Looking at the spatial variations in summer revenue (Table 6), the north-central, Texas, and the west regions have provided high payoffs. In the southeast, profits were smaller even though forecast skill (Table 3) was relatively high. This is probably due to the relative lack of natural temperature variability in this region. It is commonly said, "the larger the variability, the larger the potential returns, and the larger the risk". The northeast, on the other hand, suffered from poor overall skill and this resulted in monetary losses. This result for the northeast is surprising since the research has shown that it is one of the more predictable regions (as seen in Table 3).

Table 5 is a consolidation of the results shown in Table A2. Table 5 assumes swaps were bought at the 10-year average with a 0.4-degree bid–ask spread and a payoff of US$10,000 per degree away from the strike. The first number in each column is derived from WSI's forecasts and the second number is derived from CPC's forecasts. Note that WSI did not perform city-specific forecasts for the summer of 2000.

Conclusion

This chapter has focused on the usefulness of seasonal forecasting to the weather derivative market. The chapter presented many approaches used in seasonal forecasting and states that there is an expectation of skill in seasonal forecasts for all seasons and geographic regions covered in the chapter. This expectation relies on optimally blending the different forecasting techniques.

The expectation of skill was evaluated through an initial view of operational validations covering approximately a year and a half from two organisations: the Climate Prediction Center (CPC) and WSI Energycast Trader. Some of the expectations in the chapter have already been borne out (ie, consistent skill in the summer), but due to the small number of forecasts made per year, it will take more time for the rest of the expectations to come to fruition (ie, consistent skill in the winter).

A technique was presented to derive a hybrid probability distribution of the forecast and climatology for incorporation into pricing models. This technique allows weather traders start to incorporate seasonal forecasts into their traditional pricing models by weighting the forecast distribution with their degree of confidence. As the confidence grows, the weights can be adjusted.

In the final section of this chapter, a simulated trading exercise was presented. This exercise directly relates the accuracy of the forecast to the potential profit of a weather speculator. The exercise evaluated 19 cities using a three-month swap priced at the 10-year average with a payoff of US$10,000 per degree. The average payoff per city per forecast was US$4,300 amd US$2,200 fpr WSI and CPC, respectively. The net payoff for WSI forecasts for the period of September 2000–November 2001 was US$1.4 million and the net payoff for CPC's forecasts for the period of April 2000–November 2001 was US$1.7 million. It is also noted in the chapter that the season to season volatility is very high. The largest one-season gain was US$1.8 million and the largest one-season loss was US$1.1 million.

This chapter has provided both theoretical and operational information showing that seasonal weather forecasts can be a value to weather traders and hedgers.

Table 5. Trading payoff by region on a per-city basis (WSI/CPC)

US$ (000s)	NE	NC	SE	TX	MTN	WEST
Average	−29/8	−8/−38	99/37	85/71	88/146	161/206
Summer 00	−/78	−/8	−/−8	−/28	−/35	−/50
Winter 00–01	93/132	160/124	102/100	85/53	92/55	62/74
Summer 01	−24/−4	46/16	23/1	9/−3	17/14	59/65
Winter 01–02	−99/−197	−215/−185	−26/−56	−9/−8	−21/42	39/17

THE ACCURACY AND VALUE OF OPERATIONAL SEASONAL WEATHER FORECASTS IN THE WEATHER RISK MARKET

Table A1

	BOS	LGA	PHL	DCA	ORD	CVG	ATL	MIA	MSP	DFW	HOU	DEN	LAS	PHX	LAX	SAC	SFO	PDX	SEA	Average
MJJ 00	–/1	–/1	–/1	–/1	–/0	–/0	–/1	–/1	–/1	–/1	–/0	–/1	–/1	–/1	–/0	–/0	–/0	–/1	–/1	–/0.68
JJA	–/1	–/1	–/1	–/1	–/0	–/0	–/1	–/1	–/1	–/1	–/0	–/1	–/0	–/1	–/0	–/0	–/0	–/1	–/1	–/0.63
JAS	–/1	–/1	–/1	–/1	–/0	–/0	–/0	–/1	–/1	–/1	–/0	–/1	–/1	–/0	–/1	–/1	–/1	–/1	–/1	–/0.68
ASO	–/0	–/1	–/1	–/1	–/1	–/0	–/0	–/1	–/0	–/1	–/0	–/1	–/1	–/1	–/1	–/1	–/1	–/1	–/1	–/0.74
SON	–/0	–/1	–/1	–/0	–/1	–/0	–/1	–/1	–/1	–/0	–/0	–/0	–/1	–/1	–/1	–/1	–/1	–/1	–/1	–/0.68
OND	–/1	–/1	–/1	–/1	–/0	–/0	–/1	–/1	–/1	–/0	–/1	–/1	–/1	–/0	–/1	–/1	–/1	–/1	–/1	–/0.79
NDJ	1/1	1/1	1/1	1/1	1/1	1/1	1/1	1/1	1/1	1/0	1/1	1/1	1/1	1/0	1/1	1/1	1/1	1/1	1/1	1.00/0.89
DJF	1/1	1/1	1/1	1/1	1/1	1/1	1/1	1/1	1/1	1/1	1/1	1/1	1/1	1/1	1/1	1/1	1/1	1/1	1/1	1.00/0.95
JFM 01	1/1	1/1	1/1	1/1	1/1	1/1	1/1	1/1	1/1	1/1	1/1	1/1	1/1	1/1	1/1	1/1	1/1	1/1	1/1	1.00/1.00
FMA	1/1	1/1	1/1	1/1	1/0	0/1	1/1	0/0	1/1	1/0	0/1	1/1	1/1	1/1	0/1	1/0	1/0	1/1	1/1	0.84/0.79
MAM	1/1	1/1	1/1	0/1	1/1	1/1	1/0	1/1	0/1	0/0	0/1	1/1	1/0	1/0	0/1	1/0	0/0	1/1	1/1	0.68/0.68
AMJ	1/0	0/0	0/0	1/0	1/1	0/1	1/0	1/1	1/1	1/1	1/0	1/1	1/0	1/1	0/1	1/1	1/0	1/1	1/1	0.79/0.58
MJJ	0/1	1/1	1/1	0/1	0/1	1/0	1/1	1/1	1/1	0/0	0/1	1/1	1/0	1/1	1/1	1/1	0/1	1/1	1/1	0.68/0.84
JJA	1/0	0/0	0/0	0/1	1/1	1/0	1/0	1/1	1/1	1/0	1/1	1/1	1/0	0/1	1/1	1/0	1/0	1/1	1/1	0.79/0.53
JAS	0/0	1/0	0/1	0/1	1/1	0/0	0/0	1/1	1/0	0/0	1/1	1/1	0/0	0/1	1/1	1/1	1/0	1/1	1/1	0.63/0.63
ASO	0/1	0/0	0/0	1/1	0/0	0/0	1/0	1/1	1/1	0/0	1/1	0/0	0/0	0/0	0/1	0/0	0/0	1/1	1/1	0.58/0.58
SON	1/1	0/0	0/0	1/1	1/1	1/1	0/1	1/1	0/0	1/0	1/1	0/0	0/0	0/0	0/1	1/0	0/0	1/1	1/1	0.58/0.58
OND	0/0	0/0	0/0	0/0	1/1	0/1	0/0	0/0	0/0	0/1	0/0	0/0	0/0	0/1	1/1	0/0	0/0	0/0	0/0	0.11/0.26
NDJ	1/0	0/0	0/0	0/0	0/0	0/0	1/0	1/0	0/0	0/1	0/0	1/1	0/0	0/1	1/0	0/0	0/0	0/0	0/0	0.26/0.16
DJF	0/0	0/0	0/0	0/0	1/1	1/1	0/0	0/0	0/0	1/1	1/1	1/1	1/1	1/0	1/0	1/1	1/1	0/0	1/1	0.47/0.37
JFM 02	1/1	1/1	1/1	1/1	1/1	1/1	1/1	1/1	1/1	1/1	1/1	1/1	1/1	1/1	1/1	1/1	1/1	1/1	1/1	0.58/0.42
FMA	0/1	0/1	0/1	0/1	0/1	0/0	0/1	0/1	0/1	0/0	0/0	0/1	0/1	0/1	0/1	0/1	0/1	0/1	0/1	0.47/0.58
MAM	1/1	1/1	1/1	1/1	1/1	1/1	1/1	1/1	1/1	1/0	1/1	1/1	1/1	0/0	1/1	1/1	1/1	1/1	1/1	0.58/0.32
Average	0.65/0.52	0.47/0.52	0.41/0.57	0.53/0.65	0.59/0.61	0.41/0.39	0.71/0.52	0.82/0.70	0.59/0.65	0.59/0.43	0.65/0.57	0.76/0.83	0.53/0.48	0.59/0.70	0.82/0.78	0.76/0.61	0.76/0.57	0.82/0.87	0.88/0.91	0.65/0.62
Summer 00	–/0.60	–/1.00	–/1.00	–/0.80	–/0.40	–/0.00	–/0.80	–/1.00	–/0.80	–/0.80	–/0.00	–/0.80	–/0.60	–/0.80	–/0.60	–/0.60	–/0.40	–/1.00	–/1.00	–/0.68
Winter 00–01	1.00/1.00	1.00/1.00	1.00/1.00	0.80/1.00	1.00/0.67	0.80/0.83	1.00/0.83	0.80/0.83	0.80/1.00	0.80/0.33	0.60/0.83	1.00/1.00	1.00/0.83	1.00/0.33	0.80/1.00	1.00/0.83	0.80/0.83	1.00/1.00	1.00/1.00	0.91/0.85
Summer 01	0.50/0.50	0.33/0.17	0.17/0.33	0.50/0.83	0.67/0.83	0.50/0.33	0.67/0.33	1.00/1.00	0.83/0.67	0.67/0.17	0.83/0.83	0.83/0.83	0.50/0.00	0.33/0.83	0.83/1.00	0.83/0.67	0.83/0.50	1.00/1.00	1.00/1.00	0.68/0.62
Winter 01–02	0.50/0.00	0.17/0.00	0.17/0.00	0.33/0.00	0.17/0.50	0.00/0.33	0.50/0.17	0.67/0.00	0.17/0.17	0.33/0.50	0.50/0.50	0.50/0.67	0.17/0.50	0.50/0.83	0.83/0.50	0.50/0.33	0.67/0.50	0.50/0.50	0.67/0.67	0.41/0.35

THE ACCURACY AND VALUE OF OPERATIONAL SEASONAL WEATHER FORECASTS IN THE WEATHER RISK MARKET

Table A2

	BOS	LGA	PHL	DCA	ORD	CVG	ATL	MIA	MSP	DFW	HOU	DEN	LAS	PHX	LAX	SAC	SFO	PDX	SEA	Average
MJJ 00	–/21	–/14	–/19	–/14	–/6	–/6	–/0	–/4	–/4	–/8	–/8	–/22	–/0	–/24	–/5	–/1	–/2	–/2	–/13	–/118
JJA	–/24	–/23	–/28	–/31	–/14	–/18	–/0	–/5	–/3	–/17	–/6	–/24	–/21	–/7	–/1	–/9	–/8	–/6	–/14	–/105
JAS	–/24	–/25	–/34	–/40	–/5	–/21	–/0	–/2	–/11	–/31	–/7	–/19	–/5	–/9	–/6	–/22	–/0	–/19	–/21	–/208
ASO	–/13	–/10	–/0	–/0	–/15	–/2	–/0	–/6	–/0	–/33	–/6	–/6	–/6	–/10	–/16	–/22	–/0	–/11	–/15	–/129
SON	–/10	–/8	–/17	–/0	–/6	–/4	–/0	–/9	–/0	–/5	–/0	–/0	–/22	–/0	–/28	–/34	–/13	–/19	–/17	–/155
OND	0/31	0/31	0/34	0/38	0/–44	0/–42	0/38	0/17	0/34	0/–22	0/27	0/0	0/11	0/0	0/23	0/24	0/14	0/15	0/9	0/239
NDJ	31/31	35/35	43/43	48/48	55/0	59/59	53/53	35/35	32/32	42/–42	0/42	50/50	9/9	22/0	18/0	18/18	13/13	19/19	9/9	590/454
DJF	29/29	32/32	35/35	38/38	59/59	48/48	40/40	17/17	64/64	29/29	32/32	41/41	2/2	10/0	18/0	9/0	8/8	11/11	9/9	530/493
JFM 01	15/15	13/13	16/16	14/14	23/23	18/18	21/21	7/7	34/34	37/37	24/24	36/36	8/8	15/15	23/23	7/7	10/10	10/10	15/15	344/344
FMA	11/11	8/8	4/4	0/3	8/–8	–8/0	5/5	–15/–15	42/42	14/–14	–1/–1	23/23	7/7	7/7	30/30	15/15	16/16	18/18	25/25	209/178
MAM	0/5	3/3	3/3	–5/5	8/8	9/9	13/–13	6/6	–3/3	–6/–6	0/0	0/3	25/–25	22/–22	–15/15	0/–15	–1/–1	8/8	18/18	86/5
AMJ	14/–14	–11/–11	–11/–11	4/–4	13/13	–14/14	9/–9	10/10	16/0	6/6	7/0	8/8	32/–32	29/29	–10/10	16/16	1/0	14/14	23/23	156/63
MJJ	0/0	1/1	3/3	0/17	0/1	0/–8	20/20	15/15	21/21	–1/–1	–6/–6	13/13	34/–34	32/32	0/1	23/23	–10/0	14/14	25/25	185/149
JJA	1/–1	–6/–6	–5/–5	–14/14	1/1	0/–8	0/–19	5/5	31/31	2/–2	7/7	23/23	0/–16	0/17	12/12	3/3	4/–4	19/19	26/26	105/92
JAS	–5/0	0/0	0/0	–21/0	1/1	–8/–8	0/0	8/0	26/0	0/–8	10/10	29/29	0/–16	–22/22	0/22	16/0	13/0	14/14	0/24	61/90
ASO	–13/13	–14/–14	–12/0	2/0	–4/–4	–3/–3	19/–19	5/0	10/–10	–20/–20	0/0	18/18	–27/–27	–28/28	24/24	7/7	9/0	8/8	18/18	0/19
SON	19/19	–20/–20	–20/0	14/0	23/23	0/13	–5/0	9/0	–44/–44	6/–6	4/0	–17/–17	–36/–36	–39/0	20/20	–3/–3	1/0	3/3	11/11	–74/–38
OND	–30/–30	–37/–37	–37/–37	0/–33	40/40	–31/31	–24/0	–5/–5	–67/–67	–9/9	–6/–6	–6/–6	–21/–21	–20/20	19/0	–19/–19	–9/–9	–2/–2	0/0	–265/–174
NDJ	44/–44	–51/–51	–46/–46	0/–45	–67/–67	–46/–46	37/–37	6/–6	–101/–101	–19/19	–21/–21	3/3	–7/–7	–10/10	18/0	–8/–8	–5/–5	–4/–4	–1/–1	–278/–457
DJF	–44/–44	–48/–48	–41/–41	–35/–35	–48/–48	–32/–32	–7/–7	–11/–11	–79/–79	5/5	3/3	10/10	11/11	3/–3	11/–11	3/3	1/0	–3/–3	6/6	–295/–325
JFM 02	40/–40	–39/–39	–34/–34	26/–26	–24/24	–14/–14	9/9	9/–9	–28/–28	21/–21	13/13	42/42	–19/19	5/5	–5/–5	–19/–19	6/–6	12/12	27/27	28/–90
FMA	–25/–25	–30/–30	–31/–31	–24/–24	–6/6	–9/9	6/–6	17/–17	6/6	–10/–10	–6/–6	–26/–26	–2/–2	22/22	7/7	17/17	4/4	11/11	30/30	–49/–1
MAM	10/–10	9/–9	15/–15	13/–13	–16/–16	–4/–4	–2/–2	21/–21	–39/–39	–1/–1	11/–11	–25/–25	–2/2	–21/21	15/15	19/–19	10/10	23/23	35/35	71/–79
Total	96/–7	–154/–60	–117/16	59/83	64/8	–36/–16	194/74	138/53	–80/–84	97/37	72/105	221/348	15/–145	29/234	184/230	101/114	70/52	175/247	275/388	1403/1676
Summer 00	–/47	–/81	–/98	–/86	–/4	–/51	–/0	–/27	–/19	–/84	–/27	–/71	–/1	–/32	–/43	–/69	–/3	–/57	–/80	–/715
Winter 00–01	86/122	92/123	101/135	94/146	152/38	125/91	132/143	49/66	169/209	115/–18	55/125	150/153	51/12	76/0	73/91	50/50	46/60	67/82	76/85	1758/1713
Summer 01	16/18	–51/–51	–44/–12	–15/27	33/34	–25/0	43/–27	51/29	60/–3	–6/–30	23/23	74/74	3/–160	–27/128	46/90	59/41	18/–4	71/71	103/127	433/374
Winter 01–02	–5/–194	–195/–213	–175/–205	–20/–176	–121/–61	–136/–56	19/–42	38/–69	–308/–308	–13/1	–6/–16	–3/–49	–40/2	–21/75	65/6	–8/–46	6/–6	37/37	96/96	–788/–1126

Appendix

Each row in Table A1 represents a different season in which a forecast was made. The last five rows summarise the results. Each column represents a city for which the forecast was made. The last column provides the average of the columns. The first entry in each column –, 0 or 1 designates if WSI's forecast was not made, was wrong, or was right. The second entry is for CPC's forecast.

The summary of these results can be seen in Table 3.

Each row in Table A.2 represents a different season in which a forecast was made. The last five rows summarise the results. Each column represents a city for which the forecast was made. The last column provides the average of the columns. The first entry in each column shows the resulting payoff from WSI's forecast (or if none was made). The second entry is for CPC's forecast.

1 *See American Meteorological Society, 2001.*
2 *Numbers are based on the 1990–9 ATL cooling degree-day data and a fictional forecast.*

BIBLIOGRAPHY

American Meteorological Society, 1998, "AMS Policy on Weather Forecasting", *Bulletin of the American Meteorological Society*, 79, pp. 2161–3.

American Meteorological Society, 2001, "Statement on Seasonal to Interannual Climate Prediction", *Bulletin of the American Meteorological Society*, 82, p. 701.

Charney, J. G., 1975, "Dynamics of Deserts and Drought in the Sahel," *Quarterly Journal of the Royal Meteorological Society*, 101, pp. 193–202.

Houghton J. T, L. G. Meira Filho, B. A. Callander, N. Harris, A. Kattenberg and K. Maskell (eds), 1996, *Climate Change 1995. The IPCC Second Scientific Assessment*, (Cambridge University Press).

Lorenz, E. N., 1963, "Deterministic Nonperiodic Flow", *Journal of Atmospheric Science*, 20, pp. 130–41.

Kiehl, J. T., J. J. Hack, G. B. Bonan, B. A. Boville, D. L. Williamson, and P. J. Rasch, 1998, "The National Center for Atmospheric Research Community Climate Model: CCM3" *Journal of Climate*, 1131–49.

Moran, J. M., and M. D. Morgan, 1997, *Meteorology: The Atmosphere and the Science of Weather*, (Upper Saddle River, New Jersey: Prentice Hall).

Karl, T. R., P. D. Jones, R. W. Knight, G. Kukla, N. Plummer, V. Razuvayeve, K. Gallo, J. Lindseay, R. J. Charlson and T. C. Peterson, 1993, "A new perspective on recent global warming: Asymmetric trends of daily maximum and minimum temperature", *Bulletin of the American Meteorological Society*, 74, pp. 1007–1023.

Santer, B. D., K. E. Taylor, J. E. Penner, T. M. L. Wigley, U. Cubasch and P. D. Jones, 1995, "Towards the Detection and Attribution of an Anthropogenic Effect on Climate", *Climate Dynamics*, 12, pp. 77–100.

Spencer, R. W., and J. R. Christy, 1990, "Precise Monitoring of Global Temperature Trends from Satellites", *Science*, 247, pp. 1558–62.

10

Use of Meteorological Forecasts in Weather Derivative Pricing

Stephen Jewson, Christine Ziehmann and Anders Brix

Risk Management Solutions

Historical data can be used to estimate the distributions of payoffs of single weather derivatives and weather derivative portfolios (see Chapter 8). These distributions can then be used to calculate appropriate prices. When there are no relevant meteorological forecasts for the contract period, this is a reasonable approach. However, as the contract period approaches, relevant meteorological forecasts do become available and it is important to factor these forecasts in to the pricing algorithms.

Introduction

This chapter focuses on the forecasts that can be used in the pricing of weather derivatives. We will argue that the key to successful application of forecasts is to merge them with historical data in appropriate ways and in appropriate proportions. The first section will explain how seasonal and weather forecasts are made. We will then describe the differences between single and ensemble forecasts, and between model and site-specific forecasts. There will also be some discussion of why forecasts go wrong.

Subsequent sections will investigate how weather and seasonal forecasts can be incorporated into pricing for both index- and daily simulation-based pricing models.

Types of forecasts

There are two types of forecast used in weather derivative pricing: weather forecasts, and seasonal forecasts.

WEATHER FORECASTS

Weather forecasts are based on attempts to predict the development of individual weather systems as they move around the planet. They are usually considered to be useful up to around 10 days, but beyond that contain little information due to a combination of errors in the forecasting models and errors in the forecast's initial conditions.

SEASONAL FORECASTS

Seasonal forecasts are based on attempts to predict changes in the temperature and circulations of the ocean and the impact these changes can have on the atmosphere. The principal ocean phenomenon that leads to seasonal predictability is the El Niño–Southern Oscillation (ENSO), which is a pattern of variability of the Pacific Ocean and atmosphere near the equator. ENSO causes warming in some places and cooling in others, or rain in some and drought in others. Seasonal forecasts are, at present, to all intents and purposes, forecasts of ENSO and its effects. Forecasts as long as 10 months may be useful especially when El Niño conditions are developing, due to the slow rate of changes in the ocean. The content of a seasonal forecast is a statement about shifts in the averages of weather over a season, such as a prediction of a warmer than normal summer or winter, rather than a statement about individual weather events themselves.

Whereas weather forecasts are available for all parts of the globe, seasonal forecasts are only really useful in the continents surrounding the Pacific Ocean, such as the Americas, Japan, Australia and, to a certain extent, some parts of South-East Asia and Africa. For weather contracts based on locations in these areas, ENSO forecasts, or seasonal forecasts based on ENSO, should be used to complement

---

PANEL 1

BEATING THE 10-DAY BARRIER IN EUROPE

There is very little forecast skill beyond 10 days in Europe. Although longer forecasts do exist, it is practical at present to ignore them. There are, however, some ideas that may lead to weak but useful forecasts in the future. We review some of these ideas below.

TIME SERIES-BASED FORECASTS
Statistical analysis of European temperatures has shown that there are significant local autocorrelations of temperature out to 40 or 50 days (See Caballero *et al.*). These can be used rather easily to make statistical ensemble forecasts this far into the future. However, such forecasts are very weak, showing only very small deviations of the temperature from the climatological (or baseline) distribution.

EFFECTS OF ENSO
It is just about possible to detect the effects of ENSO in Europe in some locations, by comparing long records of ENSO state with long records of European climate. As the forecasting of ENSO and its impacts improves, these effects might be part of a useful, if weak, seasonal forecast for Europe. (See Fraedrich, 1990, for a discussion of the impacts of ENSO on European weather).

EFFECTS OF THE ATLANTIC OCEAN ON SEASONAL TIMESCALES
It is likely that the surface temperatures of the Atlantic Ocean, especially near the equator where water temperatures are highest, have some impact on the atmosphere over Europe. The surface temperatures of the Atlantic are weakly predictable a few months in advance – partly because they respond to ENSO, and partly because changes only decay slowly back to normal after being perturbed by storms. Combining a forecast for the ocean temperatures and an understanding of the impact of these temperatures on European weather could lead to a European seasonal forecast.

EFFECTS OF THE DEEPER OCEAN
The deep ocean circulation is slow compared to the circulation near the surface. Some of these slow circulations, taking perhaps 10 years or more, could possibly be predicted. If they cause changes in the surface temperatures, and if these changes affect the atmosphere, then this could lead to predictions of subtle shifts in the weather years in advance. However, such forecasts are unlikely to be very skilful, and are not being produced at present.

historical pricing at least six months before the contract begins. Neither weather nor seasonal forecasts are ever exactly correct, and both, especially at long lead-times, can have very low levels of skill. However, even in these cases, they may still contain information. When using forecasts to price weather derivatives, the challenge is to extract this information and make the best use of it.

Many years of scientific research have gone into trying to find a basis for seasonal forecasting in Europe, but with almost no success. There are, however, a few leads and some of these are outlined in Panel 1. For the time being at least, these forecasts are sufficiently weak that ignoring them completely would not seem to be particularly disadvantageous for the pricing of weather contracts in Europe.

Sources of forecasts

The previous section divided forecasts into weather forecasts (0–10, or 0–16 days), useful everywhere, and seasonal forecasts (0–10 months), useful mainly in the US, Australia and Japan (as indicated above). We will now proceed to explain in more detail where these forecasts come from and what form they take.

SOURCES OF WEATHER FORECASTS

Almost all weather forecasts are based on output from computer models of the circulation of the atmosphere known as Atmospheric General Circulation Models (AGCMs) (see Panel 2).

The main forecasts produced in this way are:

❏ US NCEP MRF (US National Centre for Environmental Prediction Medium Range Forecast): The US NCEP MRF is produced by the US National Weather Service. It is, in the authors' experience, the most widely used of the model forecasts. It also

---

PANEL 2

ATMOSPHERIC GENERAL CIRCULATION MODELS

AGCMs are computer programs that attempt to model the circulation of the atmosphere. They divide the atmosphere up into many boxes and assume that the conditions within each box are constant. The boxes then interact with each other according to approximations of the equations of motion and thermodynamics. The largest of these models use millions of boxes: however, the planet is also large, and as a result the boxes are still typically something like 50 km square, and 500 m deep, which means that weather on scales smaller than this is not represented at all in the models.

To make weather forecasts, these models begin with an estimate of the current state of the atmosphere and run for a few days of simulation. To make seasonal forecasts, these models have an additional ocean component that allows them to capture effects like ENSO: they then begin with an estimate of the current state of the atmosphere and ocean and run for a few months. To make climate forecasts, these models are run 50 or 100 years into the future, and the amount of carbon dioxide and other anthropogenic greenhouse gases is adjusted to reflect likely future changes. Different AGCMs attempt to solve the same equations, but do so in many different ways. For instance, differences between models would include:

❏ how clouds are represented (the equations for the large-scale behaviour of clouds are not known);
❏ the model resolution (how many boxes are used in the horizontal and vertical); and
❏ which variables are used to write the equations.

PANEL 3

FORECASTS, TRENDS AND ANOMALIES

It is extremely important to take trends into account when analysing historical meteorological data. Historical temperature data for most stations shows warming trends over the last 50 years, and some of these trends are extremely large. Much of this warming appears to be due to urbanisation effects. When pricing a weather contract, one has to decide whether to extrapolate (extend) such trends forward to the contract period. Back-testing analysis shows that, over the last 20 years, extrapolating trends would have worked marginally better than not extrapolating, on average over many contracts.[1] Such extrapolation could be termed a forecast: however, for the purposes of this article we prefer to consider trends and forecasts separately. Trends are discussed in more detail in Chapter 8.

Temperature forecasts can be presented either in terms of the actual value, or in terms of 'anomalies', a value relative to normal temperatures for that time of year. Because of the strong trends mentioned above, however, the use of anomalies for surface temperature forecasts can be confusing. Consider an anomaly forecast relative to the mean temperature of the last 30 years: if there has been a warming trend for whatever reason, then the weather is almost definitely going to be warmer than this average, and if the anomaly forecast just says 'warm', that contains no useful extra information beyond that contained in the trend. Clearly a better way to give anomaly forecasts would be to present values relative to the extrapolated trend itself rather than the 30-year mean.

---

extends the furthest into the future (out to 16 days: although it may not be much use beyond around 10). The US NCEP MRF is available for free on the Internet, and consists of an ensemble of nine separate forecasts (the interpretation of such ensemble forecasts is discussed later).

❏ ECMWF (European Centre for Medium Range Weather Forecasts): The ECMWF weather forecast is made using a larger and more sophisticated model than the US NCEP MRF, and consists of an ensemble of 51 members. However, it only runs out to 10 days in the future, is much less widely available than the US NCEP MRF forecast (it is expensive and difficult to obtain), and is consequently less widely used.

❏ DWD (Deutscher Wetter Dienst), JMA (Japanese Meteorological Agency), MF (Meteo-France), UKMO (UK Meteorological Office), etc. Most countries have built their own AGCMs and use them to make forecasts. These models tend to be smaller and less sophisticated than the US NCEP MRF and ECMWF models, and are not usually run as ensembles. They are used principally by the forecasters of the local weather services, and seldom by the commercial forecasting community. In some cases they have been adjusted to perform particularly well for forecasting in the country in which they were developed.

The output from these models is not particularly useful for the purpose of pricing weather derivatives because it is presented on a large scale mathematical grid that covers the globe but is not directly related to individual locations on the ground. The process of converting this output to a useful site-specific forecast, known as downscaling, is discussed later.

The main forecasting models are all reasonably close in terms of performance. The benefits for pricing weather derivatives come more from the accuracy of the methods used to downscale and correct the model output and to incorporate the forecast into the pricing algorithms than from which model is used.

SOURCES OF SEASONAL FORECASTS

Seasonal forecasts are produced using a variety of models, ranging from simple statistical models to complex models of the ocean and atmosphere. As with weather forecasts from AGCMs, most of these models produce forecasts of large-scale features that are not directly useful for weather pricing and must first be downscaled to site-specific values.

Some examples of the types of models used are:

❏ Purely statistical time-series models based on, for example, Pacific Ocean temperatures, winds and sea surface height.
❏ Mixed statistical-dynamical models that represent ocean behaviour using simple differential equations and the wind response to ocean temperatures using statistical models.
❏ CGCMs (coupled general circulation models): these are complex computer models of the atmosphere and ocean.

Most large-scale seasonal forecasts are given in the monthly Experimental Long-Lead Forecast Bulletin, freely available on the Internet (see http://www.grads.iges.org/ellfb). The only exception is the ECMWF seasonal forecast. This is partly available on the ECMWF website, and partly available commercially via the European National Meteorological Services.[2]

As with weather forecasts, the most widely used of the large-scale seasonal forecasts are those from the US National Weather Service, of which there are several.

Site-specific, single and ensemble forecasts

SITE-SPECIFIC FORECASTS

Both weather and seasonal forecasts appear in many forms. For example, weather forecasts can consist of:

1. Maps of raw AGCM output.
2. Weather maps drawn automatically from AGCM output.
3. Weather maps drawn manually from AGCM output by a trained and experienced meteorologist.
4. Site-specific forecasts derived from AGCM output.

All of these are of some relevance to the weather market: traders may well look at the first three in order to have an intuitive understanding of how the weather in a region is likely to develop. However, for the purpose of this chapter, which is to develop numerical algorithms that give objective estimates of the distribution of outcomes of a derivative contract, only the fourth of these forecasts is of use. These site-specific forecasts are made in a two-step procedure known as "downscaling":

1. Using past observational data and past forecasts, a statistical model is derived which attempts to relate large-scale AGCM forecasts to the weather at the site in question.
2. Forecasts are then fed into this statistical model and converted into the site-specific forecast.

The use of a statistical downscaling model attempts to capture local effects that are not simulated by the AGCM. In general these local effects can be very complex, including the influence of atmospheric turbulence on all scales, from the scale of the model grid (typically 50 km) down to the scale of the measuring instrument (typically a few millimetres). The combined effects of such complex processes are themselves likely to be very complex, and any downscaling model will only be a very simple representation of physical reality. For instance, the relationship between modelled

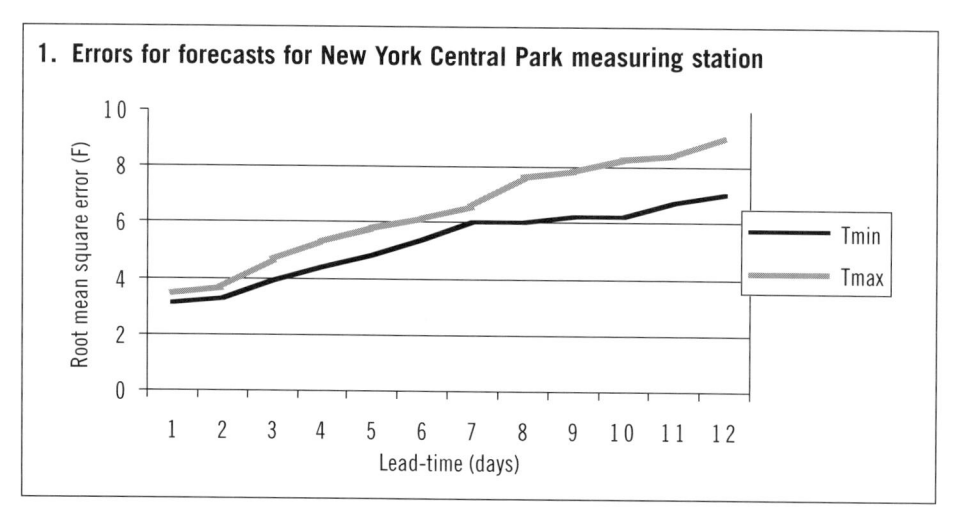

1. Errors for forecasts for New York Central Park measuring station

and site-specific temperatures is highly non-linear, but it is common to assume a linear relationship for the sake of convenience. To the extent that local conditions depend on factors that are not included in the computer model, these will simply add to the uncertainty in the site-specific forecast.

As we have seen, the computer model forecasts usually come from the US National Weather Service. The downscaling to individual locations is done by private companies (such as Earth Satellite Corporation, Weather Services International or Weather News), and the final site-specific forecasts are available commercially from these private companies.

Before moving on to describe how site-specific forecasts can be incorporated into pricing models, we will first introduce single and ensemble forecasts, and discuss their pros and cons.

SINGLE FORECASTS

Single forecasts may be either "damped" or "undamped", and it is extremely important to understand whether any particular forecast is one or the other, or in between.

An undamped forecast has the defining characteristic that the variance of the forecast at any lead-time is the same as the variance of the actual forecasted variable for that day. In this sense, undamped forecasts are realistic. A single integration of an AGCM produces an undamped forecast. The potential disadvantage of undamped forecasts is that the absolute forecast errors could be very large: imagine that the temperature for a certain day of the year has a range from –10 to 10 degrees, with an average of zero. An undamped forecast for that day could have any value in this range. Thus in the worst case, the forecast could be 10 degrees while reality comes out at –10 degrees, giving an error of 20 degrees. Compare this to the simplest possible single forecast, consisting of just the average value every day (in this case zero degrees), which can never have an error of more than 10 degrees.

If this potentially large absolute error is a problem in a particular situation then it may make more sense to use a damped forecast. A damped forecast is an optimal (mean square error minimising) combination of an undamped forecast with the climatological mean, where the relative contribution of the two depends on the skill of the forecast. It represents the best estimate of the mean temperature. If the undamped forecast has high skill, then we should follow it fairly closely: if not then we should move much closer to the climatological mean. Most of the time, somewhere in between is most appropriate. The advantage of damped forecasts is that the absolute errors are much smaller than for undamped forecasts. The disadvantage is that the variance of the forecast is now smaller than reality.

So which should we use: damped or undamped forecasts? The answer depends on the application. If, in the example above, the only thing that matters is whether or not the temperature is going to reach 10 degrees, then we should use an undamped forecast, because a damped forecast will never predict 10 degrees. If, on the other hand, the important factor is to be generally close to reality on average over many predictions, then we should use a damped forecast. Single site-specific forecasts from the commercial forecasting companies are typically damped forecasts. Figure 1 shows the size of error of such a forecast, based on New York Central Park measuring station. The errors grow gradually over the 12 days of the forecast and we see that, in this case, the errors in the maximum temperature are greater than those in the minimum temperature.

We have reviewed the different kinds of single forecast available and discussed the pros and cons of damped and undamped forecasts. For many applications, however, and certainly for weather option pricing, it is actually possible to do much better than either a damped or an undamped forecast using ensembles.

ENSEMBLE FORECASTS
Single forecasts are very easy to communicate. However, they contain no indication of the uncertainty in the forecast, or the range of likely outcomes. This is inadequate for the pricing and risk management of most weather contracts. To overcome these problems ensemble forecasts are used: an ensemble forecast is just a number of different forecasts giving a range of possible outcomes.

Statistical ensembles
One way to construct an ensemble forecast is to analyse past single forecasts, estimate their typical errors, and use these errors to create a distribution around future forecasts. Since forecast errors typically vary in size during the year, the model for forecast errors should reflect this.

Numerical (AGCM) ensembles
Numerical modellers (especially NCEP or ECMWF) run their models in such a way as to create not one but many large-scale forecasts from which an ensemble of site-specific forecasts can be created.[3] The range of forecasts that results is roughly equal to the range of possible outcomes of the atmosphere over the forecast period. If one is prepared to assume that the individual ensemble members are equally likely then these forecasts can be used to estimate the probability distribution of future outcomes.

The potential advantage of AGCM ensembles over statistical ensembles is that the spread of the ensemble can vary with state of the atmosphere. Certain meteorological situations are harder for the numerical models to forecast than others, and in these situations the spread of the ensemble would often (but not always) be large. Other situations are easier to forecast, and in these cases the spread of the ensemble might be smaller. It should be noted, however, that this variation in predictability from one atmospheric state to another is rather small and that the potential advantage of using AGCM ensembles over statistical ensembles is small compared with the advantage gained by using ensembles at all, and small compared to the other sources of uncertainty in weather pricing. An example of an ensemble forecast is shown in Figure 2. The figure shows the NCEP ensemble forecast made on December 29, 2001 for London. This example was chosen because of the dramatic warning that was predicted too early by all of the ensemble members.

Incorporation of weather forecasts into pricing models

Having discussed the various forecasts available, we now move on to describe how these forecasts can be incorporated into pricing models. We will focus exclusively on ensemble forecasts rather than single forecasts since, as explained above, ensemble forecasts give a better indication of the range of outcomes. We start by considering

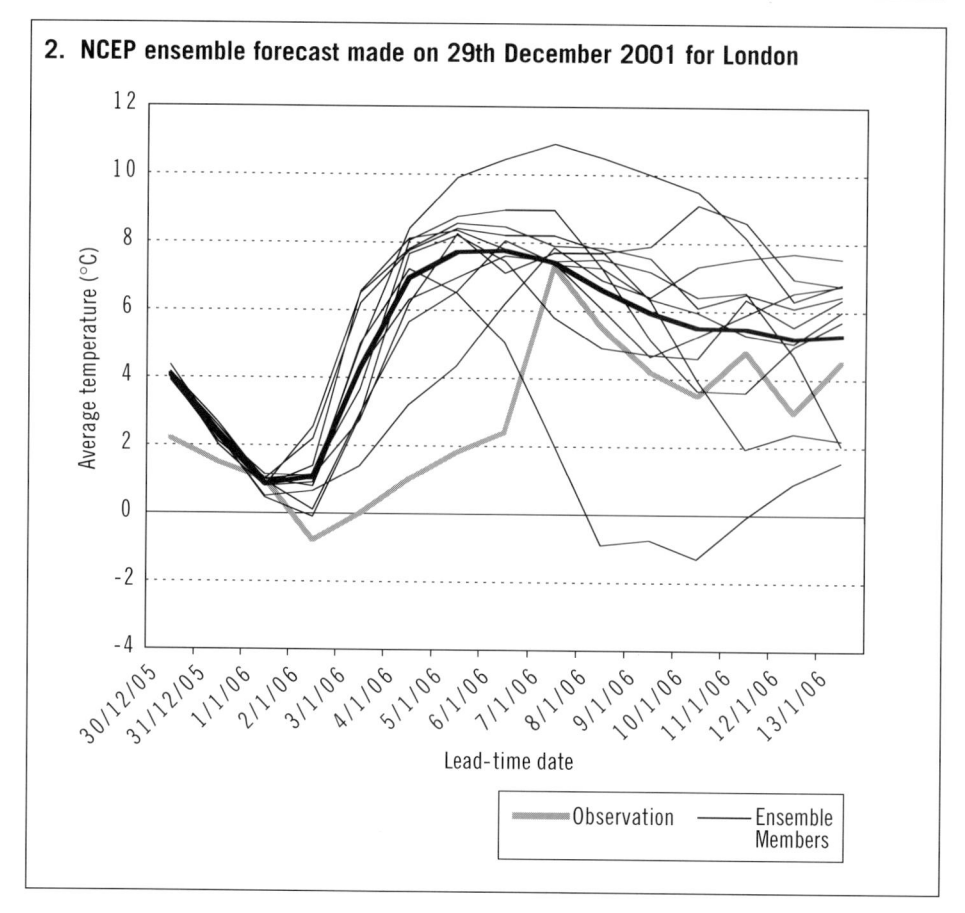

2. NCEP ensemble forecast made on 29th December 2001 for London

how forecasts can be incorporated into pricing methods based on historical or modelled indexes, and then move on to consider how forecasts can be used in daily simulation models.

WEATHER FORECASTS AND INDEX-BASED PRICING
Index-based pricing approaches are those methods that attempt to estimate the distribution of the settlement index (the index on which a weather derivative is settled), from detrended historical index values. Burn analysis (looking at how the contact would have performed in the past), parametric and non-parametric index modelling are all examples of index-based pricing. These methods do not lend themselves very readily to the accurate incorporation of forecasts because forecasts are generally given in terms of the basic daily variable (eg, daily temperatures) rather than weather indexes such as aggregate degree-days.

When using forecasts in index-based pricing models, it is common practice to proceed by splitting the remaining part of the contract period into two parts: the forecast period, during which forecasts are to be used, and the rest. It is then usually assumed that the weather variability in these two periods is uncorrelated. This is clearly incorrect as the last day of the forecast period will be highly correlated with the first day of the remaining period, but making this assumption allows the whole contract to be modelled – including the effects of forecasts – relatively quickly and simply. The effect of this approximation is that the standard deviation of the final index is likely to be underestimated. This will be most important for short contracts where the underestimation could be a significant proportion of the actual standard deviation, and could lead to under-pricing of options. In these cases, it is probably wise to use a more sophisticated model such as the "pruning method" (discussed later) for use with daily simulation models.

The steps for incorporation of weather forecasts into an index model are:

> **PANEL 4**
>
> **WHY DO FORECASTS GO WRONG?**
>
> There are two sources of forecast error in computer model predictions: errors in the initial conditions of the forecast and errors in the computer model that creates the forecast. As to which of these errors is most important is actually a source of considerable controversy. During the 1990s, encouraged by chaos theory and increases in computer power, it was thought by some that the model error had become small relative to initial condition error. If this were true it would justify the use of ensembles in which the ensemble is created by perturbing the initial conditions to try and cover the range of possible values, as is done for the US NCEP MRF and ECMWF forecasts. However, a recent scientific paper (Orrell *et al.*, 2001) which actually attempted to evaluate the relative sizes of model and initial condition error concluded that model error was also very large. This somewhat invalidates the use of initial condition ensembles, and suggests that, rather than one of the members of the ensemble being close to correct, they are all likely to be wrong. Alternative types of ensemble, which create additional spread either by using many different models, or by using one model with stochastic forcing terms, are now coming into use. These kinds of ensemble represent both model and initial condition error more comprehensively.
>
> The upside of large model error in GCM forecasts is that there is still a lot of room for improvement. Thus users of these forecasts can look forward to better products in the years to come.

Step 1. Split the remainder of the contract into the forecast period and the post-forecast period.

Step 2. Use an ensemble forecast to estimate the distribution of index values for the forecast period.

Step 3. Use historical data to estimate the distribution of index values for the post-forecast period.

Step 4. Combine the forecast and historical distributions under the assumption that they are independent.

It would be possible to go beyond this simple method and try to introduce correlations between the two periods by using simulations for the second period that are conditioned on the values in the first. However, rather than pursue this possibility, we will now consider a much more natural way to mix forecasts and historical data based on daily modelling.

WEATHER FORECASTS AND DAILY MODELS
We now consider how statistical daily weather simulation models can be merged with weather forecasts. First, we note that the models themselves can be used to create a forecast, simply by feeding them with data from the recent past and integrating the models forward in time. Such forecasts, however, are generally nowhere near as good as weather forecasts based on computer models of weather dynamics. The main reason for this is that the predictability of weather comes from complex propagating and evolving processes. These can be captured reasonably well by the numerical models, which inherently 'understand' how weather systems move and grow, but are not modelled well by statistical models. A statistical model with enough parameters to capture the behaviour of the atmosphere would be extremely complex, and it is likely that the available historical record would not be sufficient to estimate the parameters with any confidence.

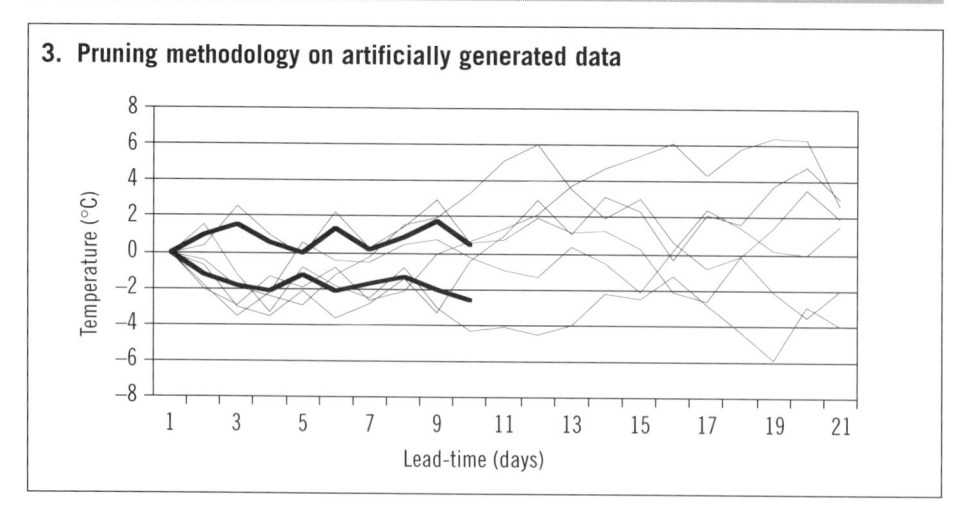

3. Pruning methodology on artificially generated data

For short contracts (less than 16 days in length) it is possible to price using forecasts alone, as long as the forecast has been downscaled to exactly the right variable at exactly the right location relevant for the contract. For longer contracts, however, the forecast stops mid-contract. If one is not prepared to make the assumption of independence between the forecast and post-forecast periods, then the challenge becomes one of modelling the correct dependencies between these periods. We describe a method known as "pruning" for how this can be achieved (see Jewson, 2000).

Pruning works as follows:

1. A daily weather simulation model is fitted to historical data.
2. The model is used to simulate a large number (say 100,000) of possible future scenarios for the weather variable(s) during the contract period. If the statistical model is a reliable one, these simulations will represent the behaviour of the daily weather process well, but will contain little forecast information.
3. An ensemble forecast is compared to the simulations. Some of the simulations will be highly consistent with the ensemble forecast, and others not. Each of the simulations is given a weight according to the level of consistency with the forecast and the climatological distribution. The simulations, along with their weights, are used for derivative pricing.

This method has the following advantages:

❑ The statistical structure of the daily simulations is preserved. In particular the correlations between the forecast and post-forecast periods are captured.
❑ As many simulations can be used as is desired, giving arbitrarily accurate convergence.
❑ The simulations can be extended arbitrarily far into the future.

An illustration of the pruning methodology is shown in Figure 3. The two thick lines represent the range of values from a 1- to 10-day ensemble forecast. The thin lines indicate simulated temperature tracks from a statistical model. The vertical axis is temperature in degrees centigrade, and the horizontal axis is forecast lead-time. Pruning involves weighting the simulated tracks according to their consistency with the forecast before using them to price weather contracts.

Incorporation of seasonal forecasts into pricing models

We now move our discussion of how to use meteorological forecasts on to the question of how to use seasonal forecasts. Again, we look at this in the context of index models and daily models separately. One complication is that many different kinds of seasonal forecasts are available, for instance:

> **PANEL 5**
>
> **METEOROLOGICAL INDEXES**
>
> There are a number of indexes in common use in climate research. The SOI (Southern Oscillation Index) is a measure of atmospheric pressure differences across the Pacific ocean, and, on monthly timescales, is a reasonable indicator of the state of ENSO. The "Niño3" Index, which consists of oceanic surface temperatures averaged over a large region of the central Pacific, is, however, a more useful indicator of ENSO because it is not as strongly affected by small-scale atmospheric processes. The SOI and Niño3 are very highly correlated on long timescales. The PDO (Pacific Decadal Oscillation), is, as its name suggests, a measure of long-term variability in the (mainly north) Pacific. Unlike the ENSO indexes it is not clear that the PDO can be predicted in advance, and so it is less useful. In the Atlantic there is the NAO (North Atlantic Oscillation). This is a measure of the pressure difference between Iceland and the Azores, and, as such, is simply an index of European weather; statements such as "positive values of the NAO are causing strong winds in Europe" are effectively equivalent to saying "strong winds in Europe are being caused by strong winds in Europe". The NAO is almost completely unpredictable – although it may be that on decadal timescales it is weakly influenced by the ocean. It is not of any particular relevance to weather pricing. Other popular indexes are the PNA (Pacific North American) pattern and the AO (Arctic Oscillation). A recent study (Ambaum *et al.*, 2001) has suggested that the latter may not even exist in a physical sense, and was simply an artefact of the use of singular vector analysis for studying meteorological data.

❏ Single forecasts of Niño3 temperatures.
❏ Ensemble forecasts of Niño3 temperatures.
❏ Site-specific single forecasts.
❏ Site-specific ensemble forecasts.

As before, we will only consider the ensemble forecasts. Single forecasts can be converted into ensemble forecasts using statistical methods based on past forecast errors (as with weather forecasts).

NIÑO3 FORECASTS AND INDEX MODELS
First we consider an ensemble forecast of sea surface temperatures in the Niño3 region. One simple model for pricing that can incorporate such a forecast works as follows:

1. A statistical model is built that relates Niño3 temperatures to the contract index using historical values for Niño3 and historical index values. The historical values for Niño3 should be calculated over the contract period, since the effect of Pacific Ocean temperatures on weather patterns is effectively instantaneous relative to the length of contracts. (The effects propagate through the atmosphere as large-scale disturbances known as Rossby waves: see, for example, Holton, 1993.)
2. The ensemble forecast of Niño3 is fed into this statistical model, which produces a distribution of index values as an output.

Ideally such a model would be built by relating past forecasts to actual index values, since this is the relationship we are trying to capture. However, since few past forecasts values are available, it is common to use past values of Niño3 instead.

The question of what statistical model to use is a difficult one. Typically one might use either a linear relationship (ie, least squares minimising linear regression) or binning (dividing the total range of Niño3 values into sub-ranges and associating a fixed response with each sub-range).

SITE-SPECIFIC SEASONAL FORECASTS AND INDEX MODELS

Seasonal forecasts often come in the form of site-specific values. These forecasts may contain more than just the effects of ENSO: for instance the forecaster may have included their view of the trend in the forecast too, and may have taken a view on the likely future state of the Pacific Decadal Oscillation. It is important to understand what has been included in a forecast and what has not, otherwise there is a danger of including effects of trends more than once, for instance.

Incorporating such site-specific forecasts into index-based pricing models is difficult: the forecast will typically be in terms of temperature, not index. The forecast may be for changes in the mean of the temperature only, even though changes in the whole temperature distribution are important if we want to quantify changes in the distribution of the index. In order to get past these problems one has to make some (often rather unsatisfactory) assumptions. A typical assumption would be that a forecast for the change in the mean temperature could be interpreted as being a shift in the whole distribution of temperature with no change of shape. This can then, with some work, be converted into a shift in the distribution of the index.

If site-specific probabilistic index forecasts are available, then the process is slightly simpler; the forecast index distribution and the historical index distribution can be combined by summing the pdfs with weights. The weights determine the extent to which one believes the historical data or the forecast, and can be estimated either using back-testing or from intuition (see Shorter *et al.*, 2002).

NIÑO3 FORECASTS AND DAILY MODELS

Niño3 forecasts can be incorporated into daily models in much the same way that they can be incorporated into index models. A statistical model is made of the causal relationship between Niño3 and US temperature or rainfall, and the latest forecast is fed in. Again, the challenge is to design a good statistical model. Since Niño3 temperatures are only available on a monthly basis, they would usually have to be interpolated to give daily values before such a model can be built. An added complexity of building this model at the daily basis is that the effects of ENSO on weather vary throughout the year. Ideally one would build a seasonally varying model to capture this effect. In practice, however, this is likely to be very difficult given the limited amount of data available.

SITE-SPECIFIC SEASONAL FORECASTS AND DAILY MODELS

In the long run most pricing will probably be done with daily models, because these make better use of the available historical data, and most seasonal forecasts will probably be presented as site-specific ensemble forecasts, as these are the only forecasts that specify the range of possible outcomes at the correct location. Fortunately, combining these two can be done reasonably simply using the pruning method described above. Many site-specific seasonal forecasts will likely be of monthly averages as we assume in the example below:

1. A daily weather simulation model is fitted to historical daily data.
2. The model is used to simulate a large number (say 100,000) of possible future scenarios for the weather variable(s) during the contract period. If the statistical model is a good one these simulations will represent the behaviour of the daily weather process well, but will contain little forecast information.
3. A site-specific ensemble seasonal forecast is compared to the monthly mean temperatures from the simulations. Some of the simulations will be highly consistent with the ensemble forecast, and others will not. Each of the simulations is given a weight according to the level of consistency with the forecast and the

climatological distribution. The simulations, along with their weights, are used for derivative pricing.

Once more, pruning has the benefits that it preserves statistical structure over the whole contract period, allows very large ensembles to be generated and can be extended as far into the future as is necessary.

Conclusion

This chapter has argued that during certain periods weather contracts cannot be priced using either historical data or forecasts alone, but must be priced using a combination of the two. We have described the relevant forecasts and how they are produced, and we have looked at various ways that historical data and forecasts can be combined. In general, incorporating forecasts into index-based pricing models is difficult, and some approximations must be made that do not work well in all cases. For this reason, working with daily models is much more satisfactory, giving algorithms that can work in all cases.

Appendix

Measuring the skill of forecasts

Producing numerical weather forecasts is a complex task and so is their proper validation; Alan Murphy is a crucial name in this area of meteorological endeavour, devoting his professional life to verification and evaluation of weather forecasts, probability weather forecasting, and decision theory (see Murphy, A., 1993). Forecast quality can be described by a set of attributes like the bias, association, accuracy, skill, reliability and sharpness. Which of these attributes is most important is defined by the specific application. The related measures can be derived from the joint distribution of forecasts and observations, both typically represented by high dimensional arrays in space and time. Traditionally, most verification techniques have been developed for single deterministic forecasts of continuous predictands (eg, temperature) or categorical predictands (eg, the presence or absence of rain).

Here we describe the three most commonly used verification measures for single forecasts of a continuous variable like temperature:[4]

1. The mean error

$$ME = <e> = <f> - <o>$$

   shows the mean difference between the forecast and the observation and thus indicates whether the model has a bias. Note that $<>$ means averaging over the number of pairs of observations and forecasts.

2. The root mean square error

$$RMSE = \sqrt{<(f-o)^2>}$$

   describes the overall accuracy of the forecast and is a measure of the amplitude of the model error. Root mean square errors for a forecast for Central Park, New York are shown in Figure 1.

3. The anomaly correlation (AC) measures whether the variability in the forecast matches the real variability:

$$AC = <((f-<f>)(o-<o>))> / \sqrt{<(f-<f>)^2><(o-<o>)^2>}$$

   An AC near to one indicates a small phase error between forecast and observation. An AC below 0.6 is typically considered a useless forecast.

DOES A FORECAST SHOW SKILL?

A forecast is said to have skill if the accuracy exceeds that of one produced with a simple strategy such as climatology, persistence or random forecasts. Medium-range and seasonal forecast skill is measured against climatology while short-term forecasts should be more accurate than persistence.

1 *Anders Brix, private communication.*

2 *http://www.ecmwf.int.*

3 *There has been considerable debate for many years among the computer modellers about exactly how an ensemble of AGCM forecasts should be generated. A rigorous statistical methodology appears not to be possible because of inaccuracies in the models, and consequently a number of different ad-hoc methods are in use. US NCEP MRF uses 'breeding' (Toth and Kalnay, 1993) while ECMWF uses 'singular vector perturbations' and stochastic forcing (Palmer et al., 1992). Disadvantages of the ECMWF approach are that it does not tend to give reasonable results for spread in the first one or two days of a 10-day forecast, and that the differences between ensemble members during this early period are local rather than global. One disadvantage of the US NCEP MRF ensembles is that they do not account for model error as well as the ECMWF model does and so the spread later in the forecast period is generally too low. This suggests that it might make sense to use the US NCEP MRF model for the first few days, and the ECMWF model after that. For less discerning use of the forecasts, the actual methodology is not particularly important; it is sufficient to know that the methods have been adjusted so that the results look fairly reasonable in the sense that the spread of the forecasts roughly matches the typical forecast errors. But it should be remembered that a rigorously precise interpretation of the meaning of the ensembles in terms of probabilities is not possible, and probability distribution functions generated from these ensembles may need correcting.*

4 *For ensemble forecasts more complex evaluation techniques need to be applied, for example the Relative Operating Characteristic (ROC), that evaluates the extent to which the ensemble forecast produces the correct probabilities of events.*

BIBLIOGRAPHY

Ambaum, M. H. P., B. J. Hoskins and D. B. Stephenson, 2001, "Arctic Oscillation or North Atlantic Oscillation?", *Journal of Climate*, 14, pp. 3495–507.

Experimental Long Lead Forecast Bulletin: http://www.grads.iges.org/ellfb.

Fraedrich, K., 1990, "European Grosswetter During the Warm and Cold Extremes of the El Niño/Southern Oscillation", *International Journal of Climatology*, 10, pp. 21–31.

Holton, J, 1992, *An Introduction to Dynamic Meteorology*, (San Diego: Academic Press).

Jewson, S., 2000, "Use of GCM Forecasts in Financial-Meteorological Models", *Proceedings of the 25th Annual Climate Diagnostics and Predication Workshop*, US Department of Commerce.

Murphy A., 1993, "What is a Good Forecast? An Essay on the Nature of Goodness in Weather Forecasting", *Weather and Forecasting*, 8, pp. 281–93.

Orrell D., L. A. Smith, T. Palmer and J. Barkmeijer, 2001, "Model Error and Operational Weather Forecasts", *Nonlinear Processes in Geophysics*, 8, pp. 357–71.

Palmer, T., F. Molteni, R. Mureau, R. Buizza, P. Chapelet and J. Tribbia, 1992, "Ensemble Prediction", Technical Report, Research Department Technical Memo 188.

Shorter, J., T. Crawford, R. Boucher and J. Burbridge., 2002, "Skillful Seasonal Degree-Day Forecasts and their Utility in the Weather Derivatives Market", AMS Annual Conference Abstracts.

Toth, Z., and E. Kalnay, 1993, "Ensemble Forecasting at NMC: The Generation of Perturbations", *Bulletin of the American Meteorological Society*, 74, pp. 2317–30.

11

The Weather in Weather Risk

John A. Dutton[1]

The Pennsylvania State University and Weather Ventures Ltd

*In constant change, the weather and climate
create risk for some, opportunity for others.*
Weather Ventures

The always-changing weather is one of the certainties in life. We expect and depend upon atmospheric change, even though the consequences are sometimes threatening or catastrophic. We have learned much about predicting weather and climate, including something about the limitations that will forever frustrate our efforts, even as we continue to bend an incredible array of technology to the task.

We link all of the adverse possibilities owing to weather events and climate variability together under the name of "weather risk". We seek to manage or mitigate weather risk on various temporal and spatial scales with strategies ranging from flight before the storm to transferring risk to others through sophisticated financial manoeuvres.

Introduction

Weather risk thus brings together the enterprises at risk, the atmospheric sciences, finance, and the strategies of risk management. In this chapter, we will explore some basic aspects of the weather in weather risk. First, we look at some of the consequences of weather and climate variability and then examine the observational capabilities and theoretical concepts that are the basis for weather and climate prediction, with a special emphasis on the differences between short-term forecasts and longer-term seasonal outlooks. We review statistical methods used for representing seasonal variability and then develop a new approach for estimating the value of climate forecasts in hedging risk.

Weather and climate risk at three timescales

We encounter weather and climate risk on three timescales: hours to days, seasons, and decades to centuries.

HOURS TO DAYS
The primary weather focus of governments, the private sector and people the world over has been on phenomena that take lives, produce injuries and damage buildings, crops and other property. A study of worldwide disasters for the period 1960–89 found almost 40,000 deaths owing to storms, floods, drought and famine, and that some 64 million people had been affected, far more than any other form of disaster (Bruce, 1994). Fatalities and damage in the US owing to severe weather during the past six years are summarised in Table 1. A 60-year record of some of the most

THE WEATHER IN WEATHER RISK

prominent causes of weather fatalities is shown in Figure 1, which also demonstrates the success of atmospheric science and modern observations and models in mitigating the effects of severe weather through forecasts and warnings.

There is considerable interest in the apparent long-term increase in the societal effects of extreme weather and in whether such trends are a consequence of social and economic changes or increased severity of weather and climate events. A comprehensive review (Kunkel, Pielke and Changnon, 1999) concludes that vulnerability has increased, but the frequency of severe events has not. A global survey is available in Downing, Olsthoorn and Tol (1999).

SEASONS

Interest in variations from normal climate, long-focused on those that disrupt agricultural activities, has now expanded to other components of the economy, most notably energy, and are a major focus of this book and this chapter. The effects of climate variability can be as significant, as shown in Table 2 for those owing to flood or heat and drought in 1986–2001.

DECADES TO CENTURIES

Long-term climate change has shaped human migration, contributes to geographic variations in societal vitality and economic prosperity, and is now an urgent issue with the possibility that human activities are significantly altering the climate with unknown consequences for the biosphere, agricultural productivity, sea level and the spread and control of disease.

The most dramatic global long-term climate change involves an oscillation between glacial periods when the polar icecaps expand into middle latitudes and the relatively warmer interglacial periods when glaciers retreat. (The warm period in

Table 1. Weather-related fatalities and property damage in the US, 1995–2000

Weather-related fatalities	1995	1996	1997	1998	1999	2000	Average
Convective storms	153	101	146	216	169	119	151
Extreme temperatures	1,043	98	132	184	572	184	369
Flood	80	131	118	136	68	38	95
Coastal storms, tsunamis	1	38	22	23	28	29	24
Tropical cyclones, hurricanes	17	37	1	9	19	0	14
Winter storms and avalanches	17	91	94	77	67	57	67
Strong winds	46	31	38	24	33	26	33
Other	5	13	49	18	15	23	21
	1,362	540	600	687	971	476	773

Weather-related property and crop damage (US$ million)							
Convective storms	2,638	2,013	1,420	4,889	3,020	1,345	2,554
Drought	63	640	277	2,182	1,333	2,439	1,155
Extreme temperatures	1,090	283	308	683	161	9	422
Flood	1,251	2,535	7,028	2,643	1,792	1,934	2,864
Coastal storms, tsunamis	2	40	1	117	1	0	27
Tropical cyclones, hurricanes	5,932	1,787	875	4,128	5,609	8	3,057
Winter storms and avalanches	111	311	774	528	62	1,036	471
Strong winds	121	222	44	84	92	50	102
Other	175	145	58	856	184	2,128	591
	11,383	7,975	10,786	16,111	12,253	8,950	11,243

Source: National Weather Service, National Climatic Data Center
Notes: Fatalities includes only those directly attributable to weather events.
Convective storms includes lightning, tornadoes and thunderstorms, and hail.
Other includes dust storms, dust devils, rain, fog, water spouts, fire weather, mud slides, volcanic ash and miscellaneous and for fatalities, drought, although none were reported in this compilation.

1. Weather-related deaths in the US

Source: National Weather Service

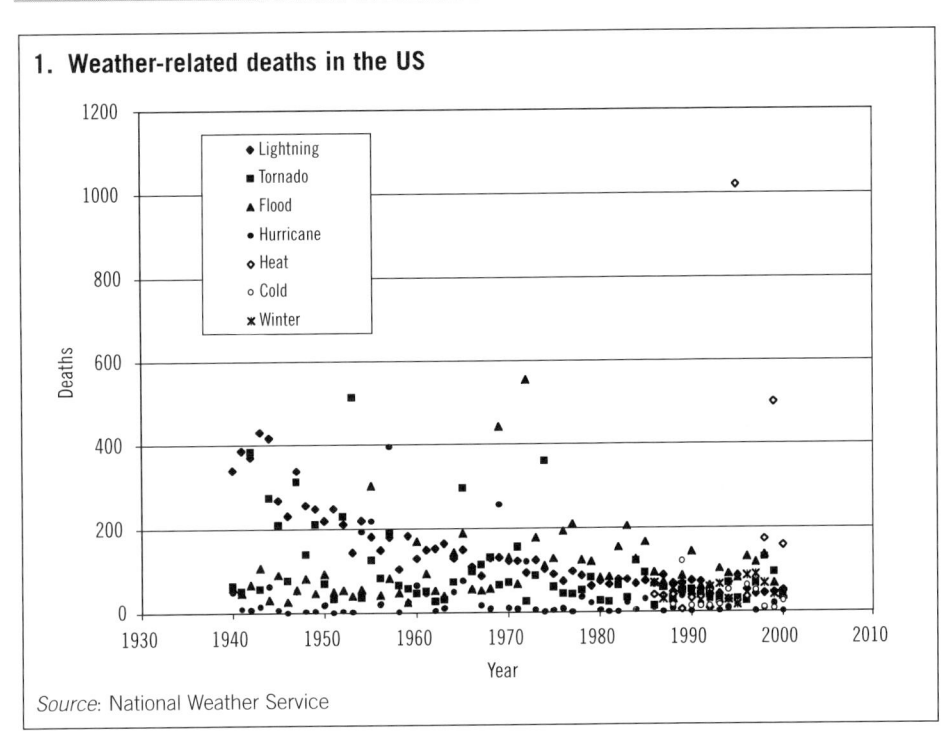

which we live began about 10,000 years ago.) But even in this current interglacial period there have been climate variations with considerable human significance; a period warmer than now enabled the Norsemen to inhabit Greenland, but a cooling period brought their occupation to an end. Today the evidence that human activities are changing the climate through accelerating release of carbon dioxide from combustion is becoming incontrovertible (Houghton, 2001). The record warm years of the past decade, whether caused by this global change process or not, have been a significant climate risk for some industries, as shown in Chapter 1.

The forces that drive atmospheric flow

The atmosphere is a relatively shallow layer of gas covering a rotating, spherical planet. The sphericity mandates ceaseless atmospheric motion; the rotation modifies it profoundly.

The global thermodynamics of the atmosphere is quite straightforward: the Earth is heated by a parallel stream of solar radiation falling on its spherical surface. Thus the energy received decreases in proportion to the cosine of latitude, while the outgoing longwave or infrared radiation emitted from the Earth's surface and the atmosphere can flow to space equally over the entire sphere. As a consequence, the equatorial regions receive more net radiant energy than the polar regions, creating a poleward decrease in temperature. This thermodynamic imbalance forces the atmosphere and ocean into motion, transporting heat toward the poles in an attempt to reduce the gradient, to reach a balance than can never be attained. Indeed, we can

Table 2. Effects of floods, heat and drought in the US, 1986–2001

Disaster	Property damage (US$ billion)	Deaths
13 Floods (some with tornadoes)	48	441
6 Heat and drought events	55	More than 6,000

show quite rigorously (Jeffreys, 1925 and Dutton, 1995, p. 192) that the atmosphere must remain in motion as long as there are horizontal temperature gradients.

The spin of planet Earth is significant for atmospheric dynamics because it creates an accelerating coordinate system and a force proportional to the rate of spin. As a consequence, the winds flow around centres of pressure, rather than from high pressure to low. Thus the Earth's spin is imparted to the air, and when we make charts of atmospheric systems or visualise them from space we see their rotating structure, the hurricane being a prime example.

Some of the most significant processes and events in the atmosphere involve the phase changes of water. Thermal energy at the surface evaporates water and is later released when the vapour condenses to form a cloud, a rain shower, or a line of severe thunderstorms. While we know the circumstances in which such events occur, we cannot predict the details in any specific instance. No matter how fine the grid on which we observe and analyse atmospheric events, there will always be the seeds of future events hidden in the smaller scales. Some of these seeds will grow through hydrodynamic instabilities to become notable features of the flow, others will vanish quietly.

Today we know that we shall never be able to forecast atmospheric events precisely beyond limits that depend on the scale of the features – hours for thunderstorms, perhaps a week for extratropical cyclones such as winter snowstorms, and perhaps tens of days for the very largest global features. The core of the difficulty is that atmospheric processes are non-linear, and as a consequence, the effects of small disturbances will eventually have chaotic consequences at the largest scales.

These limitations on predictability are incontrovertible. We can improve the forecasts of weather systems with new and more accurate observations, but the improvement will be incremental, and there are limits we cannot pass.

Identifying weather and climate risk

Risk is the possibility that adverse results or events will occur – that things will not turn out as we expect or hope. The management of risk involves identifying the possibilities of adverse outcomes, developing and comparing strategies for avoiding or mitigating the consequences of such outcomes, and selecting an approach that seems advantageous. We usually hope, but rarely prove, that an optimum solution exists and can be found.

For the purposes of this chapter, we follow Dutton (2002) to define weather and climate risk as the possibility of injury, damage, or adverse financial results as a consequence of atmospheric events or processes at any timescale. Various quantitative definitions of risk appear in the financial and insurance literature, but here we shall say that the risk associated with a certain event is given by

Risk = (probability of occurrence) × (cost or consequence should the event occur)

Since probability is dimensionless, the units of risk in this chapter will be those of the cost. The three timescales of weather and climate risk, some examples and the common mitigation strategies are given in Panel 3.

SENSITIVITY OF THE ECONOMY TO WEATHER AND CLIMATE RISK

There have been few comprehensive attempts to estimate the cost of weather and climate events to business and national economies (eg, Chapman, 1992). Because of the value of reliable and representative benefit–cost information in designing and justifying improvements in atmospheric information services, the National Research Council (NRC) (1998) recommended that the "atmospheric science community ... initiate interdisciplinary studies of the benefits and costs of weather, climate, and environmental information services".

PANEL 1

WEATHER AND CLIMATE RISK CATEGORIES AND STRATEGIES

Risk	Examples	Mitigation strategies
Severe weather or unusual weather events	Tornadoes, hurricanes, thunderstorms, intense extratropical cyclones, snowstorms	Avoidance through flight or by seeking refuge in response to forecasts or warnings Use of short-term trades to improve flexibility or resilience Preparation through design and building codes Insurance against loss
Unusual seasonal variation	Winter or summer temperatures above or below normal Precipitation anomalies, including drought Anomalies in seasonally available wind power El Niño and La Niña events	Hedges of risk through forward contracts on commodities Options with payoffs contingent on weather variables (weather derivatives) Long-term risk transfer or risk sharing arrangements Insurance
Long-term climate variation	Decadal or longer changes in spatial or temporal distributions of climate variables, such as temperature or precipitation, owing to natural causes or human activities	Modelling to determine boundaries of probable or extreme change Reduction of adverse forcing Development of strategies for reducing sensitivity of important activities Diversification of activities

Source: Adapted from Dutton (2002)

An alternative approach is to estimate the sensitivity of various segments of the economy to weather and climate risk. Table 3 thus assesses the sensitivity of each component of US GDP and, by this means, the evident sensitivity of industries such as agriculture, construction, hospitality and recreation, energy, catastrophe insurance and retail merchandising is aggregated into an estimate of a national total. The data available on manufacturing does not allow identification of those components – manufacture of snow blowers, for example – that do have some weather and climate sensitivity.

WEATHER AND CLIMATE CONTINGENCIES

The decision to act in the face of a weather or climate threat will usually depend on the relationships between potential loss, cost of avoiding loss and forecast skill.

A contingency table, as shown in Table 4, is often used to illustrate the issues involved in assessing the value of categorical forecasts (Katz and Murphy, 1997, Richardson, 2000 and Livezey, 2002). For simplicity, C is the cost of action that completely mitigates the loss, L, that would otherwise occur in adverse weather.

To see the effect of risk mitigation, let the probability of adverse weather over a

Table 3. Components of the US economy sensitive to weather and climate risk

Industries (1987 Standard Industrial Classification)	GDP components (US$ billion)	Weather sensitive components (US$ billion)
Agriculture, forestry and fishing	136	136
Farms	79	79
Agricultural services, forestry and fishing	57	57
Mining	127	110
Coal mining	10	10
Oil and gas extraction	100	100
Other mining	18	0
Construction	464	464
Manufacturing	1,567	
Transportation and public utilities	825	787
Railroad transportation	23	23
Local and interurban passenger transit	19	19
Trucking and warehousing	126	126
Water transportation	15	15
Transportation by air	93	93
Other transportation	39	0
Communications	281	281
Electric, gas and sanitary services	230	230
Wholesale trade	674	
Retail trade	894	894
Finance, insurance and real estate	1,936	379
Security and commodity brokers	144	144
Insurance carriers	168	168
Insurance agents, brokers and service	67	67
Other	1,557	
Services	2,165	261
Hotels and other lodging places	86	86
Auto repair, services and parking	94	94
Amusement and recreation services	81	81
All other services	1,903	
Statistical discrepancy	−130	
TOTAL FOR PRIVATE INDUSTRY	8,657	3,030
Federal government	387	
State and local government	829	829
TOTAL GROSS DOMESTIC PRODUCT	9,873	3,859

Source: GDP data in columns 1 and 2 for 2000 from the Bureau of Economic Analysis, Department of Commerce.

Table 4. Costs and losses of weather risk mitigation strategies

Action	Adverse weather Occurs	Adverse weather Does not occur
Mitigation action taken	C	C
No action taken	L	0
Probability of event	w	1−w

suitable ensemble of cases be w. Then if action is taken in every case, the average expense will be $E_M = C$ and if no mitigation is ever taken, the average expense will be $E_{NM} = wL$. If $C < wL$, then the advantage lies with mitigation. But suppose that the decision maker had access to a talented meteorologist and a perfect forecast in every case and thus could take action only when adverse weather would actually occur. The mean expense with a perfect forecast would then be $E_p = wC$. We can now consider $L > C$, because if it is not, we can simply take the loss and be ahead even if we had a perfect forecast. The expected expense owing to the climate is $E_c = E_{NM} = wL$ and we can estimate the value of a less-than-perfect forecast E_f as

$$V = \frac{E_c - E_f}{E_c - E_p} \tag{1}$$

THE WEATHER IN WEATHER RISK

Table 5. Forecast success and false alarm rates

	Adverse weather	
Forecast	Occurs	Does not occur
Wx will occur	S w	F (1–w)
Wx will not occur	(1–S) w	(1–F) (1–w)
Probability of adverse weather	w	1–w

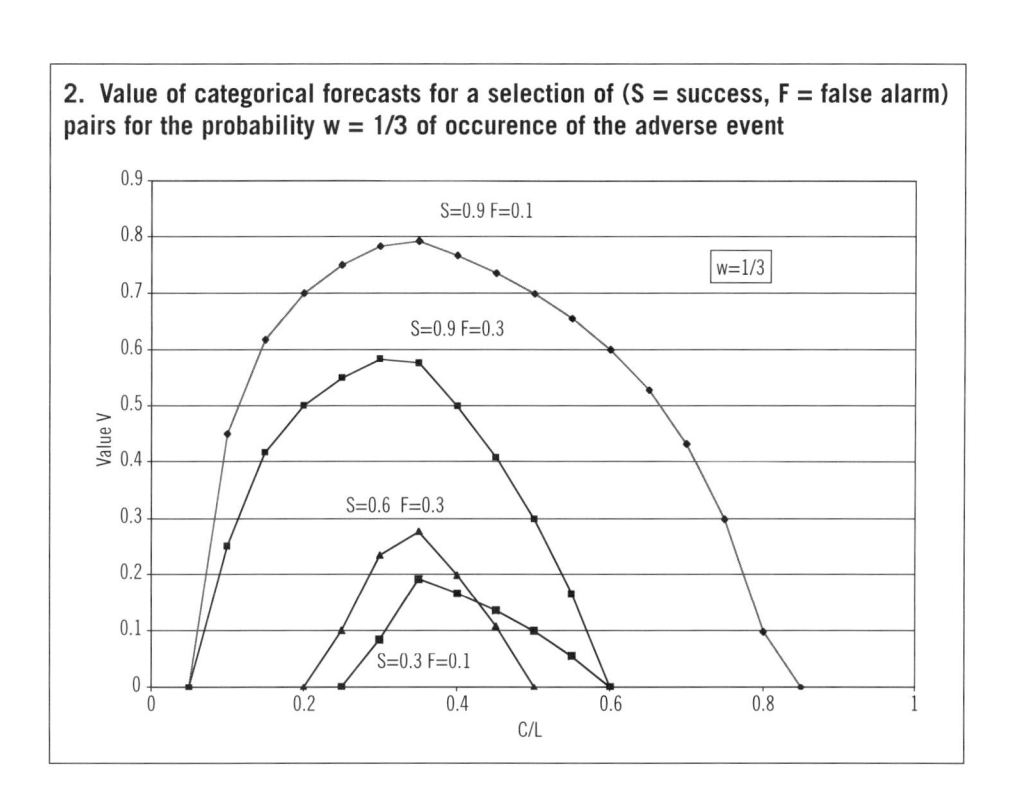

2. Value of categorical forecasts for a selection of (S = success, F = false alarm) pairs for the probability w = 1/3 of occurence of the adverse event

The means of classifying the historical accuracy of forecasts is described in Table 5, where the success rate, S, is the fraction of adverse weather events that were predicted correctly, and the false alarm rate, F, is the fraction of good weather events that were predicted to be adverse. Wx is the abbreviation for "weather" from teletype days.

To construct the table, let the numbers of cases in the first row (Wx will occur) be A and B and the numbers of cases in the second be C and D. Then the total number is $N = A + B + C + D$, the probability w of adverse weather is $w = (A + C)/N$, and for $S = A/(A + C)$, we have $A = SwN$. Converting to a probability by dividing by the number of cases gives the entry Sw in Table 6. The other entries follow similarly, given $F = B/(B + D)$.

The combination of Tables 4 and 5 gives the estimates

$$E_c = \text{Min}[C, wL]$$
$$E_p = wC \quad (2)$$
$$E_f = (Sw + F(1 - w)) C + (1 - S)wL$$

and thus we find that

$$V = \frac{\text{Min}[w, C/L] - (Sw + F(1 - w))C/L - (1 - S)w}{\text{Min}[w, C/L] - wC/L} \quad (3)$$

The value of forecasts for some (S, F) pairs is shown for $w = 1/3$ in Figure 2 as a function of C/L, which demonstrates that even forecasts of apparently moderate skill can provide significant protection or be of considerable economic value. The proviso is that $S > F$, which will usually be true for severe weather because contemporary observational technology and short-term numerical forecasts are increasingly effective in identifying severe weather and projecting its evolution. The situation may be different for seasonal forecasts, as we shall see later.

Observing the weather and climate

Management of a specific weather risk begins with atmospheric observations and forecasts. An apparent contradiction long-present in atmospheric observations is still with us today: advancing technology produces more observations than we can process and use, and yet we have insufficient observations to adequately describe the complexity of the atmosphere.

The modern, quantitative study of meteorology began in the 17th century with the invention of the thermometer and the barometer and the discovery of the relationships between pressure, temperature and density in gases. Advancing understanding of fluid dynamics and thermodynamics coupled with Newton's calculus led, early in the 20th century, to the partial differential equations that describe the evolution of atmospheric conditions. Since then, notable progress has included: the invention of radar and the creation of a global observation network in World War Two; the first numerical forecast with a highly simplified version of the equations of motion in 1951 on the ENIAC (the Electronic Numerical Integrator and Computer, the first large-scale digital computer); and the launch of the first meteorological satellite in 1960. Today, progress in atmospheric analysis and prediction involves a complex interaction among improvements in observational technology, rapid progress in computing capability, and advances in theoretical understanding (for strategies and recommendations, see NRC, 1998).

Table 6. Average daily data assimilation at the US National Centers for Environmental Prediction, April 2002

Observations	Number of observations assimilated	Percent of total
Land surface	161,297	5.26
Marine surface (including fixed and drifting buoys)	20,448	0.67
Land soundings (including radiosondes, profilers, NEXRAD winds)	8,964	0.29
Aircraft data and soundings	111,726	3.64
Total surface-based observations	**302,435**	**9.86**
Geostationary satellite soundings	75,641	2.47
Satellite cloud winds	118,679	3.87
Satellite surface observations	988,364	32.22
Polar orbiting vertical soundings	1,582,594	51.59
Total satellite observations	**2,765,278**	**90.14**

Source: National Centers for Environmental Prediction

There are distinctions between the purposes for which observations are acquired – weather prediction and analysis of the climate require somewhat different emphases on accuracy, resolution and timeliness. Thus in some countries, advanced technology is used to obtain the observations needed for weather prediction while a much denser network of manual observations is used to develop the climatological record.

While surface observations are critical for describing the weather and climate we experience in our activities, the radiosonde soundings of the atmosphere – made by ascending balloons carrying instrument packages and tracked by radar – provide essential information for defining the vertical structure for the computer forecast models. They are increasingly augmented by observations made with instruments on transport aircraft, in the climb, cruise and descent phases of flight, leading to quite dense data collections in the developed parts of the world. Today the dominant flows of data into the major, national forecast centres are observations collected from both polar-orbiting and geostationary satellites. Table 6 summarises the data ingested at the US National Centers for Environmental Prediction (NCEP) as part of the numerical prediction process, showing the average daily total of the observations in each category made worldwide at six-hour intervals beginning with midnight Greenwich Mean Time. The totals are typical of the major forecast centres; the British Met office provides on its Internet site a daily, graphical portrayal of the data it assimilates.

The contemporary data assimilation process incorporates new observations into a numerical model of the atmosphere as it runs forward in time. The motivation is that the computer model contains a wealth of physically consistent information about the state of the atmosphere, accumulated over the past few days from ingested observations, that is mixed into a comprehensive portrait with the equations of motion. So rather than throwing away this knowledge and starting over at each observation time, the atmospheric state in the model is adjusted at the new observation time to be consistent with the new observations, after they have been subjected to rigorous quality control procedures. The evolving trajectory of the computer model is forced, so to speak, to pass through the eye of a needle defined by the new observations.

As shown in Table 6, about one tenth of the observations used in numerical prediction are from surface-based systems and nine-tenths from space, with nearly 3,000,000 satellite observations being assimilated each day. Nevertheless, an overwhelming flood of data from space has just begun. The Joint Center for Satellite Data Assimilation created by the National Aeronautics and Space Administration (NASA) and the National Oceanic and Atmospheric Administration (NOAA) estimates that some 10^{11} observations per day will be available from space in 2020, not counting satellite-based laser wind profilers that may then be deployed.

So the data processing in the atmospheric sciences and services will accelerate from millions of observations today to hundreds of billions. The observations available each day will be equal to what would be available in more than 250 years at the present rate. If we can invoke sufficient computer power to take advantage of these data, then we will validate the observation made by *The Economist* (Morton, 1991):

Satellites and computers are a natural partnership; one provides the data, the other makes sense of it.

And once they have, the requirements for management of information related to weather and climate risk will accelerate, as always, in tandem with the growing flow of data.

THE WEATHER IN WEATHER RISK

Atmospheric prediction and predictability

In both economics and atmospheric science, we are interested in future events. In financial transactions, we are concerned about future rates and prices; in managing weather and climate risk we would like to know the future state of the atmosphere tomorrow, next week, next season and maybe even next year. Many schemes have been developed to predict the future values of economic or financial time-series: most depend upon empirical linear regression equations to extrapolate from the present into the future.

Atmospheric prediction is fundamentally different, as we base our predictions on the laws of physics and well-understood external forcing. Mathematical relationships describe the behaviour of the atmosphere over an incredibly wide range of temporal and spatial scales, millions (soon to be billions) of observations are recorded every day, and astonishingly powerful computers convert the observations into forecasts using the governing equations.

NUMERICAL WEATHER PREDICTION

The algorithm for calculating atmospheric predictions is based on momentum conservation specified by Newton's second law, the conservation of energy specified by the first law of thermodynamics, and the conservation of mass of dry air and of water in its three phases. Atmospheric processes are described by three wind velocity components (taken as positive to the east, north and upward), by pressure, temperature, density and a measure of the amount of water. We take account of three spatial dimensions with $\mathbf{x} = (x_1, x_2, x_3)$ and for our collection of atmospheric variables we write $\phi_i = \phi_i(\mathbf{x}, t)$, $i = 1, 2, ..., 9$, with the last three variables in the set representing the phases of water. As with the spatial coordinates, we shall also write $\phi = (\phi_1, \phi_2, ..., \phi_9)$. Now the laws of physics describing this system can be summarised as (eg, Dutton, 1995)

$$\frac{\partial \phi_i}{\partial t} = \Phi_i(\phi, \mathbf{x}, t) \quad i = 1, 2, ..., 9 \tag{4}$$

in which the non-linear operator Φ specifies the spatial derivatives and products of the variables that sum to produce the local acceleration on the left of Equation (4). With enough observations of the atmosphere, we can calculate the quantity on the

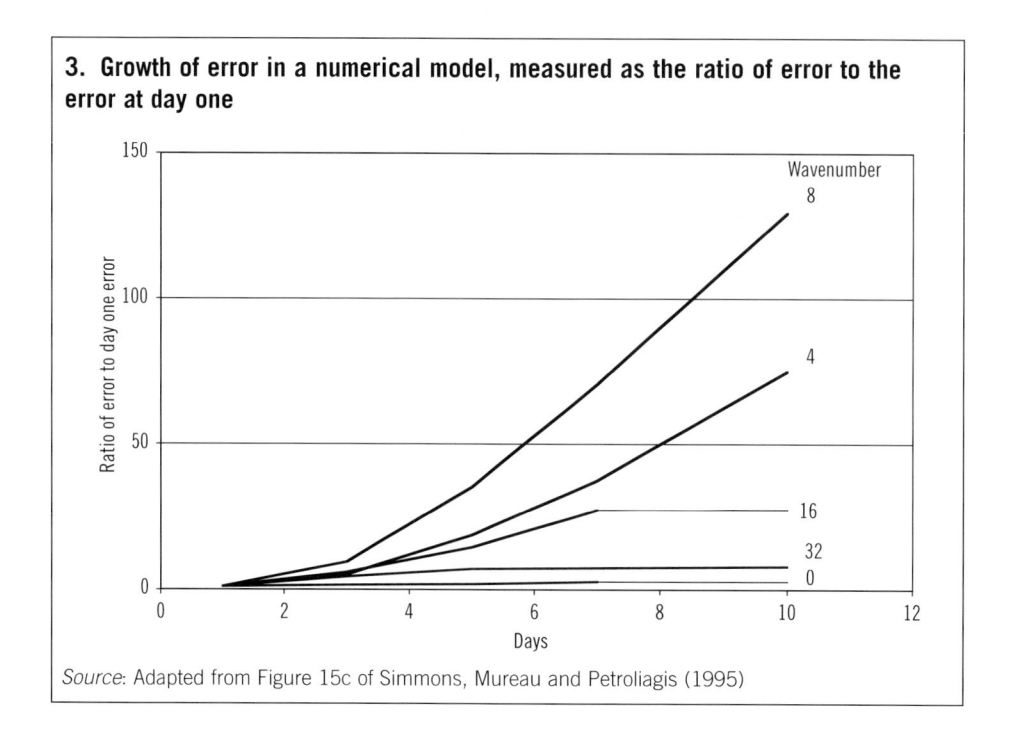

3. Growth of error in a numerical model, measured as the ratio of error to the error at day one

Source: Adapted from Figure 15c of Simmons, Mureau and Petroliagis (1995)

Table 7. The mass and thermal energy of the atmosphere, ocean and land surface

Component	Specific heat (10^3 J/kg K)	Density (kg/m^3)	Mass (kg/m^2)	Energy E/T (10^7J/m^2 K)
Atmosphere	1.0	1.2	10^4	1.2
Ocean	4.2	10^3	$10^3 d_0$	0.42 d_0
Land	2.0	1.5 10^3	1.5 $10^3 d_L$	0.3 d_L

Source: Specific heat and density for land are within a range for different soils given by Peixoto and Oort (1992), the other estimates are conventional

right of Equation (4) for each variable for every point on a coordinate grid, and so – in principle – we can write for some small increment Δt of time

$$\phi(\mathbf{x}, t + \Delta t) = \phi(\mathbf{x}, t) + \Phi(\phi, \mathbf{x}, t)\Delta t \qquad (5)$$

Upon performing these computations, we have our variables at every point at the new time, and we can return to Equation (5), fill in the data on the right side, and increment the process once again, thus stepping forward into the future an indefinite number of times.

Unfortunately, this elegant scheme may be foiled by observational error, inadequate representation of processes at scales smaller than the computational grid, or numerical instability. But even if such difficulties are overcome, the vista of long-term, *deterministic* prediction will be blocked by the limitations of chaos arising from the non-linearity of the system represented in Equation (4).

Here, "deterministic" refers to predicting (in the sense of Equations (4) and (5)) the numerical value of the variables at a specific time and place. Predictions about climate variability are a different enterprise, because we seek averages rather than specific values.

Chaos arises because of non-linearity on the right of Equation (4) in terms such as $u\partial\phi/\partial x$, where u is a velocity component.[2] The consequence of this non-linearity is that solutions to the initial value problem of Equation (4) depend sensitively on initial conditions. Atmospheric flows emanating from slightly different initial

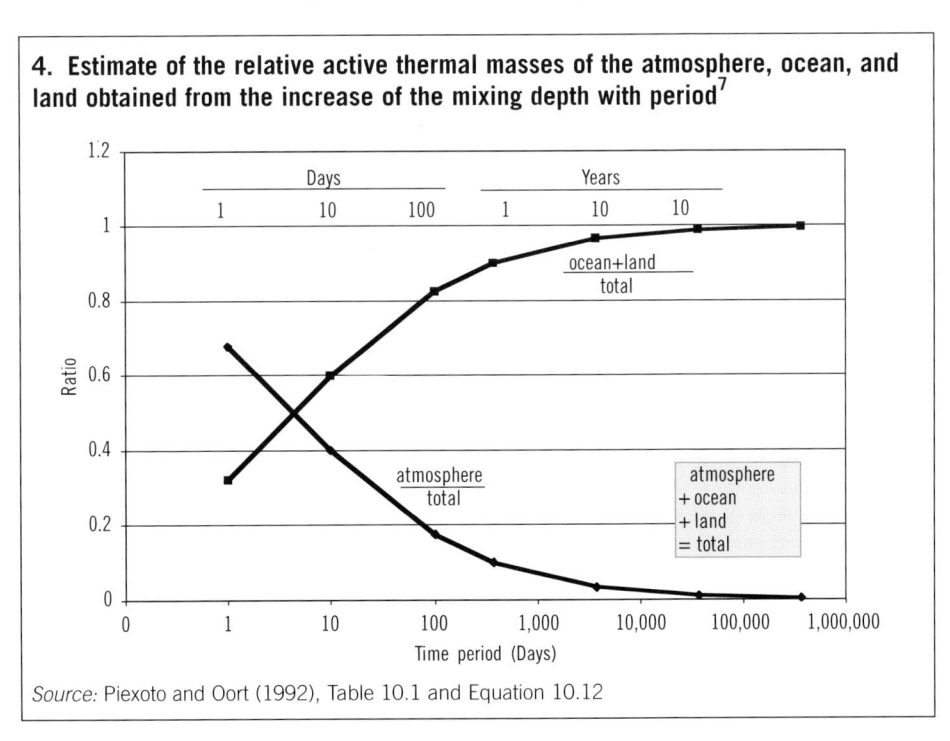

4. Estimate of the relative active thermal masses of the atmosphere, ocean, and land obtained from the increase of the mixing depth with period[7]

Source: Piexoto and Oort (1992), Table 10.1 and Equation 10.12

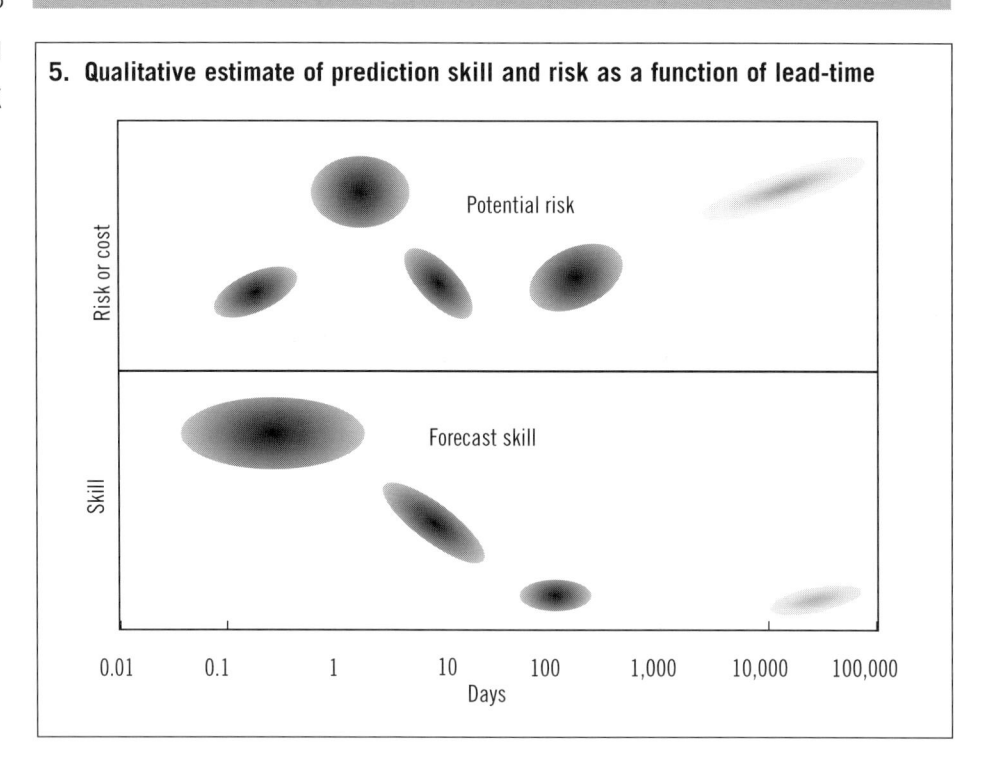

5. Qualitative estimate of prediction skill and risk as a function of lead-time

conditions will eventually become quite different. In fact, errors in numerical atmospheric models double in somewhat less than two days. The growth of errors is demonstrated in Figure 3.

Today, we try to obtain an estimate of predictability along with the deterministic numerical prediction. On the assumption that the difference or similarity of forecasts from a collection of similar initial states will indicate the degree of predictability, the major forecast centres calculate ensembles of numerical predictions – not one

Table 8. Verification scheme for categorical forecasts

		Observed result		
		Above	Normal	Below
Forecasts	Above	h_{11}	e_{12}	e_{13}
	Normal	e_{21}	h_{22}	e_{23}
	Below	e_{31}	e_{32}	h_{33}

Table 9. Fraction correct and Heidke skill score

Fraction correct	Heidke skill score
1	1.000
3/4	0.625
2/3	0.500
1/2	0.250
1/3	0.000
1/4	−0.125

Table 10. Forecast results for CPC seasonal outlooks, 1995–2000

December, January, February (DJF) (1995–2000)						June, July, August (JJA) (1995–2000)				
TEMPERATURE										
	Observations						Observations			
Forecasts	B	N	A	CL		Forecasts	B	N	A	CL
B	4	5	28	0	37	B	122	148	125	0	395
N	0	0	22	0	22	N	22	7	46	0	75
A	447	786	1,421	0	2,654	A	401	551	1,010	0	1,962
CL	219	571	1,552	0	2,342	CL	488	747	831	0	2,066
	670	1,362	3,023	0	5,055		1,033	1,453	2,012	0	4,498
	HSS 28.8% [15%]						HSS 20.3% [12%]			
PRECIPITATION										
	Observations						Observations			
Forecasts	B	N	A	CL		Forecasts	B	N	A	CL
B	228	130	130	0	488	B	79	44	65	0	188
N	0	0	0	0	0	N	0	0	0	0	0
A	233	289	498	0	1,020	A	49	21	47	0	117
CL	963	1,165	1,419	0	3,547	CL	1,503	1,164	1,526	0	4,193
	1,424	1,584	2,047	0	5,055		1,631	1,229	1,638	0	4,498
	HSS 22.0% [6%]						HSS 12.0% [1%]			

Source: Climate Prediction Center

forecast but many (eg, Kalnay, Lord and McPherson, 1998). Where they are similar, predictability is judged to be high. This same idea is being extended to find regions in which special observations would improve the forecast. Carefully targeted observations from satellites and from aircraft, perhaps remotely piloted, might thus contribute to forecast improvement in critical cases.

LONGER-RANGE PREDICTION

The preceding paragraphs have described the solution of an initial-value problem involving a system of differential equations and the initial values of the variables. But as we look ahead for periods longer than a week or so, the exchanges of energy across the boundary between the atmosphere, the oceans and the land surface, become increasingly important. Indeed, we might say that weather forecasting is an initial value problem and that climate forecasting is a boundary value problem.

To the first order, the Earth as a thermodynamic system is a body of water 4 km deep. The fact that the ocean is the thermally dominant mass in the Earth system is critical to climate prediction. The total thermodynamic and potential energy (to be called the thermal energy) of the atmosphere, ocean or land, is given by $E = cMT$, where c is the specific heat, M is the mass of the system, and T is the Kelvin temperature. Table 7 gives the active mass for a depth d_0 and d_L of the portion of the ocean and land interacting with the atmosphere and shows that the ocean will dominate the thermal proceedings when the energy is being transferred over ocean depths of 10 or more meters.[3] As the forecast or simulation period increases, the fraction of the ocean and land involved increases, and the atmosphere becomes a smaller fraction of the active thermal mass, as shown in Figure 4.

For longer timescales, the atmosphere drives the ocean and the ocean drives the atmosphere. So the actual physics mandates a computer model of the Earth System that calculates the interactions of the atmosphere, ocean and land simultaneously and predicts how all three will evolve. This is an area of active research; computing the interactions of processes with very different timescales has proved to be challenging.[4]

THE WEATHER IN
WEATHER RISK

SEASONAL OUTLOOKS

Various major forecast centres and private firms are experimenting with forecasts of seasonal departures from normal conditions, and some are making them available to users or the public, often with the caveat that they are to be regarded as experimental.

The Climate Prediction Center (CPC), of the US National Weather Service, has explored the possibilities of monthly prediction for 60 years and has issued seasonal outlooks for more than 30 years; it now regularly offers forecasts for periods extending from a week to seasonal outlooks a year in advance. In preparing the seasonal outlooks, the CPC uses (in order of the CPC assessment of perceived reliability):

❑ *ENSO analogs*: climate variations in the US during El Niño Southern Oscillation (ENSO) events are predicted when El Niño events are imminent or in progress from averages of conditions during previous events. Strong El Niño events are significant drivers of seasonal climate anomalies.
❑ *Trend analysis*: the rates of change of average temperatures in some parts of the US in some seasons have been large enough to provide some forecast skill in otherwise quiet periods.
❑ *Dynamical models*: the atmosphere–ocean interaction process described earlier in this section is approached with a two-step process. First an ensemble of coupled atmosphere–ocean models is used to predict sea surface temperatures. The resulting ocean temperatures are then used as boundary values for calculations with the NCEP climate model run in an ensemble mode to detect variations from normal conditions.
❑ *Statistical methods*: various methods are used to take advantage of statistical relationships between sea surface temperature and various atmospheric variables to infer that climate anomalies might be likely.
❑ *Soil moisture*: statistical methods are used to predict soil moisture for the summer months when it shapes the surface energy budget and hence is critical to temperature and precipitation forecasts.

This collection of material is available to the CPC forecasters who apply their collective experience to integrate it and produce a probabilistic estimate of how conditions will vary from the seasonal normals.

FORECAST SKILL

Figure 5 shows a subjective estimate of the relative skill of weather forecasts on the scale of days to seasons and longer.[5] Here the term skill has been intentionally left undefined and the skill axis is not labelled quantitatively. The important point is that contemporary observations and computer models combine to produce quite skilful forecasts of the development and motion of atmospheric systems for ranges from a few days to about a week, and that then the skill drops off rather quickly.

The upper part of Figure 5 is a similar subjective assessment of how weather and climate risk varies with period of predictability. Loss of life dominates the risk on the order of days while the serious economic consequences of long-term events such as drought become important on the seasonal scale.

Long-range forecasts are cast today in categorical terms. The CPC, for example, uses the climate record to determine the boundaries of lowest third, middle third and upper third of the average temperature or precipitation in each climate forecast region. The seasonal outlook then specifies in which of the three categories the average for the forecast period is expected to be. A fourth category is climatology (CL): the probabilities will be the same as those in the climate record with the average equally likely to fall in the lower, middle and upper third.

The basic scheme for measuring skill is shown in the contingency Table 8, in

> ## PANEL 2
>
> ## A FORECAST EXPERIMENT
>
> The scores reported in Table 10 for skill in seasonal outlooks might be considered disappointing. We might wonder whether statistical methods would give better results. After all, we often try to predict the evolution of social, economic, or financial variables by assuming that the record of the past contains sufficient information to foretell the future.
>
> We might ask for a method that gives the best possible prediction of future values that can be obtained as a linear combination of values in the immediate past. For statistically stationary series, the theoretical and numerical solution to this problem is known as "optimal linear prediction" and determines coefficients in the linear combination of past values through a non-linear process depending on the autocorrelation function.
>
> The forecast experiment was based on the 50-year record of daily average temperatures at Madison, Wisconsin, and used routines given by Press et al. (1988). The daily forecasts were made from previous daily temperatures, the three-month seasonal forecasts from previous seasonal averages. As shown in the table below, skill depends entirely on whether we are predicting the annual cycle or deviations from the annual cycle. The results confirm what we know: it will be much warmer in Madison in July than January; the real question is whether the next July will be warmer or cooler than average. On that issue, the statistical scheme showed no skill at all.
>
> **Heidke skill scores for statistical forecasts for Madison average temperatures**
>
> | Lead interval (days or seasons) | Heidke skill for daily forecasts — With annual cycle | Heidke skill for daily forecasts — Annual cycle removed | Heidke skill for seasonal forecasts — With annual cycle | Heidke skill for seasonal forecasts — Annual cycle removed |
> |---|---|---|---|---|
> | 1 | 78.5 | 41.4 | 73.5 | 1.2 |
> | 2 | 70.8 | 17.1 | 75.3 | 1.2 |
> | 3 | 68.6 | 7.4 | 70.0 | 2.9 |
> | 4 | 66.8 | 2.8 | 75.3 | −2.4 |
> | 5 | 65.8 | 0.7 | 66.5 | −0.6 |
> | 6 | 64.3 | 0.3 | 66.5 | 1.2 |
> | 7 | 63.9 | 0.1 | 70.0 | 2.9 |
> | 8 | 63.2 | 0.1 | 68.2 | −2.4 |
> | 9 | 62.1 | 0.1 | 66.5 | −2.4 |
> | 10 | 61.5 | 0.0 | 66.5 | −2.4 |

which the number of forecasts that fall in each of the nine categories are identified (the h_{ii} are correct forecasts (hits); the e_{ij} are forecast errors). Those along the diagonal are successful forecasts, the others are in error.

With the observations divided into thirds, N random forecasts would produce fractions $N/9$ in each box in Table 8. Thus the hits or successes along the diagonal would sum to *N/3*. Using this observation, the "Heidke skill score" (HSS) is defined to be zero if the sum of the diagonal is *N/3* and 1 if it is *N*. Thus we have

$$HSS = \frac{H - N/3}{N - N/3} \qquad (6)$$

Table 9 converts fraction correct into Heidke scores;

The CPC outlooks are now issued individually for each of 102 US climate divisions, an increase from the 59 divisions used prior to August 1999. The results of CPC outlooks for all lead-times for December, January, February (DJF) and January, February, March (JFM) 1995 for 1995–2000 are shown in Table 10.[6] The contingency table entries are sums of the data appropriate to the two different configurations of climate divisions. The data in Table 10 demonstrate clearly that temperatures observed in 1995–2000 were quite different from the 1961–90 statistics used to set the class limits. The preponderance of above-normal temperature observations is consistent with the unusually warm conditions in the 1990s in the US. A consequence is that the Heidke skill scores computed from Table 10 are not strictly correct according to Equation (6) because the observations are not divided into thirds.

CPC compilations of skill scores for all seasons by lead-times from one to 13 months show averages that are similar to the skill scores shown in Table 10. Thus temperature forecasts (without the CL row in Table 10) are correct about half the time, precipitation somewhat less than half.

Statistical approaches to weather and climate risk management

At periods longer than a few days, the management of weather and climate risk depends on ascertaining the probabilities that adverse events might occur and then offsetting them by various strategies, which will themselves depend on the probabilities. As shown by Figure 6, we necessarily depend increasingly on climatology as the time period of interest increases and we move out of the range of reliable weather forecasts.

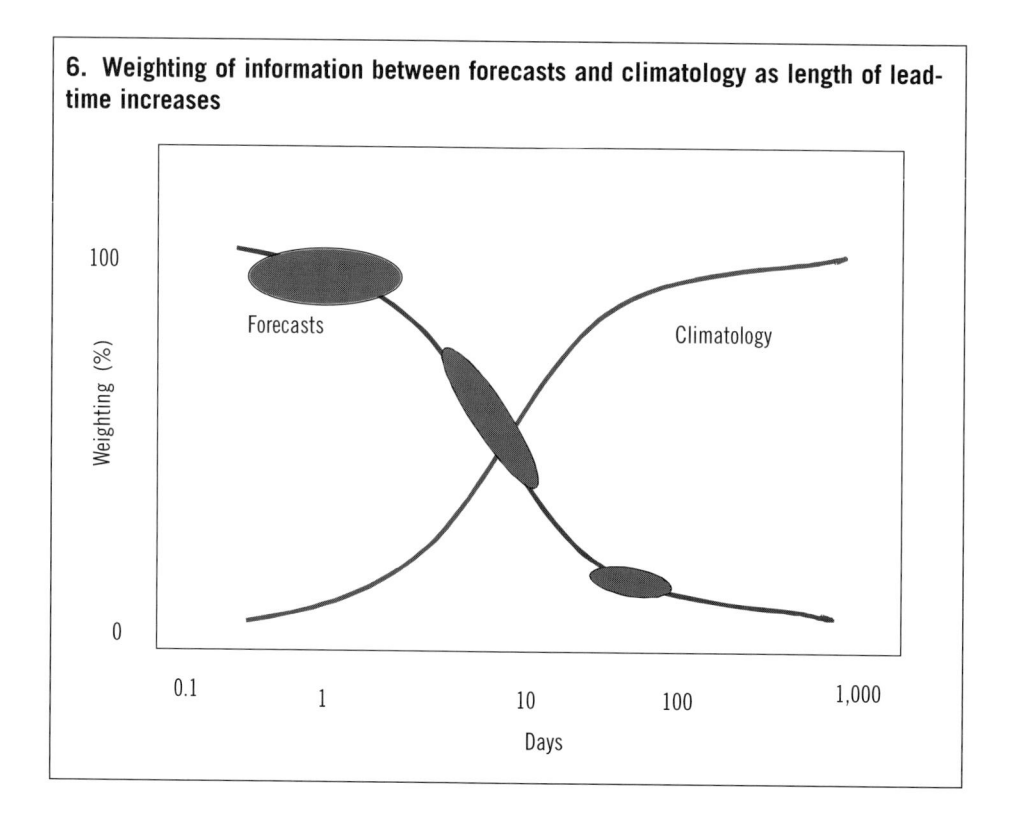

6. Weighting of information between forecasts and climatology as length of lead-time increases

TEMPERATURE, PRECIPITATION AND WIND

The three main atmospheric variables of interest in weather and climate risk – temperature, precipitation and wind – are usually measured simultaneously, but have very different properties. Temperature (a measure of molecular kinetic energy) is essentially continuous and usually varies smoothly in time and space. Precipitation comes as rain and in several frozen forms, but can be markedly discontinuous or spotty in both time and space. The horizontal wind is usually defined by speed and direction, but can also be represented as the two components of a vector, one to the east, one to the north. The wind, too, is continuous, although it can vary rapidly.

Table 11. Three probability functions used for describing atmospheric variables

Probability function	Mathematical form	Applications
Normal	$p(x) = \frac{1}{\sqrt{2\pi}\sigma} e^{-\frac{1}{2}\left(\frac{x-\mu}{\sigma}\right)^2}$	Temperature, u and v components of the wind
Gamma	$p(x) = \frac{\lambda}{\Gamma(\alpha)} (\lambda x)^{\alpha-1} e^{-\lambda x}$ $\quad \alpha > 0, \lambda > 0$	Precipitation
Weibull	$p(x) = \lambda \alpha (\lambda x)^{\alpha-1} e^{-(\lambda x)^a}$ $\quad \alpha > 0, \lambda > 0$	Wind speed

Table 12. Examples of atmospheric statistical distributions

Location, variable, distribution	Parameters	
	μ (F)	σ (F)
Madison, WI Average daily temperature (50 yrs) Normal distribution(s)		
Spring	45.2	14.1
Summer	69.1	6.83
Autumn	48.4	13.8
Winter (single distribution)	20.5	12.0
Winter (sum of two normal distributions)		
$\quad$ Weight 0.68	17.2	12.2
$\quad$ Weight 0.32	30.1	5.90
	a	λ(mm^{-1})
Groningen Airport, Eelde, NV Daily precipitation (40 yrs) Gamma distribution		
Summer	0.461	0.161
Winter	0.442	0.115
	a	λ(kt^{-1})
O'Hare Airport, Chicago Hourly wind, Dec, Jan, Feb (5 yrs) Weibull distribution		
Wind speed	2.33	0.10
Wind speed plus gusts	1.86	0.087

THE WEATHER IN WEATHER RISK

The use of atmospheric statistics in applications is almost always facilitated by analytical representations of the probability density functions $p(x)$ and the (cumulative) probability distributions $P(x)$. Table 10 lists three probability functions commonly used in climatology. Note: The Heidke scores are given first for the data without consideration of CL and then with the CL forecasts distributed equally to each of the other three categories, thus reducing the skill score to the value shown in brackets.

Table 12 gives parameters used to fit observed distributions by minimising the mean square difference between the empirical and theoretical probability distributions.

Temperature
Figure 7 shows the seasonal probability densities for daily average temperature for Madison. Single normal distributions fit the spring, summer and autumn densities reasonably well, but a linear combination of two normal distributions was required to model the two populations that together make up the winter statistics.

In the US, temperature statistics are often expressed in terms of heating or cooling degree-days, which have the advantage of being cumulative statistics and the disadvantage of being non-linear transformations of temperature distributions that are often reasonably well approximated by normal distributions.

Precipitation
Analytical models are usually facilitated by separating precipitation events from all the cases with no precipitation. Figure 8 shows the distribution of daily precipitation at the Groningen airport at Eelde in the Netherlands.[8]

Wind
Figure 9 shows five years of winter hourly wind data at O'Hare airport in Chicago. In the figure, the hourly wind speed is well represented by the Weibull distribution, but the two populations involved when speed and gusts are combined cannot be represented adequately by one function. The increasing interest in wind power generation will presumably stimulate interest in refining wind statistics and lead to

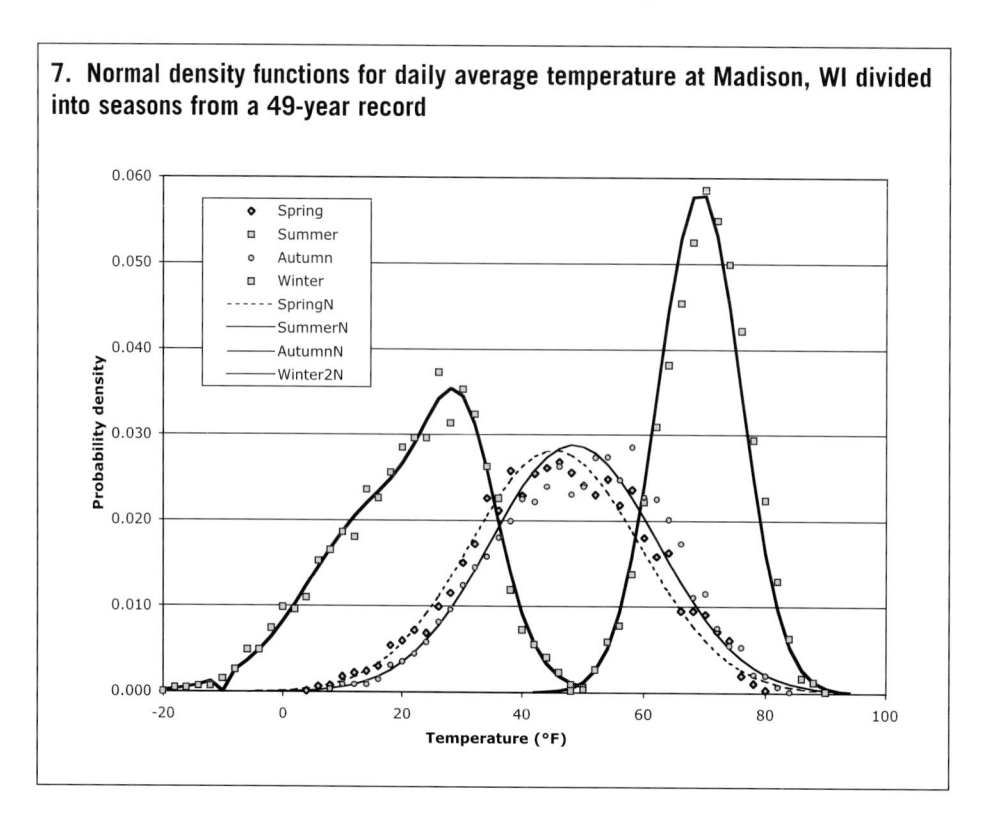

7. Normal density functions for daily average temperature at Madison, WI divided into seasons from a 49-year record

PANEL 3

THE AVAILABILITY OF WEATHER AND CLIMATE INFORMATION

The availability of weather and climate information varies widely from nation to nation.

The present US policies on atmospheric data have been formulated over the past two decades. The basic position is that the government supports a vigorous programme of atmospheric observation in order to provide forecasts, warnings and climatological information for the protection of life and property and for the management of weather and climate risk. The policy further requires that the government should not collect data unless it is needed in pursuit of those objectives, and that any atmospheric data and information that have been obtained or produced for the government's purposes should be made freely and readily available to the taxpayers who have already paid for it. The situation in some other nations is quite different: daily observations are sold by government entities at prohibitive expense, making the operation of a private weather services so costly that some firms are forced to operate private observing networks. Climatological information is only available at daunting cost. The government attempts to derive revenue from fee-based weather and climate services rather than providing a public good based on taxpayer support.

Charging citizens directly to access weather information, many would argue, makes no more sense than charging them directly for federal economic data, for crop information or forecasts, for safety information about products and medicines, or indeed, for emergency assistance or police services. It is impossible to support a contemporary national atmospheric observing programme with user charges and it is a strange policy indeed to deprive citizens and business of the very best information available about the weather and the climate.

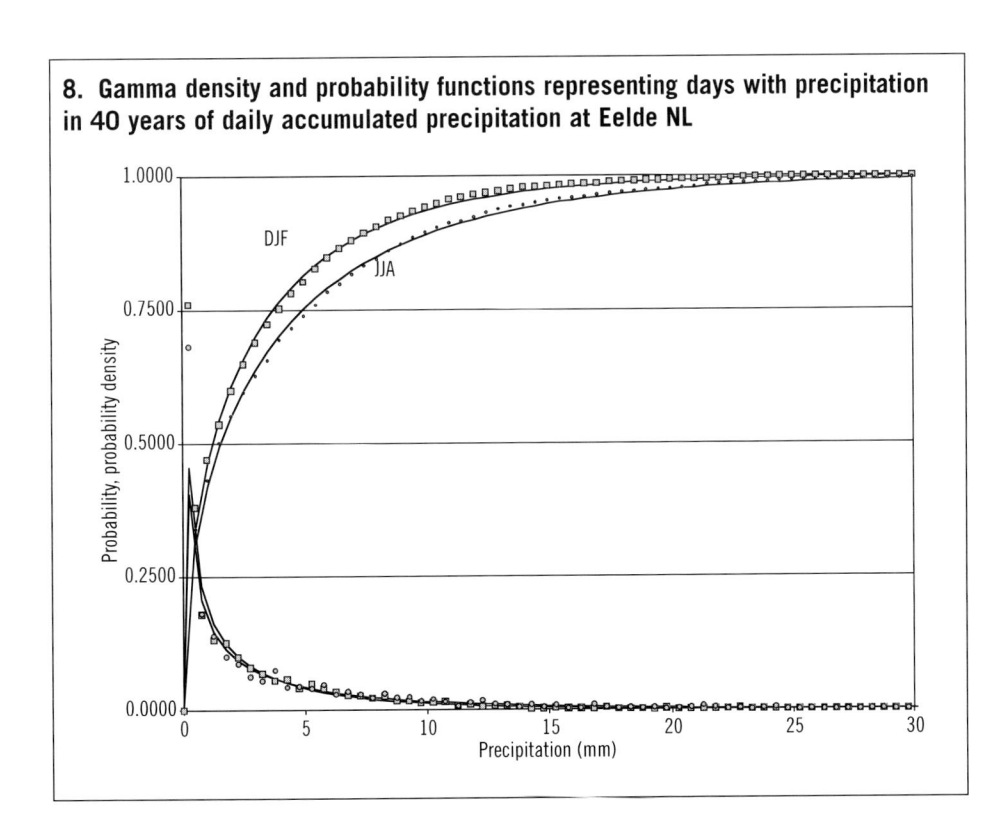

8. Gamma density and probability functions representing days with precipitation in 40 years of daily accumulated precipitation at Eelde NL

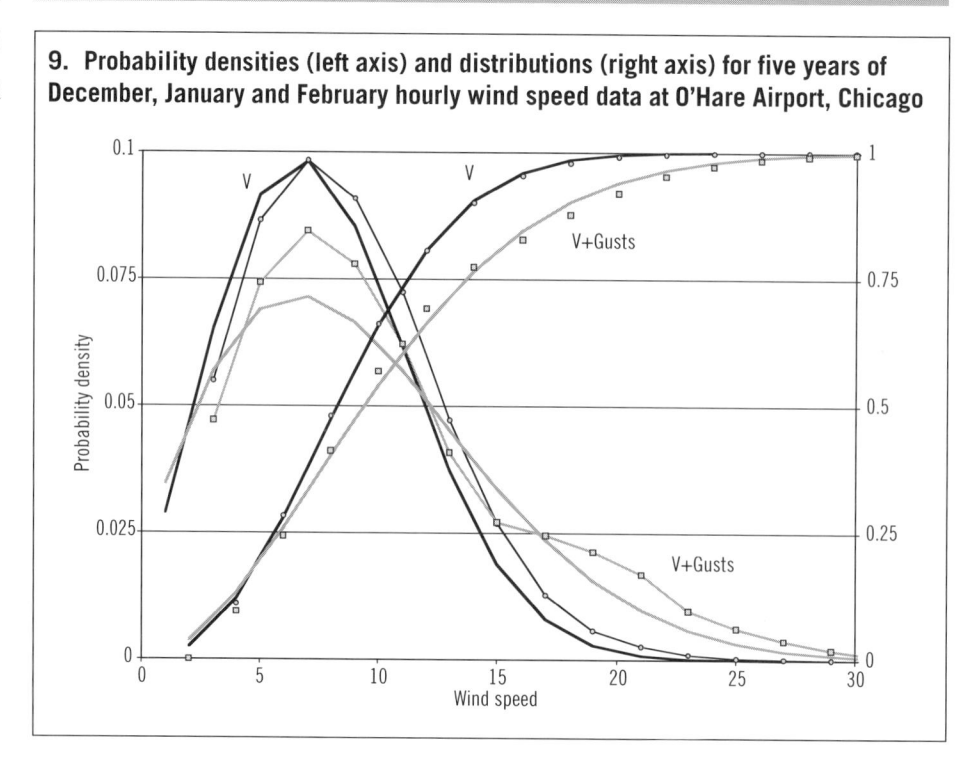

9. Probability densities (left axis) and distributions (right axis) for five years of December, January and February hourly wind speed data at O'Hare Airport, Chicago

interesting issues because the power available in the wind is proportional to the cube of the wind speed.

TRENDS IN TEMPERATURES

Long-term trends in meteorological variables create difficulties in developing statistics for climate risk management and for selecting criteria on which to base risk transfer contracts. The increase in temperature averages over the past decade is an obvious non-stationarity in the statistical record that cannot be ignored.

The US weather risk markets have responded by using degree-day statistics for the most recent 10 years. The serious disadvantage is that statistical reliability of estimates of the means and variances is rather small on such a short record.

The CPC performed a systematic study of the long-term trends in US temperatures in recent decades, treating each season separately. The results demonstrate that the trends in atmospheric temperature vary geographically and seasonally.[9]

SIMULATION OF METEOROLOGICAL RECORDS

Many weather and climate risk analyses and decisions can be based on classical statistical techniques such as integrating the product of a contract cost–payoff function and an atmospheric probability density function. For some techniques, it becomes advantageous to generate simulations of the weather variables in ensembles of records significantly larger than those in the observed record. Simulations can be used to develop prices for hedges, for evaluating a contract or portfolio by marking-to-model, or for generating examples of low-probability worst-case scenarios. They are especially useful for non-linear dependence of output upon input.

Weather and climate simulators should reproduce the observed data and model correctly the probability structure of the original ensemble and its sequential behaviour. The last point is important as meteorological records are sequentially correlated and have characteristic energy spectra and phase relationships. Thus the usual statistical generators will not produce realistic meteorological series.

A weather and climate simulator for use in the management of weather and climate risk has been developed by the Pennsylvania State University and Weather Ventures based on methods developed for simulation of turbulence for aircraft design.[10] The method uses principal component analysis to represent an ensemble of meteorological records (the deviations from the annual cycle) in the form

$$f_i(t) = \sum_{n=1}^{N} a_{n,i} \phi_n(t) \quad i = 1, 2, ..., M \qquad (7)$$

in which the principal components ϕ_n are deterministic functions determined from the data, and the expansion coefficients $a_{n,i}$ carry the statistical information. Their probability structure is determined from the observed data, and then ensembles of

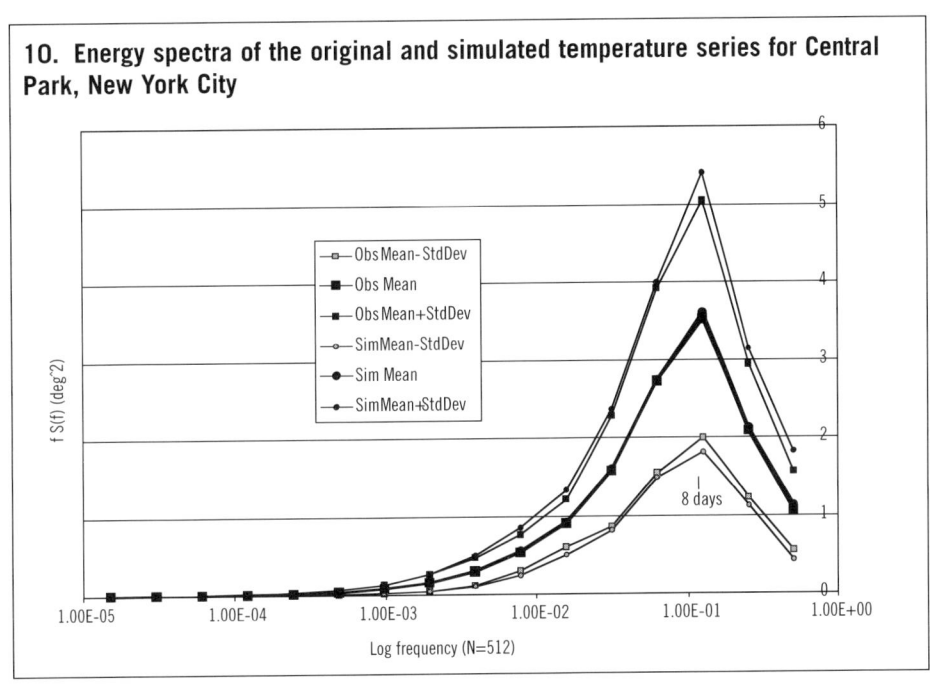

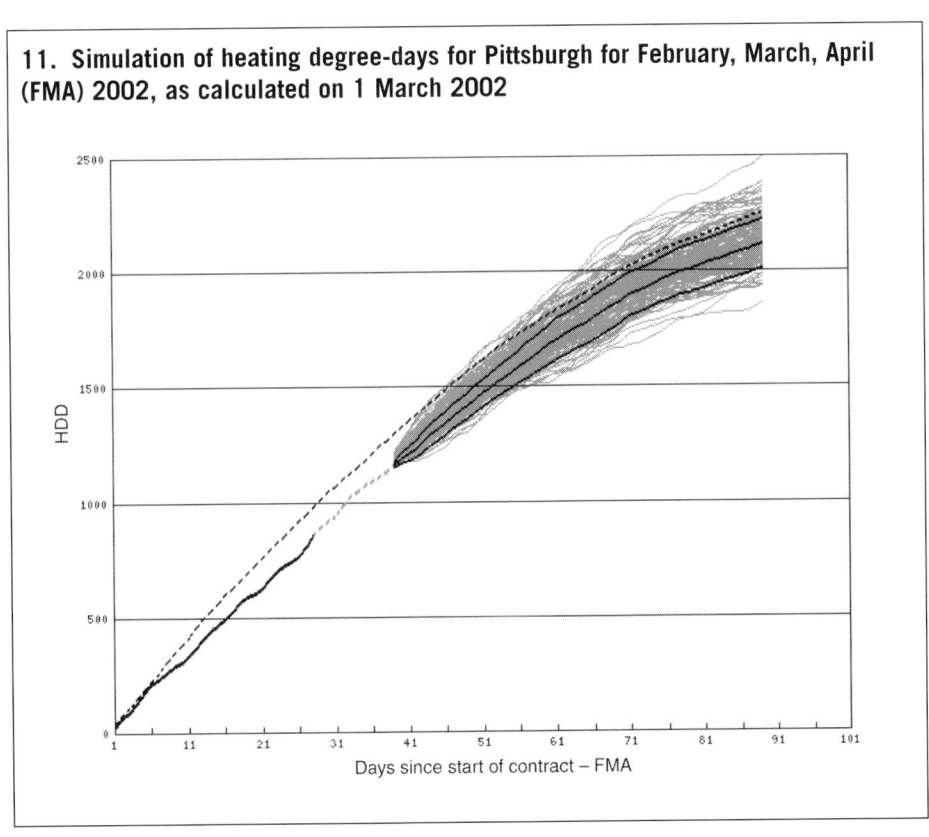

10. Energy spectra of the original and simulated temperature series for Central Park, New York City

11. Simulation of heating degree-days for Pittsburgh for February, March, April (FMA) 2002, as calculated on 1 March 2002

records can be generated by generating expansion coefficients having the correct statistical characteristics. Figure 10 compares the energy spectra of temperature at Central Park, New York City with the spectra for a simulated ensemble.

Option contracts contingent on weather variables are widely used to hedge risks related to energy usage and usually specified in terms of the seasonal average temperature or the cumulative degree-days. As the season progresses, the probability distribution of final values evolves, with both the mean and variance likely to be different from those expected at the beginning of the season.

A simulation of such a case is shown in Figure 11 for February–April 2002 for Pittsburgh. The upper dashed curve is the climatological accumulation, the lower curve, from left to right, the seasonal accumulation in 2002, the 10-day forecast represented as dashed, and then the simulation to the end of the period, with the means and boundaries of one standard deviation shown as darker lines. With an exceptionally warm winter, the actual accumulation of degree-days is significantly less than the climatological curve. At day 28 in the season, a 10-day forecast is used to extend the expected accumulation to day 38. Then a large ensemble of DFJ simulations is generated and the subset with accumulated HDD the same as that for day 38 is selected to estimate the evolution of the probability distribution from day 38 to 90. Clearly, the estimates obtained this way will converge toward the correct value at the end of the season.

Forecast skill and success in hedging weather risk

The weather risk market provides an opportunity for firms and individuals to enter into contracts whose payment terms are contingent on weather variables. These options, commonly called "weather derivatives", allow those facing weather risk to transfer the financial aspects to counterparties who accept the risk in the expectation of profit.

Since climate variability can create sizeable cost or loss of revenue, substantial amounts of money will be at risk in weather derivatives and thus the premiums for bearing the risk will also be substantial. For the market to endure, all parties involved must have a shared understanding of the advantages and cost of the risk transfer process. In this section we consider the interactions between revenue and volatility, probability of occurrence of the adverse event, and the accuracy of forecasts. For simplicity, we consider only categorical forecasts and simple contracts (calls or puts).

AVERAGE REVENUE FLOWS

The cashflows associated with possible hedging strategies are shown in Table 13, where G is the gain or revenue in normal or good weather, L is the loss of revenue if adverse weather occurs, H is the payoff of the hedge contract and P is the premium paid for the hedge.

From these definitions we can identify a number of revenue flows of interest:

Climate (no hedge) $\quad R_c \, G - wL \quad$ (8)

Always hedge $\quad R_h = G + w(H - L) - P \quad$ (9)

Return to investor or counterparty $\quad R_{inv} = P - wH \quad$ (10)

These are the average revenues over a sufficient number of events or seasons for the probability estimate w to be reliable. They constitute all of the strategies possible in a model in which we know only the climatological probability of adverse weather or climate and do not have forecasts available. It would seem that a business is not viable unless G is significantly larger than wL.

From these definitions, we have two results of significance for understanding the

Table 13. Financial flows contingent on weather or climate conditions

Risk management status		Adverse Wx	Good Wx
	Risk hedged	G + H − L − P	G − P
	No hedge	G − L	G
	Climatological probability	w	1 − w

Weather or climate condition

Table 14. Specification of revenue flows in general equation

Case	Successes	False alarms
General forecast	S	F
Climate	0	0
Always hedge	1	1
Perfect forecast	1	0
Worst possible	0	1

Table 15. The probabilities S and F (%) as determined by CPC forecast results

DJF 1995–2000 Below-normal **JJA 1995–2000**

Forecasts	Observations			Forecasts	Observations		
	B	Not B			B	Not B	
B	4	33		B	122	273	
Not B	447	2,229		Not B	423	1,614	
	451	2,262	2,713		545	1,887	2,432
	w	S	F		w	S	F
	16.6	0.9	15		22.4	22.4	14.5

DJF 1995–2000 Above-normal **JJA 1995–2000**

Forecasts	Observations			Forecasts	Observations		
	A	Not A			A	Not A	
A	1,421	1,233		A	1,010	952	
Not A	50	9		Not A	171	299	
	1,471	1,242	2,713		1,181	1,251	2,432
	w	S	F		w	S	F
	54.2	96.6	99.3		48.6	85.5	76.1

THE WEATHER IN WEATHER RISK

process of hedging risk. The first we call the fundamental theorem of weather risk hedging:

THEOREM. *If the probability of occurrence of adverse weather or climate events is greater than zero, then (without forecasts)*

$$R_h = R_c - R_{inv} \qquad (11)$$

in the long-run.

This result makes it clear that the benefit of hedging lies in reduction of volatility, because the cost of hedging mandates that long-term revenue will be less than that available by simply accepting the volatility imposed by weather or climate. A further result is:

THEOREM. *In order for a weather risk market without forecasts to endure, it is necessary that*

$$P > wH \qquad (12)$$

for the probability w of adverse conditions greater than zero.

If not, the investors absorbing the risk will be forced to leave the market.

BENEFITS OF FORECASTS

In the above model, the party at risk has only one decision to make: whether or not to hedge. In practice, forecasts of seasonal anomalies might assist in deciding when a hedge is advisable; a hedge could be set in place when the probability of adverse events is above a certain threshold probability level. Then the situation changes and the strategies become more complex. To begin, we estimate revenues if we act on forecasts at the two ends of the forecast accuracy spectrum:

Perfect forecast
$$R_p = G + w(H - L - P) \qquad (13)$$
$$= R_c + w(H - P)$$

12. Revenue as a function of volatility for w = 0.3, G = 1, L = H = G/2 and P = 1.2wH

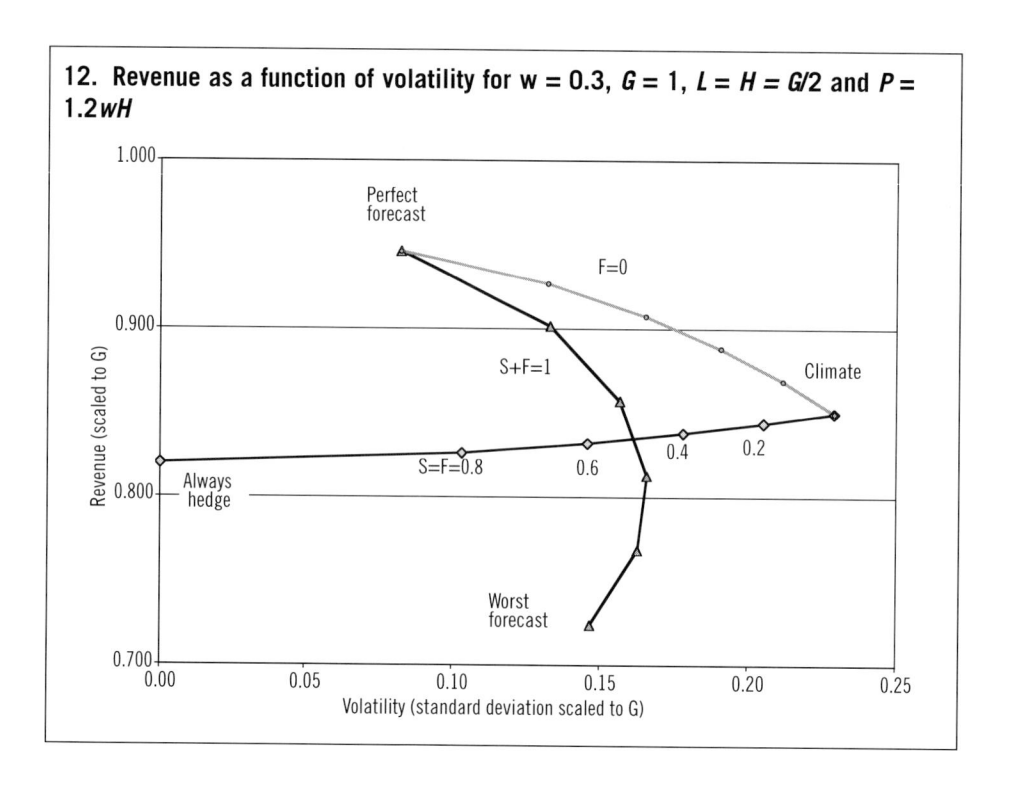

| Worst possible forecast | $R_w = G - P + (P - L)w$ | (14) |

The perfect forecast in Equation (13) (which by definition is always correct) allows us to put the hedge in place only when it is certain to be needed.

To proceed further to study the properties of the model with forecasts, we use the definitions and the scheme presented in Table 5 for measuring forecast accuracy.

The two matrices in Tables 5 and 13 are arranged so that we can calculate revenue flows as the scalar product (multiplying element by element and summing the products). Thus for the general forecast we have the revenue

$$R_f = G - wL - FP(1 - w) + (H - P)wS \qquad (15)$$

Now we see that the four revenue flows considered earlier are special cases of Equation (15). We summarise the situation in Table 14.

VOLATILITY

The purpose of hedging weather and climate risk is to provide a predictable stream of revenue – to eliminate the effects of weather and climate from the income statement. To do so requires that revenue be surrendered as the cost of reducing volatility. For some firms, it is essential to reduce volatility of earnings and protect against very large weather losses because they may have adverse effects on the market value of the business. The question is whether using forecasts can reduce the costs of hedging.

To examine the relationship between revenue, volatility and forecast skill, we calculate the variance in the usual way for each of the five cases in Table 14 using Tables 13 and Table 5.

As examples, the variances corresponding to Equations (8), (9) and (10) are

$$V_c = w(1 - w)L^2 \qquad (13)$$

$$V_h = w(1 - w)(H - L)^2 \qquad (14)$$

$$V_{inv} = w(1 - w)H^2 \qquad (15)$$

The variances for the perfect and worst forecast are similar forms, but the variance for the general forecast is too complex to write out here. Here the volatility is the standard deviation of the revenue and thus is the square root of variances such as those above.

These analytical expressions relate return and volatility to the three measures of the weather risk and forecast skill: w, S, and F. The forecast verification information in Table 10 can be converted from three categories to two in order to calculate S and F for forecasts of above-normal and below-normal conditions. The results are given in Table 15, which suggests that $S = F$ is representative of experience over a wide range of S and F values. This estimate is used to construct Figure 12, in which for simplicity we have assumed $G = 1$, $L = H = G/2$, and that $P = 1.2wH$ will be adequate return for the counterparties. As expected from Table 14, using forecasts with $S \sim F \sim 1$ will give results similar to always hedging. On the present assumptions, R_f increases with increasing S for F fixed and decreases with increasing F for S fixed, but because of our choice of P decreases very slightly as both S and F are increased together.

Figure 12 illuminates the observation made initially in this section that weather risk management and hedging techniques reduce volatility. Perfect forecasts – or even reasonably good forecasts – will give considerable gain in revenue over that expected from climatology without hedging. Indeed, forecasts with large S and small F are of considerable value. When forecasts have similar S and F, the volatility is reduced from that of climate, but without a gain in revenue (in this model, with the assumptions made).

Evidently, Figure 12 and the associated ideas may form the basis for developing a strategy for maximising the value of long-term forecasts in the management of weather risk and allowing firms and counterparties to weatherproof their profits and their business.

Conclusion

Managing weather and climate risk is an attempt to bring danger or financial volatility within acceptable bounds. Toward that end, we observe the atmosphere from the surface, the air and space; we simulate its behaviour with non-linear equations and incredibly powerful computers; we develop mechanisms for transferring the financial aspects of weather and climate risk from one party to another.

The investments in research, technology and the ongoing costs of observations and weather services have paid considerable dividends in the reduction of fatalities and property damage, in providing a basis for quantitative management of weather and climate risk.

We are in the early stages of developing the theory, the methods and the probabilistic basis of such risk management processes. Contemporary information technologies allow us to link atmospheric information with detailed enterprise performance information in order to understand risk and how to mitigate weather risk. Our capabilities and skills will evolve rapidly in the years ahead, because there are provocative intellectual issues and the allure of considerable financial rewards.

Nevertheless, as sophisticated as our science and operational and financial strategies might become, the vagaries of the atmosphere and ocean will ensure that there is always risk in the weather.

1 *A number of individuals contributed data, information, or advice during the preparation of this chapter, including James D. Laver and Huug Vandendool of the Climate Prediction Center; Robert E. Livezey, Office of Climate, Weather, and Water Services; Louis W. Uccellini, National Centers for Environmental Prediction; all of the US National Weather Service; Paul G. Knight, Department of Meteorology, The Pennsylvania State University; Harry Otten, Meteo Consult (Netherlands) and Weather Ventures Ltd; Jan F. Dutton, Weather Ventures Ltd.*

Preparing this chapter was greatly facilitated by the information made available on the Internet by various organisations of the National Weather Service, the National Climate Data Center, the British Met Office and the European Centre for Medium-Range Forecasts.

2 *For a very readable account, see Lorenz (1993).*
3 *The effective depth of the atmosphere is taken to be 10 km.*
4 *For example, see Palmer and Anderson (1994) and NRC (1998), Chapter 5.*
5 *Figure 5 was originally published in Dutton (2002).*
6 *The Heidke scores are given first for the data without consideration of CL and then with the CL forecasts distributed equally among each of the other three categories, thus reducing the skill score to the value shown in brackets.*
7 *Here we assume that the entire atmosphere below 10 km is thermally active on a scale of one day or more as a consequence of the vigorous action of weather and climate systems.*
8 *Here the probability structure of the days with precipitation is well represented except for the peaks near 0.75. Very light precipitation days should presumably be put in a separate category.*
9 *The results are available on the CPC website.*
10 *The Pennsylvania State University has applied for patent protection for this method.*

BIBLIOGRAPHY

Bruce, J. P., 1994, "Natural Disaster Reduction and Global Change", *Bulletin of the American Meteorological Society*, 77, pp. 925–33.

Chapman, R. E., 1992, "Benefit-Cost Analysis for the Modernization and Associated Restructuring of the National Weather Service", National Institute of Standards and Technology, NISTIR 4867, Department of Commerce, Washington, DC.

Downing, T. W., A. J. Olsthoorn, R. S. J. Tol, (eds), 1999, *Climate, Change and Risk*, (London: Routledge).

Dutton, J. A., 1995, *Dynamics of Atmospheric Motion*, (New York: Dover Publications),

Dutton, J. A., 2002. "Opportunities and Priorities in a New Era for Weather and Climate Services", *Bulletin of the American Meteorological Society*, Forthcoming.

Houghton, J. T., 2001, *Climate change 2001: The Scientific Basis*, (Cambridge University Press).

Jeffreys, H., 1925, "On Fluid Motions Produced by Differences of Temperature and Humidity", *Quarterly Journal of the Royal Meteorological Society*, 51, pp. 347–56.

Kalnay E., S. J. Lord and R. D. McPherson, 1998, "Maturity of Operational Numerical Prediction: Medium Range", *Bulletin of the American Meteorological Society*, 79, pp. 2753–69.

Katz, R. W., and A. H. Murphy, (eds), 1997, *Economic Value of Weather and Climate Forecasts*, (Cambridge University Press).

Kunkel, K. E., R. A. Pielke Jr. and S. A. Changnon, 1999, "Temporal Fluctuations in Weather and Climate Extremes that Cause Economic and Human Health Impacts: A Review", *Bulletin of the American Meteorological Society*, 80, pp. 1077–98.

Livezey, R. E., 2002, "Categorical Events", in: *Environmental Forecast Verification: A Practitioner's Guide in Atmospheric Science*, D. B. Stephenson and I. Joliffe, (eds), (London: John Wiley & Sons), Forthcoming.

Lorenz, E. N., 1993, *The Essence of Chaos*, (Seattle: University of Washington Press).

Morton, O., 1991, "A Survey of Space: The Uses of Heaven", *The Economist*, 15 June.

National Research Council, 1998, *The Atmospheric Sciences Entering the Twenty-First Century*, (Washington DC: National Academy Press).

Palmer, T. N., and D. L. T. Anderson, 1994, "The Prospects for Seasonal Forecasting – A Review Paper", *Quarterly Journal of the Royal Meteorological Society*, 120, pp. 755–93.

Peixoto, J. P., and A. H. Oort, 1992, *Physics of Climate*, (New York, American Institute of Physics).

Press, W. H., S. A. Teukolsky, W. T. Vetterling and B. P. Flannery, 1994, *Numerical Recipes in C: The Art of Scientific Computing*, (Cambridge University Press).

Richardson, D. S., 2000, "Skill and Relative Economic Value of the ECMWF Ensemble Prediction System", *Quarterly Journal of the Royal Meteorological Society*, 126, pp. 649–67.

Simons, A. J., R. Mureau, and T. Petroliagis, 1995, "Error Growth Estimates of Predictability from the ECMWF Forecasting System", *Quarterly Journal of the Royal Meteorological Society*, 121, pp. 1739–71.

Investor Issues

12

Weather Risk Management in the Alternative Risk Transfer Market

Julian Roberts

Aon Capital Markets Ltd

Introduction

This chapter discusses the pivotal and effective role that insurance companies and other participants in Alternative Risk Transfer (ART) have played in the market for weather risk management products over many years. The characteristics of both the demand for weather products – old and new – and those that supply them are examined in order to draw conclusions about the status of the market and its products. It is not the purpose of this chapter to provide an exposition on the workings of typical weather products – be they insurance or derivative, as these are fully described elsewhere in this title. Rather it is the intention to examine how the distinctions between the market participants compare to the constraints and drivers for its development. As a starting point, however, it is appropriate to review some of the historical context in which this market is developing.

Historical perspective

The dramatic escalation in the number and diversity of transactions that have taken place in the weather market during the years since 1997 would lead the casual observer to believe that this was truly the birth of a new risk management discipline. Arguably this is indeed the case, but its origins lie in the property insurance market – both insurance companies and Lloyd's of London. In truth the birth is actually a renaissance.

The transition from insurance to weather risk management is in fact quite a logical one; insurance contracts that have encompassed "all risks" have naturally included the risk of loss or damage as a result of the occurrence of one or more 'weather perils'. Specifically, so-called "named-peril" insurance policies have been offered to buyers of insurance across a broad range of commercial, industrial and domestic sectors as a matter of course. Even dating from as far back as the 18th century, crop hail insurance policies were issued to colonial tobacco farmers who sought to mitigate against the devastating consequences of hail damage on their crops. Since that time, weather insurances against crop damage and yield shortfall have been developed and commercialised throughout the world. In 2001, the concept of yield protection was re-introduced as one of the suite of their products by one of the leading operators in the weather protection market.

Similarly, insurance companies as well as syndicates at Lloyd's have offered so-called "pluvius" protection as an insurance product for many decades.[1] This was offered as protection against the cancellation of outdoor or other weather-sensitive

1. Weather contract quantity 1997–2001

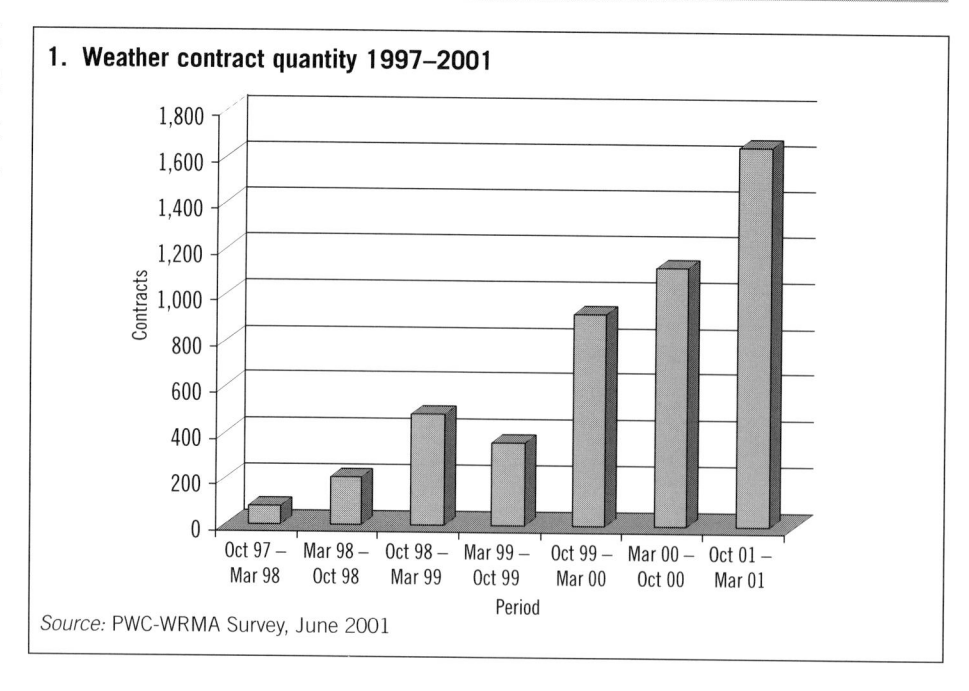

[Bar chart showing contracts by period:
- Oct 97 – Mar 98: ~100
- Mar 98 – Oct 98: ~200
- Oct 98 – Mar 99: ~500
- Mar 99 – Oct 99: ~400
- Oct 99 – Mar 00: ~900
- Mar 00 – Oct 00: ~1,150
- Oct 01 – Mar 01: ~1,650]

Source: PWC-WRMA Survey, June 2001

events (the village fête, for example) and provides indemnity against the financial impact of consequently non-recoverable fixed costs. Such simple but effective products are, in fact, the origin of today's more sophisticated weather products that have been developed to include the use of indexes or various weather parameters.

However, not only has the insurance of weather exposures been encompassed within the ambit of non-specific property insurances, but the opportunity to offer highly specific weather contracts has also been in existence for many years. A range of weather contingent products was developed in the US during the 1980s that were highly innovative and creative at the time by Lloyd's backed underwriting agents such as Goodweather Inc. For example, in a transaction that was rumoured to be closely mirrored only recently, the retailers of snow tyres were able to hedge their 'no snow money-back guarantee' with an insurance policy whose payoff was linked to a pre-agreed scale based upon the reduction in actual snowfall during the winter season. Even weather-linked, lottery-style promotional campaigns became possible with arbitrary triggers such as a "white Christmas" leading to gifts or money back. This heralded the start of far more wide-reaching weather-related products. It is therefore true to note that the insurance industry has for many years recognised the potential presented to them by underwriting weather risks.

These innovations were conceptually the vanguard of today's weather derivative market both in their simple application of a 'trigger' definition and also by virtue of the fact that their existence and their pricing were made possible by the ready and free access to meteorological data provided by the US National Climate Data Center. While these products were offered to cover event cancellation in the style of the old pluvius products, their scope was widened beyond simple precipitation measures to include temperature and wind.

The nature of demand

Oxford University's Professor John Ruskin once said "there is really no such thing as bad weather, only different kinds of good weather", which is perhaps another version of the old adage "every cloud has a silver lining". Indeed weather events influence the needs, obligations and productivity of individuals and entities in differing – and often quite complementary – ways.

That the weather generally influences the efficiency and profitability of human activity is not questioned, there are, however, numerous estimates of the degree to which the weather is the cause of lost revenue or additional expense. Meteorological

> **PANEL 1**
>
> **PLUVIUS INSURANCE**
>
> Weather risk management – in its broadest sense – has been a challenge for mankind since the dawn of time. Indeed the patterns of human settlement, the development of civilisation itself and its attendant agricultural infrastructure have emerged as a direct consequence of man's endeavours to harness and manage the weather.
>
> The most ancient sites of human habitation bear witness to man's efforts to appease the Gods that bring rain and promote fertility. One epithet of the most senior of Roman Gods, Jupiter, was "pluvius" or, literally, "the sender of rain". But the Romans progressed their weather planning beyond mere religious belief and sophisticated engineering, the Emperor Claudius even provided protection to Roman merchants for losses incurred by their trading fleet from storm losses.
>
> Subsequently, across Europe, the availability of insurance-like arrangements against weather losses – especially to farmers – is well documented from as early as the 15th century. Crop hail insurance policies were written for tobacco farmers in "The Americas" from the 18th century. Throughout the 19th and 20th centuries increasingly formal and sophisticated systems of insurance were introduced to provide protection against losses damaging weather events.
>
> Thus, the current weather market is part of a heritage that stretches back many hundreds – if not thousands – of years. The desire to control the effects of the weather remains an enduring and universal pre-occupation. Whilst the needs remain basic, however, the products have grown in complexity and usefulness. The range of weather effects that can be managed is now broader than ever. Nevertheless, even today the term "pluvius" lives on: these days not to invoke an ancient deity but to describe an insurance policy that covers the loss of income or profits caused by rain or other weather conditions.

research indicates that more than 80% of business activity is to some extent affected by the weather and that according to a study carried out by the Chicago Mercantile exchange, in the USA alone, this could amount to as much as US$9 trillion in terms of effected turnover and lost profit. Of course, such figures can be nothing more than broad approximations but they serve at least to show the potential magnitude of the issue. Figures of such scale also underlie the obvious point that the impact of such influences is truly fundamental to both commercial and social undertakings.

Businesses that are so greatly affected by the weather have taken its variability literally 'into account' and reflected the volatility of revenue or cost in their year to year (or season to season) reporting. Owners and managers of enterprises that are influenced by the weather have, in the absence of alternative strategies, adopted practical risk management measures to avoid or mitigate the adverse impact.

Such strategies have typically been physical or logistical in nature and rely upon practical knowledge and understanding. Farming, for example, adapts itself according to the combined influences of geography and market economics and the resultant pattern of cropping is determined accordingly. Under such circumstances, in a rather Darwinian sense, owners and managers who best plan their business to manage the adverse effects of the weather are the long-term survivors. Those that fail to anticipate and adapt the needs of a business that is strongly influenced by the weather and its consequences are likely to under-perform. Indeed, the need for good risk management is increasingly understood: where risks can be managed, so they should be.

So it was presumed, as financial products were introduced enabling business to

WEATHER RISK MANAGEMENT IN THE ALTERNATIVE RISK TRANSFER MARKET

hedge the adverse effects of the weather, that demand would emerge and rapidly outstrip supply. Or, put the other way around, that the mere availability of such products would stimulate the demand for an apparently fundamental product that had hitherto not been available. However, the pattern of growth and development in the reinvigorated weather risk market has not reflected the heady expectations of those early years. The reasons for this, as we will explore later, are both logical and instructional for what is needed to sustain its growth and future. But the truth is that mere supply has not created demand; had it been that simple, such products would surely have emerged from their origins much sooner.

The experience of the last five years is that demand remains confined to a fairly narrow band of business users who very clearly understand their relationship with the weather – namely the energy and utility sector. To some extent this is self-fulfilling; the momentum of the weather risk industry accelerated in 1997 with the inclusion of the energy companies and in the following years their influence in marketing and product design has dominated. The most powerful driver in that market is the negative impact upon energy sales (gas volume or electricity consumption) that result from a 'winter period' that is milder than normal. In the US, at least, the seasonal reciprocal also happens to be true, that is to say, cooler than normal summers result in less use of air conditioners and therefore less electricity consumed.

So it can be observed that the vast majority of trading, starting in the US but spreading to Europe and Japan, has been concentrated on the exposure of the energy sector to its sales in the anticipation of relatively warm winters. In many ways, this pattern is ironic as the actual experience of the past few winters has, indeed, been warmer than the long-term average in many key US and European locations (see Figures 7–9 in Chapter 1). This, in turn, has made the need to diversify into a broader range of exposures and contract types even more critical. The market otherwise faces peak risks or accumulations of exposure that are hard to manage and hinder portfolio diversity.

So what is needed to change the nature of demand in the weather market? There are probably two critical drivers that might be categorised as "push" and "pull" factors, the factors that make companies feel they *need* to be involved in the weather market and those that tempt companies to become involved, respectively.

Push factors are those that encourage entities to adopt weather risk management strategies for the first time, possibly as a result of competitor activity or qualitative evaluation by influential commentators. It is doubtful that there are yet clear cases to describe where such coercive pressures have been the specific cause of any organisation adopting a weather hedging strategy. However, the contribution of the principle credit rating agencies as well as equity analysts is likely to play an increasingly important role.

The truth is that it has been (and largely still is) quite acceptable to the investor community to tolerate worse than expected results due to adverse weather conditions. Annual reports often include such statements in which a downturn in revenue is quite justifiably blamed on "unexpected weather conditions in a key sales period". Although no one seriously expects the businesses of the future to be immune from such influences, the question is, if suitable mitigation strategies exist, how long will adverse results so reported be considered mere bad luck as opposed to bad management? It would, for example, be wholly unacceptable for a large publicly traded corporation to announce large, unhedged foreign exchange losses. Such circumstances clearly bring into perspective the relative effectiveness of a company's risk management policies and can provide invidious comparisons between corporates within the same industry sector.

Perhaps one of the most significant push factors is, in fact, fear. Fear that to leave weather risk unmanaged – essentially 'do nothing' different – will mark the company down as less well managed and prone to more volatility in its earnings. In this context, will the approach of the analysts and rating agencies be to regard the adoption of some form of weather risk management as the benchmark and penalise those who fail to reach it?

Pull factors are the other side of the market equilibrium and it is in this category that the offerings have yet to broaden their appeal. As has been observed, the market has emerged from the deregulating energy sector and its products have fitted the needs of that sector. Deregulation is mentioned as a "pull" rather than a "push" factor at this point, as it was the greater liberalisation of the energy market that exposed its participants to increased demand-driven revenue shortfalls. If a broader section of the potential weather market is to be attracted to purchase such products, then aspects of product design, pricing and delivery need to adjust in order to merit that demand. The factors leading to the availability of and demand for attractive weather risk products make up a complex matrix. We should therefore examine in further detail the constituent parts of that matrix in order to understand better the interrelation of the forces of supply and demand – the requirements of buyers and sellers.

WHAT ARE THE BROAD CHARACTERISTICS OF WEATHER THAT DRIVE DEMAND?

This process commences with a brief overview of the attributes of weather risk that characterise the demand for weather protection. It must be recognised that weather affects our lives along a continuum of severity – from a very warm day that causes discomfort, to life threatening heat spells, for example. It is also largely the frequency of such impacts that contributes to the significance of their consequences. For example, while rainfall on the wrong afternoon may lead to the cancellation of a sporting event which may, in turn, result in consequences ranging from mild irritation to pecuniary losses, vast excesses of rainfall over time (and geography) may lead to widespread flooding, the consequences of which may range from damage to property and infrastructure to loss of life. Clearly, then, the effects of weather vary fundamentally according to the frequency and severity of their occurrence and the products available to mitigate these effects need to reflect those differences. Broadly speaking, these effects can be categorised as being "near the mean" or "far from the mean".

Near the mean events

These are events that are significant to the extent that they deviate from the expected norm against which they are measured and result in appreciable – but not catastrophic – consequences. For example, strong winds from a thunderstorm, in contrast to tornado winds, occur frequently and are therefore probable and closer to the norm. Such events occur with a medium (to high) probability and are therefore of a kind that the affected activity should expect from time to time but is not necessarily fully prepared to absorb by virtue of the very unpredictability of their timing.

Often these events are not single incidents *per se* but are recognised as the accumulating impact of daily weather over a period of time – days, weeks or months. These distinctions need not be exact in scientific terms and, indeed, the semantics may just cloud the issue. The definition of an event may be one, such as the November–March heating contracts, that appears well suited to a broad universe of users. Alternatively, event definition may be very specific to the precise circumstances of the enterprise in question. Here the example of the long-established City of London wine bar chain, Corney & Barrow serves as an excellent example. They recognised that their earnings were strongly affected (for the better) by warm summer lunchtimes especially at the end of the week. Accordingly a weather product was structured so that it paid if an agreed threshold number of 'hot' Thursday or Friday lunchtimes were not reached.

The most pertinent illustration of near the mean weather risks and their associated weather products is the market dominating warm winter/cool summer products. They are not defined by objective seasonal or meteorological standards but rather by the empirical needs of the commercial market that their products serve. In other words, there is a period of time during the winter period for example, in which energy sales and ambient temperature are strongly correlated. To lengthen this period to include additional weeks would reduce that correlation and to reduce it would fail to capture the true seasonal exposure.

A challenging attribute of such significant, but moderately predictable, weather dependencies is that it is often very difficult to isolate the impact of their occurrence from the general range of non-weather 'background' variability to which businesses are also exposed. In fact very few enterprises have a simple "cause and effect" relationship with a single weather variable; contrary to popular belief it is rarely possible to attribute a strong correlation between, say, turnover and temperature or yield and rainfall. In practice critical business metrics are related to weather by somewhat more complex functions. In that regard it is logical and significant that the energy sector should have emerged as an early adopter of weather products. As is discussed below, it is this very hurdle of accurately quantifying the impact of weather that may be largely responsible for the difficulty in broadening the scope of weather products to other target sectors and end-users.

Certain weather conditions such as seasonal rainfall cannot be defined in terms of a single incident (or its even non-occurrence). Obviously, a single rain-free day does not itself constitute a drought nor would two or three such days, but there comes a point after a certain number of rain-free days at which the period could be deemed a drought, perhaps because it has become a problem, in that crops are wilting or dam levels are low. The occurrence of an event in such cases must therefore be defined as the cumulative effects of a given weather parameter over a relevant time period. For such concerns, very often the compensating effect of a reverse weather condition may largely or partially offset the accumulating condition (rainfall during a drought being the obvious example). This, however, tends to hold more true for non-extreme events – to extend the crop analogy, a plant that has suffered dehydration to the extreme of drought will never recover from subsequent rains.

Far from the mean severe events
In contrast to near to the mean events, these events are characterised as being rare in their occurrence but extremely severe in their consequences. In other words exceptional, catastrophic events that have the potential to destroy a business and even damage societies.

These events tend to be isolated, sudden and unforeseen – if not entirely unexpected. Such events are often given names and remembered by communities and those that serve them for many decades. In contrast to the accumulating effects of near to the mean events, the impact of low probability, high severity incidents is more usually experienced as a shock occurrence. They might not last more than a day or so – ice storms or hurricanes, for example – like the severe named hurricanes, such as Hurricane Andrew that caused widespread property damage and loss of life in Florida in 1992.

These are "catastrophes" and have been the domain of insurance and reinsurance products for a century or more. The worldwide insurance and reinsurance industry has underwritten and funded damage caused by hurricanes, floods, earthquakes and the like by sharing the burden among the large number of participating insurance companies worldwide. Arguably the starkest illustration of the manner in which the insurance industry absorbs an insured catastrophe event is given by the destruction of the World Trade Center – although a man-made catastrophe, it is the largest insured loss recorded to date. The consequences of this loss have been shared by means of insurance and reinsurance across the entire industry.

However, there is, in fact, an inherent difficulty in managing the effects of catastrophe on the balance sheet of an insurance company, namely that the predominantly equity-based capital used to support such risks is poorly suited to the highly skewed probability distribution of catastrophe events. The expectation of the shareholders of insurance companies for returns on their holdings implies returns of a level that are consistent with being in the risk business. Conversely, buyers of insurance against events that are extremely unlikely to occur show little inclination to pay high levels of premium. Therein lies an economic dilemma. Insurance ultimately relies upon achieving diversification of the risks accepted (in terms of hazard type, time and geography) but it is difficult, or even impossible, to achieve true diversity for a relatively small number of rare but devastating events.

Nonetheless, insurance remains the natural source of protection against such hazards and it is therefore a logical and conventional aspect of risk management to turn to such products for their management. Furthermore, it is perhaps now an embedded part of social and commercial practice for Risk and Insurance Managers, Chief Financial Officers, the Board of Directors and even shareholders to presume that such catastrophic risks that can be insured (at reasonable cost), are.

THE VARIABLE IMPACT OF TIMEFRAME
In addition to considering the impact of weather events by reference to their relationship to a long-term mean, the notion of the timeframe of weather events also has a material influence on how they affect us, are perceived and are best managed.

While consideration of the timeframe of meteorology is potentially the subject of a discourse on its own, here we should at least consider three broadly valuable distinctions for the purpose of drawing risk management conclusions: they can be categorised as the long-, medium- and short-term timeframes.

Long-term
The long-term changes in weather – or climate, as it becomes – are ultimately the fundamental drivers of weather dependency. If climate change, however it may manifest itself in a given situation, is inevitable (and, of course, there are those that would question whether this is indeed the case), then those affected should evaluate their options accordingly. Under such circumstances, risk management is a core process of selecting the right business for the right location. If long-term climate changes ("global warming", for example) are causing constraining changes to the production environment, be they drier summers or warmer winters, then it is rational to consider the long-term suitability of the activity in question to that changing environment. Indeed, new opportunities may well arise from such changes, which can be taken advantage of. There is evidence, for example, that the changing climate in certain northern hemisphere agricultural growing areas may permit a change in cropping to species that have previously been non-commercial due to their higher requirement for insolation or heating than the area could previously support. Either way, business strategies may need to alter in the face of long-term changes in the climate solely in order to adapt to and optimise the risk management consequences.

This may be easy to describe in theoretical terms but, of course, the impact of such long-term change is experienced gradually over a number of years, or rarely, suddenly or irreversibly in a single event. Hence it is immensely difficult to distinguish randomness from trend without the all-seeing benefit of hindsight.

Some long-term influences may tend to be more cyclical in nature, which may offer some scope for prediction and, therefore, risk avoidance. An obvious example of this would be El Niño/La Niña cycles that are highly documented and monitored to the extent that El Niño years are predictable to a reasonable degree of accuracy, at least in the ocean. For many, if not most activities even accurate predictions of periods of elevated risk (such as El Niño) may not permit many real risk management options. Is it feasible to relocate, sell a business or find new clients? Alternatively, do you re-engineer your assets or business in such a way as to withstand the predicted effects? These are decisions that have known costs and expected payoffs and, using such assumptions, the rational decision-maker can make long-term planning decisions that are consistent with that cost–benefit analysis. We discuss below the importance of this quantitative approach to weather risk management planning.

The influence of long-term and cyclical climate events may be predictable to some degree but typically such predictions can only be accurate on a macro scale – within wide geographic boundaries and uncertainty of timing. Some predictions indicate changes in the long-term average but of course do not permit the anticipation of specific seasonal outcomes. So while risk mitigation or avoidance strategies are logical in the long-term, the short-term influences and variability may need to be managed by financial products.

Medium-term

As "medium term" is somewhat of an arbitrary categorisation of timescale, in the context of this chapter, it is intended to refer to seasonal periodicity. Clearly, this incorporates the very contract types that have made up practically the entire weather market to date: namely: winter and summer seasons. There is compelling logic as to why this should be so. For temperature-dependent activities it is rare that winters and summers provide a natural offset to each other. For example, an ice-cream manufacturer would be unlikely to find compensation for a poor (cool) summer's revenue in a mild winter and there are generally poor correlations between summers and winters. So the most effective hedging strategy is likely to be achieved by focussing on the specific time window that corresponds with the majority of volatility in the metric of concern.

The variability of the medium-term event affects businesses and operators over the timeframe for which no reliable forecasts or predictions are possible. Despite significant research and development advances, the reliability of seasonal forecasts remains very poor. In any case, activities with weather-dependant sales volumes are highly unlikely to be able to take steps to compensate for predicted abnormal weather periods.

A somewhat hybrid consideration is the management of the combined or accumulated consequences of a series of seasonal exposures – in other words multi-year contracts. While the individual seasonal components of such contracts are obviously manageable separately (in either insurance or derivative form), there are practical and commercial challenges to combining these over long periods of time, say three or five years, as well as potential benefits such as cost reduction and business certainty. Yet, logically, weather-exposed enterprises might be more interested in smoothing the predictable effects of a single adverse season (say one out of the next three winter periods) with compensating more 'normal' seasons. Such thinking avoids the obligation to pay in full for the long-term cost of the downside without taking benefit for compensating upside of normal trading.

Short-term

The third category to consider is that of short-term weather influences. These can be thought of as either brief periods of significant weather dependency (critical days) and/or near-term occasions (eg, next weekend).

In different ways, near-term events or critically important days may be significantly affected by the weather. While the overall seasonal average for the parameter in question will be known, the key concern is: by how much will the actual weather on one particular day be different. This takes us back to the logic of the early pluvius insurance contracts. Certainly, event cancellation or impairment remains an important area for weather risk management. However, in keeping with product development in the weather market, increased data availability and more sophisticated pricing tools, short-term products are now significantly more useful than just event cancellation or pluvius insurance. Weather risk management solutions available today can address a broad and complex array of weather-related problems.

Key to the management of near- or short-term weather consequences is the role and accuracy of forecasts. Without quoting specific accuracy statistics, it is clearly the case that while one-day forecasts may be adequately reliable, five- or seven-day forecasts are considerably less so (as the general population will recognise from the forecasts upon which they base their actions). Yet typically, the forward needs of many activities are dependent upon an accurate estimate of conditions up to a week away. For example, the deployment of de-icing equipment or road gritting teams is predicated on 12- to 24-hour forecasts; whereas planning the stocking of supermarket shelves with perishable goods may need three- five-day forecasts. The consequence of forecast inaccuracy for activities that have a high degree of reliance on that accuracy may be unnecessary expenses or opportunities foregone – both result in losses or inefficiency.

Thus, it can be seen that the demand for weather risk management products manifests itself as a result of the many and diverse influences of weather. While "weather" in this context is a single word, the demand for products to manage it may vary distinctly according to the severity/frequency and/or time frame of the problem. As we will see, the differing nature of demand for these products has led, quite naturally, to a diverse range of potential suppliers who are able to offer suitable products. So, in better understanding the characteristics of both the supply and the demand sides, we are better able to understand the constraints and drivers of the market as a whole. Let us now consider the sellers of weather risk products.

The nature of supply
WHO ARE THE PLAYERS?
Weather risk management products are sold by a variety of different market participants. Each has a different role to play and a differing perspective of the most appropriate products and services they can offer.

Insurance/reinsurance companies
As discussed above, insurance (and reinsurance) companies have been involved in the weather risk market – directly or indirectly – for a very long time. Insurers of domestic or commercial property portfolios are inevitably exposed to severe weather events – widespread underwriting losses can be suffered as a result of windstorms, precipitation (flooding) and freezes. These exposures arise as a consequence of insurers' normal undertakings and are not considered to be a particular focus or specialty.

The role of insurers as pioneers of weather event insurance and weather contingent products (such as retail promotions) has also been noted above. However, there are some apparent contradictions and inconsistencies to note in their standing in this market. While it is certainly true that some insurers have indeed been highly entrepreneurial in embracing the topic of weather as a business opportunity and even a logical extension of their existing commercial operations, others have concluded that it is beyond their scope or simply not profitable and exited the market. In this, insurers face a challenging decision that is specific to their *modus operandi*; namely, that the first principle of underwriting is to achieve a 'balanced' and 'diversified' portfolio of risks. Contrast this to a risk trader who seeks to hedge his position on a dynamic basis. But, as has been widely recognised, activity to date in the weather risk market has been far from diversified, being concentrated on winter heating degree-day (HDD) contracts in a relatively small number of locations and the end-user generally wanting to purchase puts.

Conversely the question arises: if insurance companies, by virtue of their existing risk exposures carry appreciable amounts of offsetting risk positions, are they not the ideal counterparty in many transactions? Whether or not this is entirely so (and it will vary from company to company), we must also consider the normal technique adopted by insurers to lay off their risk – reinsurance. Conventionally, reinsurance contracts are contracts of indemnity and "follow the fortunes" of the insurer's original underwriting. Reinsurance is conveniently expressed in a back-to-back format, which allows the insurer to protect a given line of underwriting with a single contract. To the extent that this practice continues to be available and economic, how much less likely is it that a reinsurance buyer would seek to fragment his buying strategy into its component parts and take a modular approach? As a consequence, while insurers may indeed be the natural holders of weather risk, few find it efficient or practicable to consider managing that portion of their risk in such a specific way.

However, this does not detract from the fact that insurers are indeed 'long' (the holders of) weather exposure. This confers a potential market advantage to insurers in that they are consequently able to aggregate and warehouse risk which, in a strictly trading environment, may otherwise be hard to manage. Thereby insurers are able to seize time-critical risk opportunities without the constraint of laying off and building their position. It should be noted however that this 'advantage' can only be achieved if both the premium charged for the additional risk is correctly priced and it has a beneficial incremental impact on the diversity of the insurer's overall weather portfolio.

As participants in the weather risk market, insurers also have the benefit of their broader franchise; insurance companies value long-term trading relationships and maintaining their market share. It is natural that risk managers turn to their existing insurer or insurance broker relationships to discuss ongoing risk management issues. Their insurers are both well placed and generally willing to broaden and strengthen their client relationships; it is a relatively easier step to include a specific weather exposure into a broad spectrum of risks already underwritten to a specific client than to start from scratch.

Thus it is that insurance and reinsurance entities have a natural role to play in the weather risk market. It is therefore not surprising that during the past five years, a number of well recognised insurers have started specific weather ventures or, at least, formalised their commitment to the market by investing in the necessary personnel, data and marketing.

Energy companies/traders
It is no exaggeration to credit the recent rapid development of the weather trading market to the participation of a reasonably small number of US energy companies. Deregulation in the North American energy market permitted companies to exploit their position and market knowledge. This can be seen as an entirely rational development, as it allowed those companies both to manage their own weather exposures and to capitalise upon their expertise to provide protection to third parties – thus making a virtue out of necessity. It is ironic to observe that Enron's position as one of the leaders in that market has been cited by the less informed media as clear evidence of the follies to which it was tempted and, by inference, one of the very reasons behind its ultimate collapse. In truth, there is no evidence to suggest that this was the case.

Currently energy companies both in the US and Europe are still active leaders in the weather trading market. These participants are seeking to expand their client base and product offerings in order to encourage greater take-up by end-users.

Banks
While not the first movers in this market, a number of banks – particularly European ones – were quick to respond to the opportunity of offering weather products to their clients. This has led to the establishment of a number of weather desks within banks over the last four or five years. On the face of it, this is a logical move; such banks have sophisticated client relationship managers who can sell a broad range of

financial and risk management products, so to extend this to include weather products was a short step; however, experience has shown that their success in this area has been mixed.

As with the energy companies, regulatory constraints dictate that the weather products sold by banks are in derivative form (banks are specifically not allowed to issue insurance policies). However, a number of banks have also established insurance subsidiaries in order to 'transform' their derivative offerings into an insurance form where client requirements dictate. These transformer companies are established in domiciles that legitimately enable the creation of an insurance company by a banking parent company to enable it to deliver their product in insurance form.

Some institutions have established funds that are specifically focussed on weather products (often to include catastrophe bonds) so that risk positions can be accumulated through well priced buys in the market and issuance to clients, and not necessarily offset by selling. Weather and natural catastrophe products show great potential as an alternative, interesting high-yield investment whose performance is largely uncorrelated with that of prevailing equity or debt markets.

The weather therefore presents two main opportunities: as a risk management product and an investment.

The exchanges
Exchanges that have introduced weather contracts are a further supplier of weather risk management products – these include the Chicago Mercantile Exchange, the Deutsche Börse, LIFFE and most recently the Weather Board of Trade.

Out of necessity, exchange-traded products (in this case weather future and options) are standardised contracts in order to facilitate their efficient sale and purchase. Therefore it might be expected that the role of the exchanges will develop where demand and liquidity exists. The extent to which weather products can successfully be "commoditised" in this way has yet to be proven and low trading volumes to date would suggest that exchanges have some way to go before developing their potential (see Chapter 13).

Some degree of web-based trading is also taking place although this would appear to be an alternative delivery platform for the key existing suppliers rather than (yet) a distinctly different source or type of risk management product.

Not surprisingly perhaps, the market has already witnessed a number of players closing down their weather desks and exiting the business. The abundance of sellers and relative lack of buyers has not encouraged all of the early movers that this yet a profitable line of business. It is probably too early to draw much of significance from these early departures or, indeed, recent additions to the pack of players in this 'market' other than observing evidence of the usual ebb and flow of an emerging market sector.

Products: insurance vs derivative

So we can observe a wide range of providers and intermediaries on the supply side of the weather market. These providers, by virtue of the differing institutions they represent, broadly cleave into two categories: insurance and derivative. Demand for weather products, especially at the end-user level, is itself influenced by the structure of the products on offer. So it is appropriate to consider the distinguishing characteristics of these products and their relative advantage and disadvantages for different types of users.

The market for weather risk management products is often loosely referred to as the "weather derivative" market. However, as is clear from the foregoing review of the participants in the market, there is in fact a choice both of provider and a choice of product type. While it is not the intention of this section to describe individual

WEATHER RISK MANAGEMENT IN THE ALTERNATIVE RISK TRANSFER MARKET

products in detail (see instead Chapter 2), it is appropriate to consider the practical differences that arise as a result of selecting between insurance and derivative form.

INSURANCE

Advantages

❑ In the environment of corporate risk management, insurance products have the enormous advantage of familiarity. Risk managers, Finance Directors and Chief Financial Officers alike understand and appreciate the benefits (and potential drawbacks) of insurance. As such, an insurance contract is a well-established part of mainstream risk management culture and in this familiarity there is reassurance that a potential risk has been prudently mitigated.

❑ While specific weather contracts are more likely to be arranged separately and form distinct policies, nonetheless the insurance form retains the ability to blend weather and non-weather risks into a single contract. Indeed, historically this has been the case with weather exposure forming a part of a general "fire and allied perils" policy. Importantly, insurers are familiar with the catastrophic (or far from the mean) component of weather and do not generally exclude such extremes. By contrast, insurance against weather events with a high frequency of occurrence is likely to be inefficiently priced by insurers who will price near the mean exposures according to their high probability of occurrence. And this may effectively mean that such products are merely "dollar swapping" exercises with little commercial value – in other words, the cost of the premium may approach the expected cost of the loss.

❑ In most tax environments the premium charged for an insurance policy is treated as an allowable expense for tax purposes that, according to the actual rate of corporate taxation, is certainly advantageous. Furthermore, the treatment of insurance for accounting purposes is well understood and straightforward.

Disadvantages

❑ In order that a contract may be treated as insurance (regardless of the entity with which it is transacted), there are certain qualifying characteristics with which it must comply. Importantly, the insured must have an "insurable interest" and, a corollary of this, there must be "proof of loss" for a claim to be legitimately payable.[2] This constrains the valid uses to which insurance may be put and consequently restricts the use of insurance from objectives such as gambling or speculating. For the majority of legitimate corporate risk management requirements, this is unlikely to present a practical problem. However, it may constrain the use of insurance for protection that is triggered exclusively against a weather parameter without regard for actual loss incurred.

❑ Finally, in most countries insurance is subject to taxation at the point of sale (Insurance Premium Tax, or similar). This tax may be very significant – over 20% in some countries. This frictional cost, over and above the insurer's overhead costs may act as a very real disincentive to transacting risk products in insurance form.

DERIVATIVES

Advantages

❑ The emergence of derivative-style contracts as the predominant form for weather trading is logical considering the nature of the entities involved – energy and utility companies. Risk management and trading using International Swaps and Derivatives Association (ISDA) documented swaps and options for energy prices is the established practice and, in any case, energy companies do not routinely have the necessary insurance vehicle to trade otherwise.

❑ The relative simplicity and standardisation of weather derivative agreements is a prerequisite for a traded market. This should confer the advantage of liquidity and price efficiency.

❏ In contrast to the regulated environment of insurance and its strict requirement for an insured to have insurable interest, no such restrictions apply to derivative contracts. It is therefore possible to enter into an agreement where the cashflows bear no relationship to the direct interests of the parties involved. A weather hedging strategy that employs derivatives can therefore ignore the concept of indemnity and pure speculation is possible. This environment has facilitated the existence of heating/cooling degree-day products in which end-users chose the amount of protection they purchase to suit their particular risk appetite and hedging objectives. In practice, this may lead to a gain or a loss and the management of any such basis risk lies with the end-user.

❏ Such flexibility permits not only the free use of index-based products but also opens up the potential for end-users to take a view on the indirect consequences of weather. It may be valid, for example, to take a view on the weather conditions that influence the competitive advantage of a main competitor. This would be difficult, if not impossible, to achieve with an insurance contract.

Disadvantages

❏ In structure and function, there are few drawbacks to derivatives *per se*. As has been discussed above, they do not necessarily share the positive attributes of insurance form and perhaps the key differences arise in their accounting treatment and the nature of the basis risk. Under US Generally Accepted Accounting Principles (US GAAP), payments under a derivative are treated as an asset on the balance sheet, and changes in fair value (the value of the derivative as at the date of the accounts) must be reported as unrealised gains/losses. This is clearly an added complexity that may deter certain users from selecting derivative form over insurance.

❏ Credit risk is a serious concern for any protection purchaser and this remains the case whether the product is in insurance or derivative form. The end-user should take care to establish that the financial strength or credit quality of their counterparty is of suitable quality. In the case of swap agreements, this becomes a two-way consideration.

Constraints and drivers

NEED FOR LIQUIDITY

Regardless of the differences between the suppliers of weather risk management products and of the structural differences between those products, the market's success is constrained by its currently small trading volumes. Without critical mass, the necessary liquidity and price transparency cannot be achieved which, like a vicious circle, impedes growth in the market.

All suppliers have a role to play in the growth of the market but insurers are particularly well placed to exploit their existing trading relationships and use their balance sheets to retain risk and repackage risk for the secondary market and take advantage of such offsets (diversifiable risk) as can be achieved.

From the point of view of potential end-users, we should also consider what is preventing their wider up-take of weather products. If the case for the existence of weather risk management products seems so compelling, why is it that demand has so far failed to match supply?

There must be something in the old adage "you can take a horse to water, but you cannot make it drink". In this case, merely inventing a new risk management product evidently does not – in itself – stimulate demand. The fact is that weather has always been a major risk factor and, in the absence of such products, social and commercial activity has managed without them. As such, the purchase of weather risk protection is often merely perceived (correctly or incorrectly) merely as an additional expense although, of course, that 'expense' may turn out to be a benefit. So ultimately, the long-term cost–benefit analysis of changing weather risk management strategy needs to be evaluated. Yet for many businesses the alternative of doing nothing apparently remains most expedient – or just easiest.

A cost–benefit evaluation is therefore a critical function that needs to be performed before an entity can rationally take the decision whether to enter into a weather contract. The evaluation must quantify a relationship with one or more weather parameters – this can be formidable, to perform a probabilistic analysis of the chosen weather parameter(s) to derive an expected cost of loss. Thereafter, the price of alternative risk protection strategies (including the 'do nothing' option) can be compared objectively. The ideal decision-making process would then include an optimisation approach to select the alternative that provides the best risk weighted returns. This requires a considerable amount of data, including data both for enterprise performance and for its associated meteorology. Also it pre-supposes access to adequate analytical resource and, not least, access to the fullest range of market prices.

It is not surprising then that many potential participants are deterred by the complex and time-consuming nature of this process and choose to make their evaluation against more rule-of-thumb, not to say visceral, judgments – which usually comes down to the simple up-front cost of hedging. Indeed many potential buyers are evidently deterred by cost alone and this may be an entirely reasonable assessment if that cost is simply not containable within the margins of production or overall budget. But, of course, the 'cost' of the weather does not just go away and it must be borne within the business, which may reflect itself in terms of volatility of earnings and ultimately shareholder value. Nor need this, *per se*, be wrong: shareholders of companies that are fundamentally weather influenced may view that fact as a legitimate component of the risk and return associated with that stock.

Conclusion

Taking an informed view of the impact of weather and the alternatives that are available to deal with it is the logical and appropriate approach to weather risk management. This may indeed support the 'do nothing' approach, if the expected cost of hedging or trading exceeds the potential returns. But the availability of accessible data as well as the plethora of advisory and consultancy services on offer increasingly means that to remain ignorant of the choices available and their consequences is now poor risk management practice. The universe of risk management products and the variety of available protection providers has already started to expand in order to match this challenge.

There are many different predictions for the future of this brand of risk management. Some see it as being a short-lived fad, others believe that it is here to stay. Either way, the extent to which weather risk management techniques become fully incorporated within the mainstream of risk management is not yet known. However the rate of growth in the market since 1997 as, for example, indicated by the results of the survey carried out by the Weather Risk Management Association (June 2001) implies increasing acceptance of such products.

Insurers and reinsurers have a long-standing and important involvement with weather risk but the developing 'alternative' risk marketplace now also incorporates providers of protection from traders, banks and exchanges, each of whom provides distinctly valuable products and services to manage weather risk.

1 *Ironically, pluvius literally means "sender of rain" and was an epithet of the Roman god Jupiter who was called upon to bring rain to the crops (see Panel 1).*
2 *For an accessible explanation of the legal requirements of insurance contracts, see C. O'Connell (2002).*

BIBLIOGRAPHY

O'Connell, C., 2002, "Legal Risks Mitigating Document Risk – Some Hard Lessons Learned", in: M. Lane, *Alternative Risk Strategies*, (London: Risk Books).

13

Weather Note Securitisation

Frank Caifa

Swiss Re[1]

A weather note securitisation is a blend of weather derivatives traded in the over-the-counter (OTC) weather market that create a new security with the cashflow features of bonds with embedded options. Because this new weather security is in a familiar bond form, it appeals to investors – institutions, large mutual funds and hedge funds – who might normally be averse to accepting weather risk in an option form but might be willing to accept weather contingent returns in a portfolio of weather bonds.

Introduction

This chapter reviews the early attempts at weather note securitisation and why they were only partly successful. The main portion of the chapter discusses the characteristics of a new securitisation and offers guidelines for success.

Securitisations to date

The securitisation market in general has grown extremely strong in the last decade. Issuers of the securities successfully combined individual debts in the form of home and commercial mortgages, automobile and credit card loans – to list only a few examples – into bond-like structures in a form that appealed to investors in the capital markets. As the weather market matures, there is much interest in launching and developing a strong market for tradable weather risk issues.

The weather market could serve the marketplace as the mortgage market did; the creation of the pass-through mortgage-backed security enabled the lenders of capital to homebuyers to securitise individual loans and to sell the package to investors through private placements and openly in the capital markets. This returned to the lenders the capital they had used for lending to homebuyers, and increased the lender's ability to finance new home loans. The creation of weather notes will enable the originators of weather assets to securitise individual assets and to sell the package to investors through private placements and to the capital markets. Weather risk is thereby transferred to those investors with an appetite for it, and allows the weather asset originators to issue even more weather assets.

A weather note is distinct from other classes of risk currently available in the financial markets including weather derivatives: its risk is based on weather variations that are not correlated with other investment instruments, it carries the credit rating of the issuer not the underlying contracts, and it can be engineered to guarantee principal or interest or some measure of both.

Both issuers and investors are motivated to see weather notes succeed. The risk bearer (the issuer) would be able to offload acquired weather risk to the capital markets, and, if desired, issue more assets. The risk taker (the investor) could diversify his/her portfolio with a non-correlating new risk class. Weather assets are attractive to managers of diversified portfolios because their value does not respond to the Dow, Nasdaq or any other market segment performance, thus they will not be affected when the Federal Reserve raises or lower interest rates.

WEATHER NOTE SECURITISATION

From a seller's perspective, using the capital markets to transfer weather risk can be more efficient than using the current OTC market as it can be limiting if a risk-bearing entity wanted to deploy large blocks of weather risk; currently the capacity for OTC weather deals tends to dry up above about US$100 million. For example, it is unlikely that the current OTC marketplace would be able to absorb a large weather risk at once; costs would be large, and the issue may have to be placed in several yearly tranches, increasing issuance costs. Pricing this risk into the open market generally yields higher prices with each prospective risk taker pricing the risk separately. Some of the counterparties taking a share of the risk may not have the credit rating needed to make the risk bearer comfortable to engage in a multi-year contract with them which further limits its ability to place the risk all at once.

Deploying this risk in to the capital markets, however, can solve three concerns simultaneously. First, the capital markets can digest more than US$100 million in weather risk much more easily than any one entity currently in the weather market because of access to large investors. Second, since the issue will be sold all at once, the pricing will be determined in advance; the seller will know the entire cost of the risk transfer and not be forced to make projections or estimates of future pricing, which would increase opportunity costs. Lastly, a weather securitisation note would likely be issued for at least a three- to five-year term. This last aspect frees the ultimate risk taker from having to recreate this structure each year.

Another benefit to the seller of risk in the capital markets is the aggregation of capital that will be paid up as soon as the note is sold. Investors will be looking for an AA-rated (or better) entity to issue this note, while if this strong counterparty were to use the current OTC market to place this risk, it would be exposed to predominantly lower than A-rated credit. As a result a credit dilemma can be avoided and both parties win – the investing public has a strong note counterparty and the issuer has the capital in hand from the capital markets.

Another important reason to offer the risk to the capital markets is the fact that having a strong counterparty such as a very highly rated AA or AAA entity issuing the note significantly reduces the default risk of the underlying counterparty and the concern to the investor. Considering today's market psychology with all of the large scale bankruptcies occurring – especially Enron's – having a strong issuer is a necessity. Thus, for capital market investors who are looking for a non-correlated product to invest in, the weather note is likely to be an absolute benefit to them.

A weather note could further expand the scope of investors who would be willing to look at the existing OTC weather market. This would be a tremendous benefit to the overall growth of this industry. Packaging weather risk in a form that the capital markets currently understand and use on a regular basis should greatly ease the concerns about participating and looking at weather risk. From a weather market perspective, it allows for a different way to shed some of the risk that is being absorbed every day in the regular, OTC weather market.

A properly structured weather note would not only permit investors and speculators to participate in the weather market but should do so in a way with which they are familiar and comfortable. Two of the main requirements would be simplicity of structure and transparency. This would enable investors, rating agencies, regulators and anyone else looking into the weather market to understand and analyse the market. In this regard, the weather note should be very similar to any other note that is used to securitise debt or mortgage obligations in the present capital markets.

Weather bonds would foster the development of today's weather market in various ways. In broadening the scope of participants in this market, securitisations will begin the transformation towards greater transparency. Today's investors are demanding more disclosure than ever before. The early days of the weather market were shrouded in secrecy and unsubstantiated claims of whose model or data set was better. Better models and more complete data sets were the route to competitive advantage. These walls, including complexity, must come down if this market is to thrive. New consultants and new software have assisted in allowing newer entrants in this market to compete on a level playing field with the established players. As this trend continues, the weather industry should become a more mainstream market.

Further, weather bonds would clear a path for new end-user risk to be brought to the weather market, which would, in turn, increase liquidity in the secondary market – the more issues there are and the broader the variety of weather risk available to the investment community, the stronger the market will be. Last but not least, weather bonds provide a means for the existing weather market participants to shed some of their risk positions from the weather derivatives OTC market to the capital market. In fact, this possibility was at least part of the motivation to structure the two weather bonds discussed in the next section.

THE FIRST SECURITISATIONS

Innovators have to work harder than those who follow in their footsteps; the leader in a cross-country ski race works to "cut the trail" in the snow while those who choose to follow the leader use far less energy getting to the finish line. This phenomenon is mirrored in capital markets innovation: the first two attempts at weather securitisation met with limited success as they cut the trail for issuers who might follow. Innovation is an adventure, and while these first attempts carried some interest in the marketplace, they encountered difficulties as well.

Entergy-Koch and Enron each attempted one of the first two weather securitisations. Entergy-Koch's issued its "Kelvin" note in late 1999/early 2000, but met limited success. Enron's "weather-indexed return securities" (WINRS) note was marketed but never issued.[2]

When marketing a new product it is helpful to keep in mind that the investing public is generally looking for one of two things – a financial benefit or a compelling reason to own or use the product or service. Kelvin was the first weather note issued to the market and participants were eager to analyse its structure and assess its merits, but apparently – because of its complexity – were unable to find clear evidence of the financial benefit they expected to see in this novel product.

The Kelvin notes consisted of three tranches, each having a different level of principal guarantee and interest rate of return. The make-up of the note included some heating degree-day (HDD) indexes, some cooling degree-day (CDD) indexes, and some average temperature indexes measured at 28 US cities. Further, some of the HDD and CDD structures were based around a temperature level other than the standard 65°F threshold. Since this was the first weather note to hit the market, having such a complex structure made it very difficult to analyse the risk or understand the issuer's motivation.

Moreover, the complexity of the note made the accurate pricing and the assignment of an appropriate risk level for the various tranches and localities a challenge from an investor's perspective. Of the anticipated principal amount of the note to be placed, approximately 25% was actually placed.[3] Considering the difficulties in assessing the risks and merits of the Kelvin notes, the limited interest at that time was not surprising.

Enron's WINRS notes had a simpler structure than the Kelvin notes: Enron limited the risk to only 10 cities, and five derivatives in each of the two seasons – 4 HDD calls, 1 HDD put, 2 CDD calls, and 3 CDD puts. This made it easier to judge the pricing of this structure. As a result, preliminary market feedback indicated that investors were going to purchase about 70–85% of this risk.

It would seem that Enron wanted to sell weather transactions in the 10 cities to their customers at a mark-up, then sell it back wholesale to the investment community or capital markets via the WINRS notes. This would have had to be done at a coupon rate that would have guaranteed them a profit on the spread and eliminated most of the risk. If structured correctly, any losses on the options would

WEATHER NOTE SECURITISATION

have been absorbed by the bond buyers via loss of principal. Hence, they would have had to repay a reduced portion of principal to holders of the note if they realised losses on the OTC derivatives. Conversely, if they made profits in the OTC market, they would have no problem paying back the principal including the coupon rate to the note holders. Presumably, the profit on the bond would have exceeded the mark-up they would have made by trading only in the OTC market.

Timing is the key factor in financial investment and, unfortunately for Entergy-Koch and Enron, late 1999 was not the most advantageous time for bond issuances in general, particularly for new bond offerings such as these. Pricing also played a role, as investors were demanding rates that were uneconomic from an issuer's standpoint.

The time these notes were being offered to the marketplace was also the hey-day of the tech-stock boom and the returns investors were receiving in the tech-stock arena far surpassed (for a while) anything the weather market would have been able to offer, and weather trades were less liquid. Issuers, as the investing public has now come to learn, focused their efforts on exploiting new markets and constantly pushing pricing in their favour, whether it be loans from banks or investment banks, or offerings to the public.

PANEL 1

CURRENT SECURITISATION ACTIVITY

Since 1999, there have been no new offerings of weather notes. There are at least two reasons for this:

1. The limited success of the first two notes caused the weather market to wonder whether there really was an appetite for weather risk to be absorbed by the capital markets.
2. The focus of the issuers of weather assets so far has been to explore and exploit price inefficiencies in the OTC weather derivative markets, therefore they have not been willing to pay the premium needed to raise the interest rate paid to potential investors.

This is changing. The OTC weather market as it stands today has grown. Not only has it grown in terms of total dollar value and notional amounts of contracts traded, but in terms of the number of participants as well.[4] This points towards an increased likelihood that end-users might be more willing to transfer their weather risks to others if the pricing and structuring are right. With the increase in publicity that this weather market has now obtained, there should be a good platform to launch the next note – possibly even in the near future.

The investing community is becoming increasingly aware that the impact of weather on certain companies' financial results, and that the volatility of those financial results can be mitigated by using some of the tools that the weather market now has to offer. Large utilities and other energy-related entities that currently do not trade regularly in the OTC market are looking at this market to hedge large blocks of their exposure and eliminate some of the volatility in their cashflows. Non-energy-related end-users such as agricultural companies, construction companies and companies in the various fields relating to entertainment are taking a serious look at this market now (see for example, Chapters 16 and 17). This all points to a wider base of potential investors for the next weather note securitisation. As this base broadens and people understand more about the weather market, the likelihood of a successful placement increases.

RISK BOOKS

Initially, these notes were viewed as similar to catastrophe bonds and pricing was supposed to be in line with BB or BBB offerings. The weather note issuers failed to realise that in issuing a new risk class, especially one that few people understood, they would have to offer a rate premium to get this type of note placed. Investors, however, required a higher interest rate than the issuer was willing to offer.

Specifically, the WINRS coupon rate kept increasing to meet a rate that the market felt was reasonable, but never reached it. The pricing advantage or the arbitrage Enron thought it could make evaporated entirely. In the end, it may be that Enron felt the pricing economics from their perspective did not warrant issuing this note and hence it was never issued.

Characteristics of a successful note
UNCOMPLICATED AND TRANSPARENT STRUCTURE

As stated above, without question, simplicity is the key to a successful issuance. One characteristic required for success in securitising a weather note is to limit the structure of the note to consider no more than 8–10 cities, as was done in the Enron note, in contrast to the 28-city Kelvin note. Making things relatively easier to understand would certainly aid in the investors' knowledge of and appetite for the weather note because there would be less to understand, review and digest. A weather note will have the inherent risks of the underlying weather market as its basis and core, so a structure having only 8–10 cities (possibly even less for the first note) should lessen the necessity for investors to spend a lot of their time on review and data mining, and aid their understanding of what it is they are buying and where their potential risks and rewards lie. If reference sites were equally spread throughout the country in a handful of cities, investors would not need to consult meteorologists or doctors of climatology to truly appreciate the risks involved.

As demonstrated in Table 1, the degree-day correlation between some cities is highly positive and between others it is highly negative. It is therefore possible to reap the benefits of good diversification of cities by constructing a portfolio of offsetting correlations. All of the cities remain for the most part either positively or negatively correlated over time, regardless of the timeframe selected, and the correlation differs only slightly for each period indicated. For example, the cities of Sacramento and San Diego in California are negatively correlated to Dallas, Texas for the CDD season regardless of whether the data is for the 10, 25 or 40 years presented. If these three cities were part of a securitisation and they were all either CDD puts or CDD calls, an investor would gain a diversification benefit and would not expect to suffer losses in all three cities in the same season. Also, HDDs in some cities are negatively correlated with CDDs in other cities leading to intra-season diversification benefits and a reduction in weather risk for this type of note.

Also, in order to increase price transparency, the issuer could pick cities that have the cleanest weather data and a station history with as little change as possible in both the location and the measurement of the weather data. Further, cities should be selected where the current OTC weather market trades frequently. Liquidity in these cities would provide the opportunity to trade out of positions if necessary. This would address one of the things that is preventing non-energy end-users – and, for that matter, even some large energy end-users – from participating in the current weather market: fear of uncertainty or the fear of getting into the game and being at a disadvantage, as new entrants are unlikely to know as much as the established players.

Being open and free with information would go a long way towards eradicating or eliminating some of these fears. One way for the issuer to do this would be to simply streamline a weather note with only a handful of cities where all the data can easily be spelled out in the offering. The offerer should select major metropolitan sites from around the country, and provide data that can easily be tracked by those participating. Referring to the cities listed in the table as a representative sample, one

Table 1. Correlation matrix – raw data

10-Years RAW	1 DFW-CDD	2 Miami-CDD	3 Sacramento-CDD	4 San Diego-CDD	5 Wash., DC-CDD	6 Atlanta-HDD	7 Boston-HDD	8 Pittsburgh-HDD	9 Seattle-HDD	10 Tucson-HDD
DFW-CDD	1.0000									
MIAMI-CDD	0.6319	1.0000								
SACREMENTO-CDD	(0.4200)	(0.6255)	1.0000							
SAN DIEGO-CDD	(0.3343)	(0.5039)	0.4664	1.0000						
WASH DC-CDD	0.3214	0.6245	(0.5245)	(0.8654)	1.0000					
ATLANTA-HDD	0.6873	0.1554	0.0776	0.0521	(0.0815)	1.0000				
BOSTON-HDD	0.0168	0.5673	0.4773	0.6425	(0.5900)	0.5035	1.0000			
PITTSBURGH-HDD	(0.1383)	(0.6400)	0.5901	0.4555	(0.5037)	0.4513	0.9185	1.0000		
SEATTLE-HDD	(0.1948)	0.0106	(0.0840)	(0.5368)	0.6701	(0.2044)	(0.1877)	(0.0496)	1.0000	
TUCSON-HDD	0.6239	0.7416	(0.5476)	(0.1805)	0.3299	0.1466	(0.4111)	(0.6636)	(0.0778)	1.0000

25-Years RAW	1 DFW-CDD	2 Miami-CDD	3 Sacramento-CDD	4 San Diego-CDD	5 Wash., DC-CDD	6 Atlanta-HDD	7 Boston-HDD	8 Pittsburgh-HDD	9 Seattle-HDD	10 Tucson-HDD
DFW-CDD	1.0000									
MIAMI-CDD	0.2520	1.0000								
SACREMENTO-CDD	(0.3883)	(0.3202)	1.0000							
SAN DIEGO-CDD	(0.4117)	(0.4350)	0.3311	1.0000						
WASH DC-CDD	0.4257	0.2647	(0.1718)	(0.6205)	1.0000					
ATLANTA-HDD	0.6336	0.0421	0.0632	(0.0088)	0.0438	1.0000				
BOSTON-HDD	0.2594	(0.1922)	0.0310	0.3152	(0.4399)	0.5563	1.0000			
PITTSBURGH-HDD	0.3212	(0.1764)	0.0424	0.1136	(0.3244)	0.5862	0.9097	1.0000		
SEATTLE-HDD	(0.1568)	(0.0410)	(0.2200)	(0.3927)	0.1595	(0.3274)	(0.2602)	(0.1620)	1.0000	
TUCSON-HDD	0.1297	0.4145	0.0863	(0.2514)	0.0455	0.1018	(0.2292)	(0.1483)	0.1410	1.0000

40-Years RAW	1 DFW-CDD	2 Miami-CDD	3 Sacramento-CDD	4 San Diego-CDD	5 Wash., DC-CDD	6 Atlanta-HDD	7 Boston-HDD	8 Pittsburgh-HDD	9 Seattle-HDD	10 Tucson-HDD
DFW-CDD	1.0000									
MIAMI-CDD	0.1353	1.0000								
SACREMENTO-CDD	(0.3818)	(0.1946)	1.0000							
SAN DIEGO-CDD	(0.3149)	(0.3404)	0.3037	1.0000						
WASH DC-CDD	0.1041	0.2391	(0.0891)	(0.4964)	1.0000					
ATLANTA-HDD	0.4628	0.0883	0.0959	0.0132	0.1317	1.0000				
BOSTON-HDD	0.2141	(0.1601)	(0.0184)	0.3163	(0.2558)	0.4426	1.0000			
PITTSBURGH-HDD	0.3099	(0.0892)	0.0174	0.1806	(0.2506)	0.5610	0.8131	1.0000		
SEATTLE-HDD	(0.0184)	0.0199	(0.1996)	(0.3577)	(0.0213)	(0.2545)	(0.2122)	(0.2239)	1.0000	
TUCSON-HDD	0.0034	0.3293	0.0633	(0.3662)	0.1983	0.1778	(0.2625)	(0.2193)	0.2289	1.000

can see not only the correlation benefits but also that all of the sites are major metropolitan locations. Also from the issuer's perspective, by picking the major metropolitan sites that have clean data and a good history, it will be easier for those cities to be represented accurately when the data is presented in the offering document.

A typical investor is likely to have more appetite for a diversified weather risk than for weather at a single site during a short period of time – as is usually the case with a weather derivative. Therefore, a weather note with an appealing structure should allow the investor to participate in the weather risk of non-correlating sites and should allow both buyers and sellers of the note to participate in both winter (typically November–March) and summer (typically May–September) seasons.

Additionally, including both seasons in a note allows for semi-annual cashflows within the structure. Principal values can be re-determined and interest payments can be made around these dates. This should assist prospective institutional investor and hedge fund portfolio managers concerned with the present value of their cashflows and fluctuating principal values over time to manage this new investment and compare it to their other holdings.

Further, the note could also be split evenly between puts and calls so that the temperature swinging in one particular direction or another – in either winter or summer – does not advantage or disadvantage either of the parties involved. In other words, offering only one-sided weather risk by including all puts or all calls in a note, an investor could be adversely affected provided those seasons turned out to be, for example, particularly warm in summer or particularly warm in winter. By structuring the note in such a way that the cities are both equally split by summer and winter and relatively spread out over the geographic regions of the US, one can offer a buyer of this note the comfort that some of the OTC weather market will be going in their favour at all times, and that the issuer is not just looking to take advantage of the investor.

CREDIT OF THE ISSUER
It is imperative that the issuer of this type of product at this point in time be a company that has a strong balance sheet and also possesses a strong reputation in the current weather market. As mentioned earlier, the two companies that have previously offered notes are Entergy-Koch and Enron. Entergy-Koch is still a thriving player in this market, and is currently equivalent to an A-rated entity. Enron is – at the time of going to press – in the midst of the biggest bankruptcy filing in US history to date, which is likely to have had a negative impact on holders of their notes, which were due in 2004.

Not only should the next issuer possess a strong balance sheet – AA- or AAA-rated at least – but also the company issuing the note should be a company that the weather industry regards as being long-established in this market.

Prospective buyers would be likely to ask a company proposing a weather securitisation:

1. Why should I buy this note from you?
2. What is your business rationale for the offering?

An established company with a good reputation can legitimately maintain that it is entering into this area to broaden the market, and that it is going to be around when the note matures. Credit is a big issue in this marketplace, especially at this time in light of the large bankruptcies of 2001–2. In addition to the Enron bankruptcy, a host of energy-related companies have been downgraded by the major credit rating agencies. As a result, credit quality of a counterparty is at a premium, not only with respect to weather securitisation, but also in general. The marketplace is looking for, and more likely than not will continue to look for, an instrument that provides non-

WEATHER NOTE SECURITISATION

correlated risk, and the assurance that the counterparty will be prepared to perform (that is, pay) at the maturity of the note.

Solid credit should remove yet another barrier of entry for the institutions and prospective buyers to come in off of the sidelines.

TRANCHES, GUARANTEES AND CASHFLOW

A successful weather note securitisation should have a maturity of at least three to five years, as this is long enough for investors to receive cashflows over a few years and not have to rollover their money each and every year. With a three- or a five-year note, there would be three HDD and three CDD periods or five HDD and five CDD periods, respectively. Offering both HDD and CDD periods within a weather note should further help reduce the volatility and risk associated with such a note. Offering both a winter and summer season on the note securitisation enables its investors or buyers to participate and track the deal through successive winter and summer seasons. By including a relatively equal amount of puts and calls on negatively correlated sites in both winter and summer, the volatility is further reduced by not subjecting the buyer to any one particular type of extreme season.

The next weather note should have at least two layers or tranches to appeal to investors with different appetites for cashflow and risk. In this respect, this weather note would be similar to what was done with both the Kelvin notes and what was projected to be done with the WINRS notes.

In a two-tranche note, for example, the return of principal could be guaranteed in the most secure tranche although the initial rate of interest could be at risk of being lowered, and the rate of interest could be guaranteed in the least secure tranche although the principal could be at risk. The interest rate offered in the most secure layer might be 500 basis points (bp) above treasuries upon note issuance, while in a lower layer, the rate might be 800–900bp above treasuries.

The principal layers will be at risk based upon the resulting accumulation of HDD and CDD totals per season per year. In the event that seasons were adverse to the performance of the weather options underlying the note, the tranches in the note would not perform at their peak and maximum potential cashflows would not be realised – some cashflow erosion from the maximum is possible and probably should be expected. This is why the initial interest rates are offered higher than treasuries. It is possible; of course, that the options perform well and there is no erosion of potential cashflows in the note.

The investor could suffer losses in any one season: the losses are associated with each season and only for that particular year and/or season. The investor in the most secure tranche will always receive full principal but may only receive 200bp above treasuries or may just get the principal back if the weather is that severely adverse to the underlying options.

The investor in the least secure layer receives a higher interest rate than in the other tranches because it is here that the principal is eroded first. This interest would be calculated over the three- to five-year term based on the initial principal in the tranche regardless of whether or not the principal is eroded. For example, a guaranteed interest layer of US$50 million could be structured to be at a principal risk of a maximum of US$10 million per year over the next five years. The principal exposure could then be further limited to US$5 million for winter and US$5 million for summer. If an extremely adverse winter occurred, the investor might lose up to US$5 million of its principal in that season, but would be guaranteed the underlying or notional interest rate based on that US$50 million regardless of the principal erosion in that tranche. If the principal declined US$5 million, the investor would therefore only be entitled to repayment of the US$45 million.

Conclusion

Weather note securitisation is an efficient way to bring a non-correlated investment to institutional investors. It is also beneficial for both the buyer and seller of regular weather risk that currently operate in the weather market.

The creation of the pass-through mortgage-backed security enabled home lenders to securitise individual loans and to sell the package to investors through private placements and the capital markets. This returned to them the capital they had used for lending to homeowners and buyers, and increased their ability to finance new home loans. The creation of weather notes would enable the originators of weather assets to securitise individual assets and to sell the package to investors through private placements and the capital markets.

Innovation was shown in the early attempts at weather securitisation and lessons were learned. Providing a simple structure while taking advantage of diversity – in geography, time and exposure to above and/or below average weather conditions, and positive and negative correlation – will provide both buyer and seller of a weather note an equal chance at success. The cross-country ski tracks have been laid, so the next issuer will find it easier to cross the finish line.

Learning from these experiences, this chapter suggests that a successful weather note should have the following characteristics:

- the note structure must be simple to understand;
- the note structure should contain no more than 8–10 cities;
- the note should contain both winter and summer seasons over a three- to five-year period;
- the note should be geographically diverse;
- the note should contain both puts and calls;
- the issuer should have an established presence in weather market; and
- the issuer should be a AA-rated entity or higher.

These positive aspects should reduce the barriers of entry that keep companies and investors on the weather sidelines, and should help to draw in investors who have never seriously considered the weather market before. Focusing on these factors will help to isolate and market a risk that investors are looking to take – weather risk – a risk that is volatile enough, without adding additional uncertainties.

In simplifying the risk so that it can be more easily understood, the issuer should be able to reduce the time investors typically spend on analysing structures. If the issuer can also demonstrate that credit quality is solid and that the deal offers prospective investors an equal footing on either side of the weather as the note matures and is not simply an arbitrage play in the marketplace, then there will be a high probability of a successful launch.

Comparing this type of securitisation with catastrophe (cat) bonds, investors either win or lose based upon whether significant large-scale events occur in a cat bond. But the weather happens every day, so a weather note's market value should not be exposed to extreme movements from one day to the next. As a result, a weather bond's underlying exposure can be tracked regularly and marked like a regular equity option with the resulting time decay over the option period.

The better the credit quality of the issuer, the better the chances of a successful placement. And as the securitisation of home loans actually benefited the ultimate homebuyers by allowing for more mortgage lending, so too can the successful securitisation of weather risk ultimately benefit the weather derivative end-user, the investor and others in the market through enhanced liquidity. Considering that over US$2 trillion of the US economy's GDP is affected by weather, this enhanced liquidity is a necessity in order for this market to grow.[5]

WEATHER NOTE SECURITISATION

1 *This chapter represents views of the author and not necessarily those of the members of the Swiss Re Group.*

2 *For further information on these two notes, see http://www.casact.org/coneduc/reinsure/2000/handouts/lane.pdf and http://www.riskwaters.com/riskawards/riskawards-enron.htm. (References supplied by the editor.)*

3 *See http://www.kochind.com/newsroom/news_detail.asp?ID=228. (Editor's note.)*

4 *According to the PricewaterhouseCoopers survey commissioned by the Weather Risk Management Association in 2001, from October 1997 to March 2001, 4,800 contracts traded with a notional amount of US$7.5 billion. This timeframe represents the history to date of the weather market.*

The most recent survey released in June 2002, indicates that 3,937 contracts were written for the period ended March 31, 2002 compared to 2,759 for the prior period, a 43% increase. The survey also indicates contracts were written with notional amounts of US$4.3 billion and US$2.5 billion respectively for the periods that ended March 31, 2002 and April 14, 2001. This equates to a 72% increase.

5 *Dutton (2002).*

BIBLIOGRAPHY

Dutton, J. A., 2002, Weather Risk 2002 Conference Presentation, January 24.

14

Managing a Portfolio of Weather Derivatives

Lixin Zeng; Kevin D. Perry[1]

Willis Re Inc.; the University of Utah

The risk and return associated with weather derivatives depends solely on the variation of weather conditions. This obvious yet extremely important fact makes weather derivatives an attractive asset class for investors and speculators because their returns are generally independent of the state of the economy and the fickle shifts of the collective emotions of capital market investors. This is especially true since the globalisation of the economy and modern communication technologies have virtually linked the world's financial markets into an integrated playing field. This globalisation has dramatically reduced the diversification benefits of foreign investments such as emerging market stocks and debts because this asset class is no longer independent of the capital market in the developed nations.

By introducing weather derivatives into an investment portfolio consisting of traditional capital market instruments, the risk of the combined portfolio can be substantially reduced without necessarily reducing the return. While any asset class that generates a return comparable to or better than the capital market portfolio and is independent of the global economy can achieve this goal, the authors believe that weather derivatives are the best candidate. The reason for this optimism is that weather derivatives have the greatest potential among all available independent asset classes to become a large and liquid market.

Introduction

The primary focus of this chapter is how to best manage a portfolio composed of weather derivatives. This task consists of:

1. quantifying the risk/return profile of a weather derivative portfolio; and
2. constructing a portfolio with an optimal risk/return profile.

Although these tasks are no different than managing a portfolio of traditional capital market instruments, special attention must be paid due to the unique nature of weather derivatives. In fact, addressing the special issues related to weather derivatives and developing the special techniques required to tackle these issues are the main themes of this chapter.

Since the concept of risk/return lies at the core of the discussions here, we first define and quantify the risk/return characteristics of a portfolio of weather derivatives.[2] We then discuss the advantages and disadvantages of commonly used risk and return measures. Although a single universally applicable parameter to quantify risk and return does not exist, it is demonstrated that the probability distribution function (PDF) of the portfolio return always forms the foundation of risk/return analyses. Hence, we next introduce approaches for calculating the PDF of

MANAGING A PORTFOLIO OF WEATHER DERIVATIVES

the portfolio return. Lastly, we introduce the framework of constructing a portfolio with optimal risk/return characteristics demanded by the investor.

The risk/return profile of a portfolio

The goal of investors when buying and selling weather derivatives is to generate a positive return.[3] The return, denoted X, is defined as the relative change of the portfolio value during a period when the investor holds long and/or short positions of these contracts (the holding period):

$$X = (V_1 - V_0)/V_0 \qquad (1)$$

where V_0 and V_1 are the values of the portfolio at the beginning and the end of the holding period (eg, a year), respectively. The portfolio value is equal to the sum of the values of the individual weather derivative contracts, cash, and other assets within the portfolio. Unless the portfolio consists of only risk-free assets, it is impossible to predict the value of V_1 or X precisely, prior to the end of the period. This uncertainty leads to the *risk* in the return. Usually, the investor must rely on probabilistic methods to estimate the risk and return potential. Its general concept is briefly reviewed in the following few paragraphs.

THE RANDOM NATURE OF RETURN

The probability density function (PDF) and the cumulative distribution function (CDF) are the most common functions used to describe the randomness of the return mathematically. The PDF, which is denoted by $f_X(x)$ for a real continuous random variable X, is defined as a function such that

$$f_x(x) \geq 0, \ \forall x \in R_x$$

$$\int_{x \in R_x} f_x(x) dx = 1$$

where R_x is the valid range of X. Using this notation, the probability that the value of X is between x_1 and x_2 ($x_1 \in R_x$, $x_2 \in R_x$) is given by:

$$P(x_1 \leq X \leq x_2) = \int_{x_1}^{x_2} f_x(x) dx \qquad (2)$$

In addition, the CDF, which is denoted by $F_X(x)$, is defined as:

$$F_X(x) = \int_{\tau \leq x} f_X(\tau) d\tau \qquad (3)$$

A sample PDF and CDF of a portfolio return are shown in Figure 1. The shaded area in the PDF graph represents the probability that the return is less than –20% (ie, a 20% loss). This corresponds to an X value of –0.2. This scenario is also represented by point A on the CDF graph, which shows that the probability of a 20% loss or worse is approximately 10%. This, by definition, is also the area of shaded portion of the PDF graph.

If a random variable X is not continuous, but rather one of a set of discrete values, then the analogous PDF and CDF equations become:

$$P(X = x_i) = f_X(x_i) \qquad (4)$$

$$F_X(x_i) = \sum_{x_j \leq x_i} f_X(x_j) \qquad (5)$$

where x_i are the discrete values that X can take, and

1. A sample PDF (left) and CDF (right)

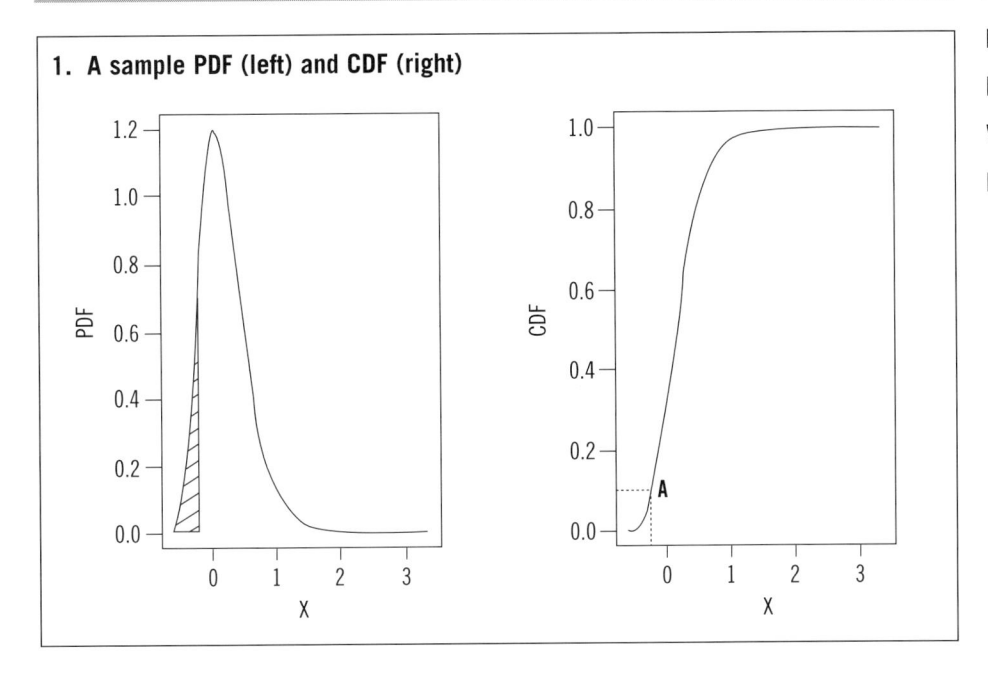

$$f_X(x_i) \geq 0, \forall x_i \in R_x$$

$$\sum_i f_x(x_i) = 1$$

There may be, in theory, a finite or infinite number of these possible discrete values (ie, x_i). Although the value of most weather indexes and the portfolio return are theoretically continuous, they are frequently approximated by discrete values in the numerical algorithms that we introduce in the remainder of this chapter.

The calculation of the PDF and CDF of the portfolio return will be covered in the next section. In the remainder of this section, we will focus on how to extract risk/return information from these probability functions.

THE NECESSITY OF QUANTIFYING RISK AND RETURN WITH ONE NUMBER
Mathematically, the PDF or CDF of the portfolio return provides a complete accounting of the random nature of the portfolio return provided that the functions are calculated accurately using high quality input data. However, in order to make prudent investment decisions, investors generally like to condense the information contained in the probability functions into *one* real number. For example, two investment strategies (A and B) lead to two different portfolios with PDFs of $f_{XA}(x)$ and $f_{XB}(x)$, respectively (Figure 2). The expected return of Portfolio B is greater than that of A, but the return of A is less likely to be negative than that of B. Nevertheless, this information does not provide the investor with a direct solution to which is the better strategy.

In fact, the answer depends on the investor's objective. We consider two extreme cases. First, if the objective is to achieve the maximum possible return without considering any uncertainty (ie, risk), then strategy B will be better. To the contrary, if the primary objective is to minimise downside risk at the expense of a lower overall return, then strategy A will be better. In reality, investors are usually willing to accept a greater degree of uncertainty in exchange for a higher overall return. Traditionally, a numerical value called the "return/risk ratio" is often used in an attempt to measure the balance between potentially positive returns and the associated uncertainty. The return/risk ratio, which is denoted by r, is defined as:

$$r = \frac{r_t}{r_k} \quad (6)$$

2. PDFs for portfolios A and B

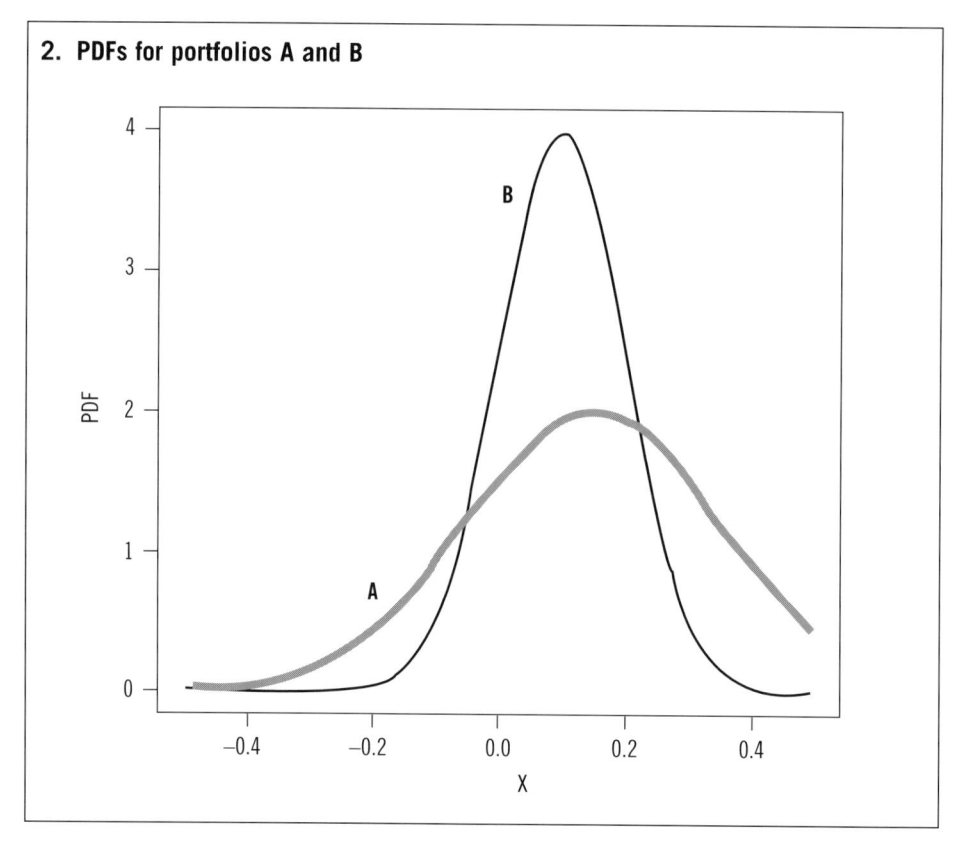

where r_t and r_k are measures of potential return and risk, respectively. The parameter r helps investors differentiate portfolios with various return potentials and uncertainties. It is commonly believed that a greater value of r implies that the portfolio has a better "risk/return profile". In reality, investors compare the r values of different portfolios or strategies in conjunction with certain other constraints.

In the example above, we shall assume that the r values for A and B are somehow arrived at to be 0.7 and 0.6, respectively. This type of one-number analysis makes it possible for an investor to make a better-informed decision. (In this case, strategy A is actually better than strategy B!) The next step in the process is to show how the values of r_t and r_k can be determined from the PDFs or CDFs.

QUANTIFICATION OF RETURN
There is generally a well-established consensus among investors that the expected return of a portfolio should be used as the measure of return. If X represents the portfolio return defined in Equation (1), then we can define r_t as:

$$r_t = EX = \begin{cases} \int_{x \in R_x} x f_X(x) dx, & \text{for continuous } X \\ \sum_{x \in R_x} x_i f_X(X = x_i), & \text{for discrete } X \end{cases} \quad (7)$$

where E denotes the operator that calculates the expected value of a random variable. EX convolves the size of the return and the probability of that return over the entire range of possible values. It can also be viewed as the average return over the holding period if the holding period were to be repeated for an infinite number of times. The EX of the sample portfolios A and B (as derived from Figure 2) are 10% and 15%, respectively, indicating that the long-term return of B is greater than that of A.

QUANTIFICATION OF RISK OR UNCERTAINTY

Unfortunately, there is no consensus as to how the risk or uncertainty associated with a portfolio should be measured. In fact, quantifying risk is one of the most intriguing challenges in financial analysis and actuarial research (D'Arcy, 1999).

The most widely adopted measures of risk are standard deviation (SD) and value-at-risk (VAR). We also introduce a less commonly known measure, namely the tail value-at-risk (TVAR). The advantages and disadvantages to an investor who uses these measures are discussed and summarised below.

Standard deviation

As the most commonly used risk measure by investors, SD measures the degree of variability of a random variable, X, around its expected value, EX. It is defined as:

$$r_k = SD = \sqrt{E(X - EX)^2} = \sqrt{EX^2 - (EX)^2} \tag{8}$$

Letting X represent the portfolio return defined in Equation (1), the most frequently used return/risk ratio based on SD is defined as:

$$r = \frac{EX - r_f}{SD} \tag{9}$$

where r_f is the risk-free rate of return (eg, the return of treasury bills or bank certificates of deposit). The square of r is known as the Sharpe ratio (Treynor and Black, 1973).

The SD for the sample portfolios A and B, shown in Figure 2, are 10% and 20%, respectively. Assuming an r_f of 3%, we arrive at the values of r for A and B (0.7 and 0.6, respectively) based on Equation (9). This indicates that A has a better risk/return profile than B. Although it appears that the problem of differentiating portfolios with various return and risk characteristics has been solved satisfactorily, several problems remain with this approach. The most detrimental one is that the measure defined by Equation (9) can sometimes run into the paradox illustrated by sample portfolios C and D below. Figure 3 shows the PDF of the returns of these portfolios.

Apparently, the return of portfolio D (denoted R_D) demonstrates a greater expected value and a greater degree of uncertainty than C (denoted R_C). However, even the lower end of the range of R_D tends to be greater than the higher end of R_C (ie,

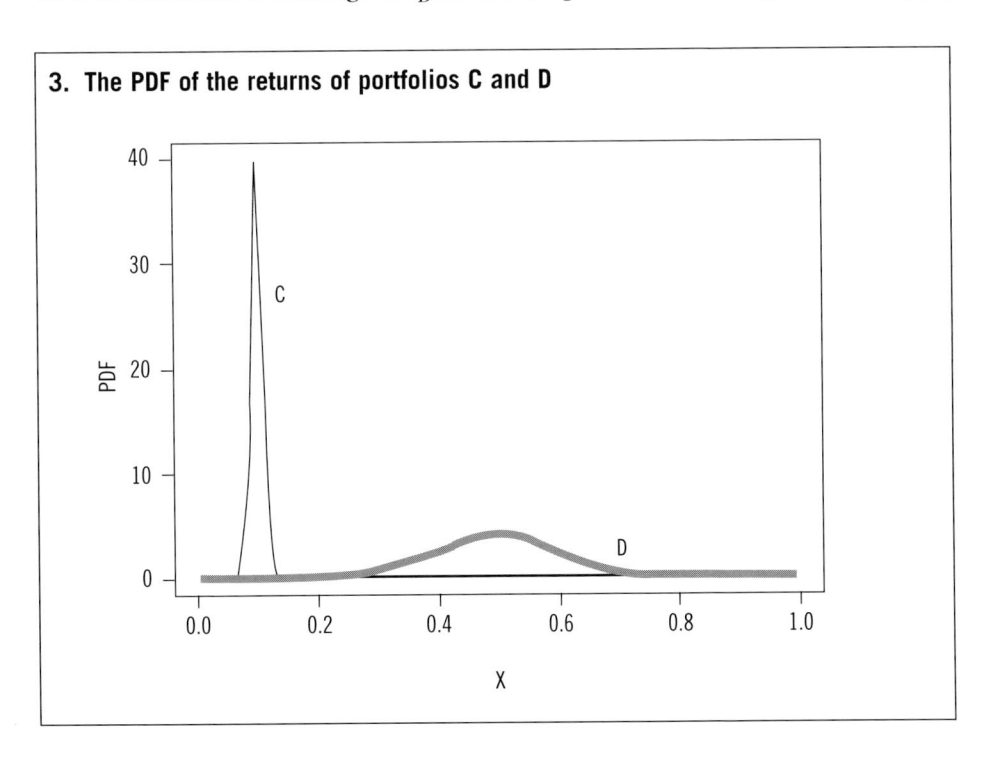

3. The PDF of the returns of portfolios C and D

portfolio D almost always delivers a higher return than C no matter how uncertain the return is). Hence, intuitively, portfolio D is a better choice and a reasonable risk measure must indicate so. Nevertheless, Equation (9) suggests that portfolio D, because of the larger standard deviation, has a lower return/risk ratio than C (Table 1). In the risk theory literature (eg, Artzner et al., 1999), this paradox is the reason why SD is not a coherent measure of risk.

Table 1. Summary statistics for sample portfolios C and D

	C	D
EX	10%	50%
SD	1%	10%
r_f	3%	3%
r	7.0	4.7

Value-at-risk

If X is the portfolio return defined in Equation (1) then VAR (denoted v_α) is defined according to the following equations:

$$F_X(x_\alpha) = 1 - \alpha$$
$$v_\alpha = -v_0 x_\alpha \qquad (10)$$

where α is known as the confidence level, v_0 is the initial portfolio value, and x_α is the return. When α is sufficiently large, the return x_α is usually negative. Thus v_α is a measure of the size of a potential loss (ie, negative return).[4] Equation (10) indicates that there is a probability of $1-\alpha$ that the investor will lose v_α or more. This is equivalent to the statement that there is a probability of α that the investor will not lose v_α or more, which is why α is called the "confidence level". For example, point A in Figure 1 indicates that there is a 90% probability that the investor will not lose 20% of the initial value or more. In this case, the VAR with a 90% confidence level is equal to the initial value, v_0, multiplied by 20%.

An attractive property of VAR lies in its direct relationship with the capital at risk. For example, if it is required that the probability of default or bankruptcy should remain 1% or lower over the holding period, then the amount of capital that the investor must hold will be at least equal to the VAR with a 99% confidence level for the same period. Hence, the return/risk ratio defined by Equation (11) effectively measures the portfolio return on the amount of capital at risk.

$$r = \frac{(EX - r_f)v_0}{v_\alpha} \qquad (11)$$

This measure is equivalent to the "risk-based return" or the "risk-adjusted return on capital" (RAROC), which are widely used by insurance and reinsurance companies to quantify the risk-adjusted return (eg, Zeng, 2001 and Nakada et al., 1999). The confidence level is usually chosen by the investor based on his/her risk tolerance and other requirements (eg, solvency requirement imposed by counterparties, rating agencies and/or regulators).

However, there still exists a paradox that hinders the universal applicability of VAR as the measure of risk. For example, assume that the return/risk ratios, defined by Equation (11), are positive and the same for two portfolios E and F. This scenario can be represented by the following equations:

$$r = \frac{a^{(E)}}{v_\alpha^{(E)}} = \frac{a^{(F)}}{v_\alpha^{(F)}} > 0 \qquad (12)$$

$$a = (EX - r_f)v_0$$

where (E) and (F) denote the variables corresponding to portfolios E and F, respectively, and a stands for the total dollar amount (ie, not percentage) of expected excess return. If these two portfolios are combined into one larger portfolio (denoted G), it is natural to conclude that, from a risk/return perspective, the combined portfolio should be better than or at least the same as portfolio E or F because of the diversification benefit offered by merging two portfolios. This conclusion, which can be written as

$$r^{(G)} \geq r \tag{13}$$

is based on the assumption that portfolio G is likely to be more diversified than either E or F. The equality of Equation (13) will hold only if the returns of portfolios E and F are always identical (ie, no diversification benefits are achieved by combining E and F). Furthermore,

$$r^{(G)} = \frac{a^{(G)}}{v_\alpha^{(G)}} = \frac{a^{(E)} + a^{(F)}}{v_\alpha^{(G)}} \tag{14}$$

The second equality is due to the fact that the expected value of the sum of two random variables (in this case, the amount of portfolio return) is equal to the sum of the expected values of the same two variables. Combining Equations (12), (13), and (14) yields:

$$\frac{a^{(E)} + a^{(F)}}{v_\alpha^{(G)}} \geq r$$

or

$$\frac{r(v_\alpha^{(E)} + v_\alpha^{(F)})}{v_\alpha^{(G)}} \geq r \tag{15}$$

or

$$v_\alpha^{(E)} + v_\alpha^{(F)} \geq v_\alpha^{(G)}$$

The last of these equations shows that, in order to be consistent with the common sense of diversification benefits by combining different risk portfolios, the VAR for the combined portfolio must be less than or equal to the sum of the VAR of the individual portfolios. However, it can be shown that this condition is not always satisfied. To illustrate this situation, we further assume that the returns of portfolios E and F are independent (CDF shown in Figure 4) with the VAR of 10% confidence level being US$11.9 million and US$12.6 million, respectively. However, the VAR with the same confidence level for the combined portfolio (G) is US$27.3 million. This value is greater than the sum of the respectively VAR for portfolios E and F (the method of calculating VAR will be introduced in the next section) and, therefore, violates the conditions specified by Equation (15).

This paradox is the reason why VAR is not a coherent measure of risk (Artzner *et al.* 1999); the major disadvantage of using VAR as the measure of risk is that it does not always recognise or reflect diversification benefits.

Another problem of VAR is that it only measures the size of the loss corresponding to a single probability on the PDF. In other words, it does not tell the investor "how bad" the return can be once the threshold v_α is exceeded.

Tail value-at-risk
To alleviate the problems associated with VAR, we introduce another risk measure, namely the tail value-at-risk (TVAR, denoted u_α), which is defined as:

$$u_\alpha = -v_0 \frac{\int\limits_{x \leq -v_\alpha/v_0} x f_X(x) dx}{\int\limits_{x \leq -v_\alpha/v_0} f_X(x) dx}, \text{ for continuous } X \tag{16}$$

4. CDF of portfolios E, F and G

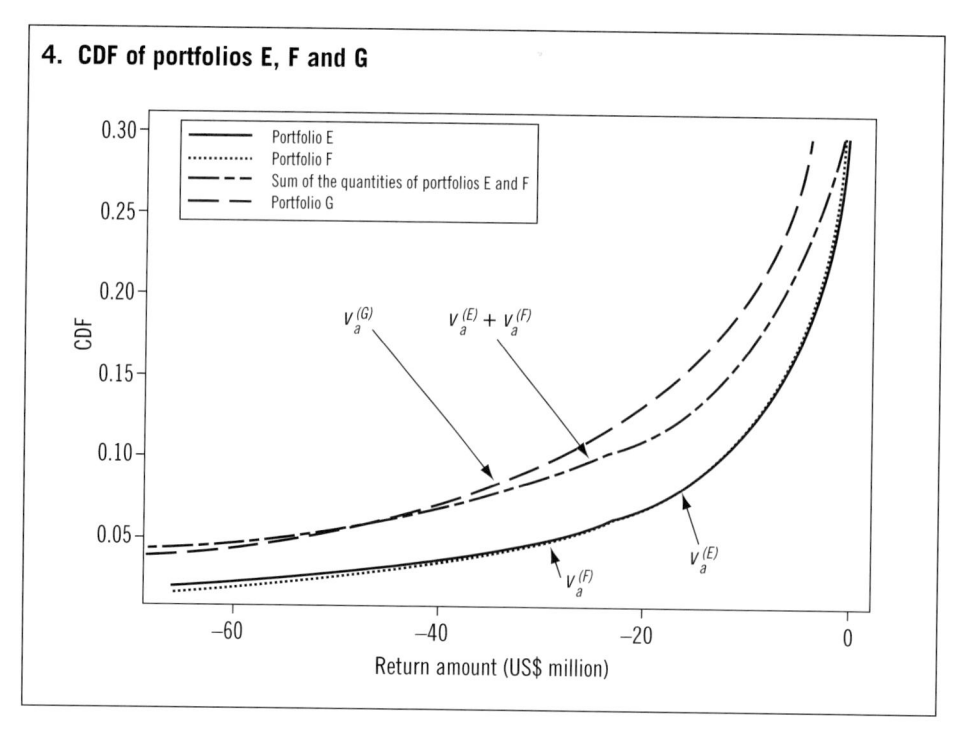

[Figure: CDF plot showing Portfolio E (solid), Portfolio F (dotted), Sum of the quantities of portfolios E and F (dash-dot), and Portfolio G (dashed). X-axis: Return amount (US$ million) from -60 to 0. Y-axis: CDF from 0 to 0.30. Labeled points: $v_a^{(G)}$, $v_a^{(E)} + v_a^{(F)}$, $v_a^{(F)}$, $v_a^{(E)}$.]

$$u_\alpha = -v_0 \frac{\sum_{x \leq -v_\alpha/v_0} x_k f_X(X = x_k)}{\sum_{x \leq -v_\alpha/v_0} f_X(X = x_k)}, \text{ for discrete } X$$

In fact, u_α is the size of the conditional expected loss in the event that the loss is worse than $-v_\alpha$. Hence, it is also known as the "expected shortfall" or "expected value-at-risk". It was initially used because it takes into account the losses worse than $-v_\alpha$ and consequently provides a more comprehensive portrayal of the downside risk. However, as shown in Artzner *et al.* (1999), an important attribute of TVAR is that it measures risk in a coherent manner in the sense that it does not suffer from the paradoxical problems associated with SD or VAR discussed above.

The advantages and disadvantages of the three risk measures discussed above are summarised in Table 2. Although each risk measure may perform best under certain special situations, we believe that TVAR is the best overall choice. In addition, since the risk-free rate, r_f, is merely a constant to be subtracted from EX, we do not

Table 2. Risk measures: a summary

Risk measure	Advantages	Disadvantages
Standard deviation	❏ Widely adopted and well understood ❏ Easy to calculate	❏ Incoherent (see Artzner *et al.*, 1999) ❏ Financial implication not explicit
Value-at-risk	❏ Directly related to risk capital and probability of insolvency	❏ Incoherent ❏ Determined by only a single point on the PDF
Tail value-at-risk	❏ Coherent ❏ Takes into account the probability and severity over a broader range of losses	❏ Sometimes difficult to calculate

explicitly include it in the remainder of this chapter to simplify the presentation of formulas. Hence, the return/risk ratio we use hereafter is:

$$r = \frac{EX}{u_\alpha/v_0} = \frac{EV_1}{u_\alpha} \qquad (17)$$

where EV_1 is the expected portfolio value at the end of the holding period. By definition, it is equal to the expected percentage return, EX, multiplied by the initial portfolio value, V_0. In addition, the confidence level α is chosen based on the risk tolerance of the investor and is usually subjectively determined.

To calculate EX and u_α, which are needed to arrive at r, we must calculate the PDF of the portfolio return. This topic is discussed in the next section.

Calculating the PDF of the portfolio return

The portfolio return, X, depends on the values of the portfolio at the beginning and the end of the holding period, V_0 and V_1, respectively. In general, a portfolio includes cash, weather derivatives and traditional assets. Equation (1) can be rewritten as:

$$X = \frac{V_1}{V_0} - 1 = \frac{c_1 + S_1 + W_1}{c_0 + s_0 + w_0} - 1 \qquad (18)$$

where c_0 and c_1 refer to the values of cash, s_0 and S_1 for other assets, and w_0 and W_1 for weather derivatives. 0 and 1 refer to the beginning and the end of the holding period, respectively.

The values of cash (c_0 and c_1) can be easily determined. In addition, it is assumed that the traditional assets are frequently traded. Thus, s_0 can be determined reliably based on market prices, assuming we analyse the portfolio after the beginning the holding period. The value of S_1 can be estimated probabilistically, but is not of primary interest in this section (see subsection "A portfolio with cash, weather derivatives and traditional assets"). To keep the equations mathematically simple, we ignore the interests generated by holding cash. The terms reflecting such interests can be added back on to the equations without any fundamental difficulty.

However, the value of weather derivative contracts cannot be marked-to-market because they are usually traded infrequently, if at all. In fact, the only reliable way to arrive at the value of a given contract is based on its settlement value, which is known only after the date on which the contract ends. In order to use the settlement values to represent W_1, we assume the contract terms of all weather derivatives end prior to the end of the holding period. Furthermore, we assume that a portfolio always starts without any weather derivative position or the value of the weather derivative positions can be marked-to-market at the beginning of the holding period (ie, w_0 is equal to zero or a constant). This allows v_0 to be treated as a constant.[5]

We first illustrate the calculation of the PDF of X using a simple portfolio with cash and a single weather derivative. This calculation is shown to be a generally simple procedure. However, a portfolio generally contains multiple weather derivatives. The calculation of the PDF of the total value of the weather derivatives is complicated by the fact that the payoff or settlement values of the individual contracts are usually dependent on one another. In fact, the joint PDF of all the weather parameters involved in the portfolio must be arrived at in order to complete the calculation. This is the most challenging part of the calculation due to the lack of:

1. sufficient historical data to support a thorough statistical analysis of the interdependence among the weather parameters; and
2. physics-based models that can adequately simulate such interdependence (not yet widely available).

MANAGING A PORTFOLIO OF WEATHER DERIVATIVES

We introduce several approaches commonly used to tackle this problem and discuss their strengths and weaknesses.

Lastly, we calculate the PDF of a portfolio with cash, weather derivatives and other assets. Although the procedure is relatively simple, the results strongly reinforce the view that introducing weather derivatives to an investment portfolio can greatly enhance its risk/return profile.

A SIMPLE PORTFOLIO WITH CASH AND A SINGLE WEATHER DERIVATIVE CONTRACT

We consider a portfolio that holds only cash in the amount of c_0 at the beginning of the holding period (ie, $v_0 = c_0$). Then the investor sells a swap for which the weather index is the February heating degree-days (HDD) for city B (denoted Y). It has an exercise level of 975 degree-days (denoted y_0) and a notional value of US$100 per degree-day (denoted k). Hence, the portfolio return is:

$$X = \frac{c_0 + W_1}{c_0} - 1 = \frac{W_1}{c_0} \qquad (19)$$

where W_1 is solely determined by the settlement value of the swap, which is a linear function of the weather index:

$$W_1 = -k(Y - y_0) \qquad (20)$$

Figure 5 shows the PDF of the weather index, the settlement value of the swap, and the portfolio return (assuming an initial cash holding of US$20,000).

In this example, the PDF of HDDs is based on fitting the historical data during the past 30 years (adjusted for instrument bias and trends) to the normal distribution. We believe this can be the simplest acceptable method for HDD over the period of a month or more. Other more advanced approaches, such as blending historical data with seasonal predictions, can be found in Zeng (2000).

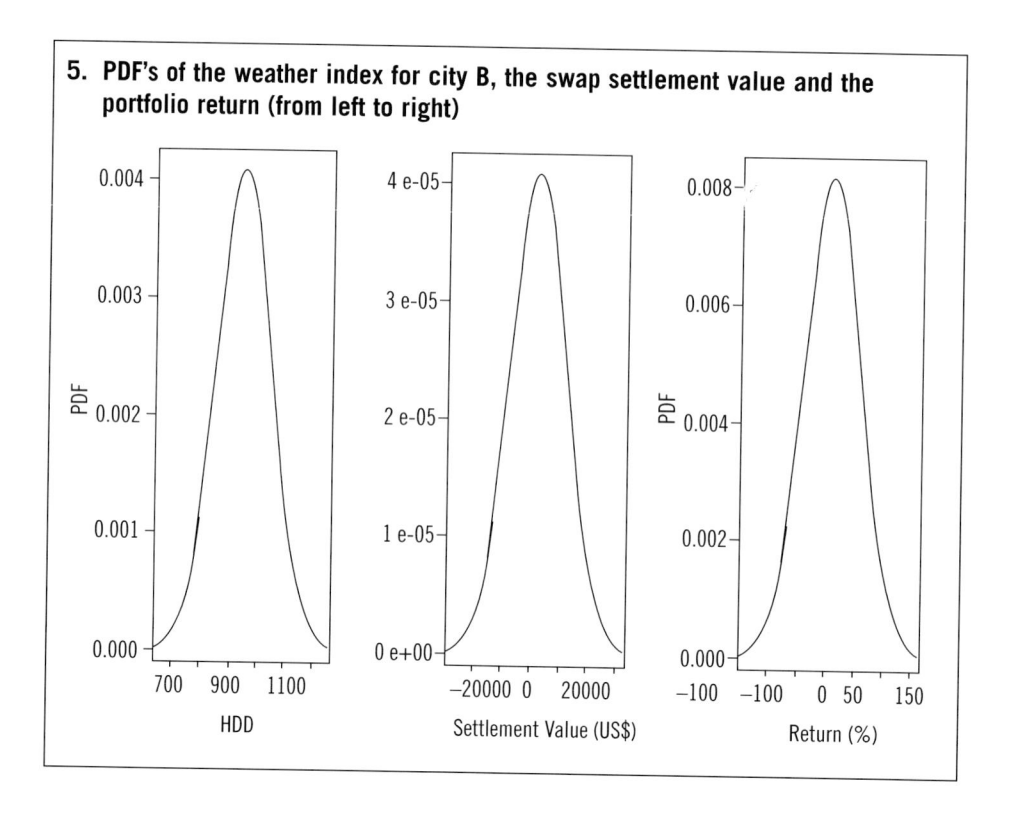

5. PDF's of the weather index for city B, the swap settlement value and the portfolio return (from left to right)

Given their linear relationship with the weather index (Equations 19 and 20), the settlement value and the portfolio return are also normally distributed. Hence:

$$\begin{aligned}
Y &\sim N(\mu_{hdd}, \sigma_{hdd}) \\
W_1 &\sim N(\mu_w, \sigma_w) \\
X &\sim N(\mu_x, \sigma_x) \\
\mu_w &= k(\mu_{hdd} - y_0) \\
\sigma_w &= k\sigma_{hdd} \\
\mu_x &= \mu_w/c_0 \\
\sigma_x &= \sigma_w/c_0
\end{aligned} \quad (21)$$

where $N(\mu, \sigma)$ denotes the normal distribution with a mean of μ and a standard deviation of σ. μ_{hdd} and σ_{hdd} are the sample mean (951 degree-days) and the sample standard deviation (98 degree-days) of the adjusted historical HDD data, respectively.

In reality, analytical solutions are usually unavailable because either the PDF of the weather index must be expressed in a non-parametric fashion or the relationship between the weather parameter and the contract settlement value is complicated. In this case, simulation-based approaches are usually used (eg, Monte Carlo simulations, bootstrapping, etc). Since this is not the main focus of this chapter, interested readers are referred to Zeng (2000).

A PORTFOLIO WITH CASH AND MULTIPLE WEATHER DERIVATIVES

In this case, Equation (19) still holds except W_1 now represents the total settlement value of all weather derivatives in the portfolio. This can be written as:

$$W_1 = \sum_{i=1}^{n} s_i(Y_{i1}, Y_{i2}, ..., Y_{in}) \quad (22)$$

where s_i is the settlement function of contract i (ie, the value of the settlement as a function of one or more weather parameters denoted by Y_{i1}, etc). To calculate the PDF of W_1, the joint PDF of **Y** must be known, where **Y** represents the vector of random variables consisting of the unique weather indexes (with a total number of n) referenced by all contracts in the portfolio:

$$\mathbf{Y} = \{Y_1, Y_2, ..., Y_n\} \quad (23)$$

Estimating the joint PDF, denoted $f_\mathbf{Y}(\mathbf{y})$, can be relatively simple or extremely difficult, depending on the characteristics of **Y** and the amount of data available. Commonly used approaches include fitting historical data to a probability model, modelling historical data dynamically, and the use of physics-based models. Detailed discussions on how to calculate $f_\mathbf{Y}(\mathbf{y})$ can be found in Appendix 1.

In general, $f_\mathbf{Y}(\mathbf{y})$ can be represented by one or more of the following three forms: analytical, discrete or simulation. A common example of an analytical representation is the multivariate normal distribution:

$$f_\mathbf{Y}(\mathbf{y}) = (2\pi)^{-n/2} |\mathbf{A}|^{-1/2} \exp[-\frac{1}{2}(\mathbf{Y} - \boldsymbol{\mu})^T \mathbf{A}^{-1}(\mathbf{Y} - \boldsymbol{\mu})] \quad (24)$$

where $\boldsymbol{\mu}$ and **A** are the mean vector and variance–covariance matrix, respectively. Each of these quantities can be derived from the historical data of weather indexes.

Alternatively, $f_\mathbf{Y}(\mathbf{y})$ can be represented by a table of discrete values. However, this method can only be used when two weather indexes are involved. For example, according to the sample distribution shown in Table 3, the probability that Y_1 is between 1050 ± 25 and Y_2 is between 260 ± 5 is 0.0006.

For the purpose of modelling weather indexes, however, the most flexible and powerful representation of $f_\mathbf{Y}(\mathbf{y})$ is using the matrix of simulated realisations (denoted $\mathbf{Y}_s$):

MANAGING A PORTFOLIO OF WEATHER DERIVATIVES

Table 3. Sample joint PDF table

	Y_2			
Y_1	250	260	270	...
1,000	0.0001	0.0002	0.0005	...
1,050	0.0003	0.0006	0.0007	...
1,100	0.0004	0.0009	0.0010	...
...	...	...	...	...

$$\mathbf{Y}_S = \begin{pmatrix} Y_1^{(1)} & Y_2^{(1)} & \cdots & Y_n^{(1)} \\ Y_1^{(2)} & Y_2^{(2)} & \cdots & Y_n^{(2)} \\ \cdots & \cdots & \cdots & \cdots \\ Y_1^{(m)} & Y_2^{(m)} & \cdots & Y_n^{(m)} \end{pmatrix} \quad (25)$$

An element of the matrix, $y_i^{(j)}$, is the simulated value of weather index i for simulation j. Each row of the matrix represents the values of the weather indexes in a single realisation of the simulation process. It is assumed that the number of simulations performed, m, is large enough for $\mathbf{Y}_s$ to properly represent the frequency of occurrence of nearly all combinations of possible values of the weather indexes. In addition, each column of $\mathbf{Y}_s$ represents the marginal PDF of the corresponding weather index. Examples of the analytical and simulation representations are illustrated in Figure 6 for a portfolio with two weather indexes. Figure 6 shows the the PDF contours (left) and simulated values for index A and index B (+s, right panel). The open circles are historical observations.

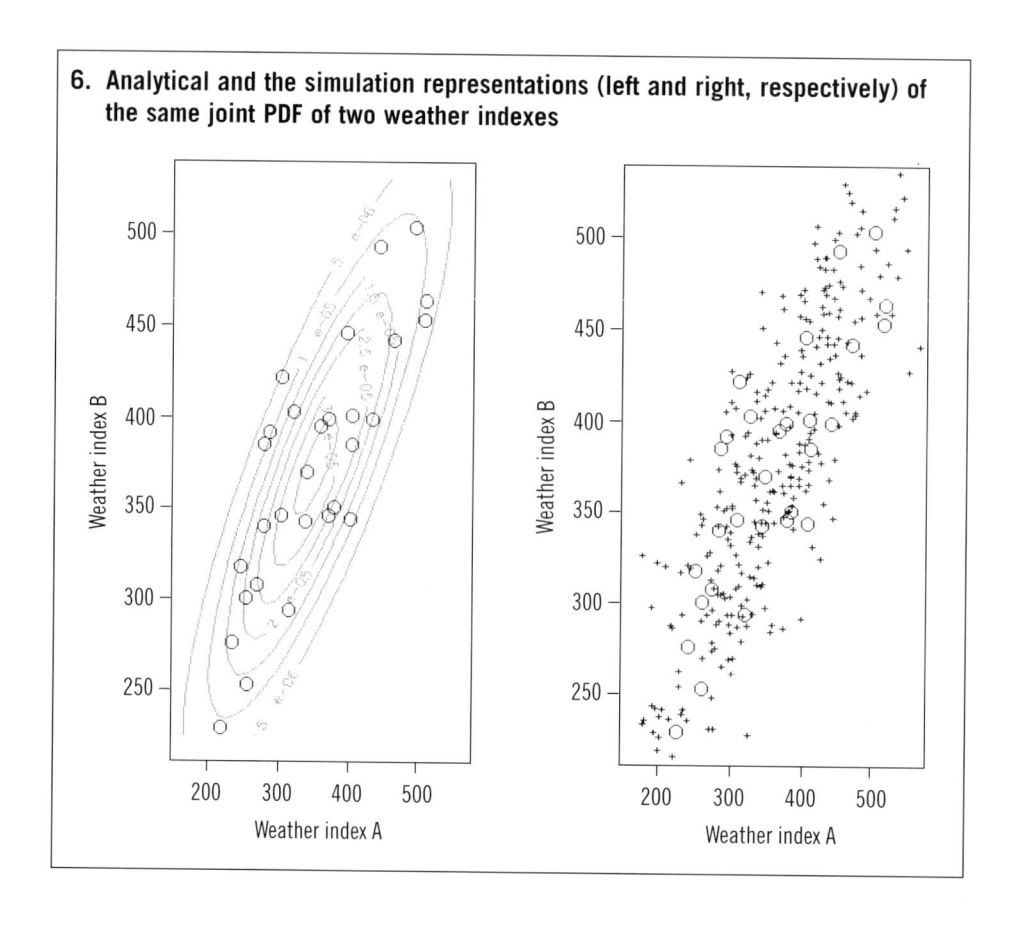

6. Analytical and the simulation representations (left and right, respectively) of the same joint PDF of two weather indexes

The unique strength of the simulation-based representation is its ability to handle those joint PDFs that cannot be adequately represented or even approximated by any analytical formula or table. The simulated realisations are usually created by data-driven and/or physics-based dynamic models. In addition, the simulations can also be created from an analytical formula or the tabulated discrete values discussed earlier (see Appendix 1 for details). In general, the simulation-based representation is the most effective approach for most weather derivatives. As a result, this type of representation will be used exclusively throughout the remainder of this chapter.

Given the joint PDF of weather indexes, represented in the form of $\mathbf{Y}_s$, the joint PDF of contract settlement values can be expressed in the same format:

$$\mathbf{S} = \begin{pmatrix} s_1^{(1)} & s_2^{(1)} & \cdots & s_p^{(1)} \\ s_1^{(2)} & s_2^{(2)} & \cdots & s_p^{(2)} \\ \cdots & \cdots & \cdots & \cdots \\ s_1^{(m)} & s_2^{(m)} & \cdots & s_p^{(m)} \end{pmatrix} \quad (26)$$

$$s_i^{(j)} = s_i(y_{i1}^{(j)}, y_{i2}^{(j)}, \ldots, y_{ip}^{(j)})$$

where $s_i^{(j)}$ is the simulated settlement value of contract i for simulation j, which is determined by the settlement function s_i (Equation (22)). The number of contracts in the portfolio (denoted p) is generally different from the number of weather indexes (n) referenced in the portfolio. This is because a contract may be based on multiple weather indexes and a weather index may be referenced by multiple contracts. The sum of a row of S is the total settlement value of all the contracts corresponding to a single realisation of the simulation process. Hence, using Equations (19) and (22), the probability distribution of W_1 and X can be expressed by the vectors $\mathbf{W}_s$ and $\mathbf{X}_s$, respectively:

$$\mathbf{W}_S = \begin{pmatrix} w^{(1)} \\ w^{(2)} \\ \cdots \\ w^{(m)} \end{pmatrix} \quad \mathbf{X}_S = \begin{pmatrix} x^{(1)} \\ x^{(2)} \\ \cdots \\ x^{(m)} \end{pmatrix} \quad (27)$$

$$w^{(j)} = \sum_{i=1}^{p} s_i^{(j)} \qquad x^{(j)} = w^{(j)}/c_0$$

Effectively, the portfolio return X is approximated by the discrete values $x^{(j)}$. Thus, the statistics vital for the calculation of portfolio return/risk ratio can then be calculated using Equations (7), (16) and (17).

A PORTFOLIO WITH CASH, WEATHER DERIVATIVES AND TRADITIONAL ASSETS
As discussed at the beginning of this chapter, the return of weather derivatives is generally independent of that of traditional assets. Given the respective PDFs of the returns associated with these two parts of the portfolio, the PDF of the combined return can be easily calculated using the approach introduced in Section A of Appendix 1. In this subsection, we focus on demonstrating the benefit of introducing weather derivatives to a traditional investment portfolio using the following hypothetical example.

Assume that the annual return of an existing investment portfolio is normally distributed with an expected return of 10% and a standard deviation of 20%. Furthermore, the investment portfolio is assumed to be the market portfolio (ie, its

MANAGING A PORTFOLIO OF WEATHER DERIVATIVES

risk can no longer be reduced using any capital market instruments without reducing the expected return). We also assume that the return of a weather derivative portfolio follows the same distribution but is independent of that of the existing portfolio. Thus, by definition, introducing weather derivatives into the portfolio will not change the expected return. However, as shown in Table 4 and Figure 7, the risk can be significantly reduced.

Building an optimal portfolio

After introducing the methods for quantifying a risk/return profile of the portfolio and estimating the probability distribution of the portfolio return, we now turn to the ultimate application of these methods: how to build a portfolio of weather derivatives with the best possible risk/return profile.

BACKGROUND, CHALLENGES AND OPPORTUNITIES

The task of forming a portfolio with a superior risk/return profile, also known as "portfolio optimisation", is thoroughly studied by participants of the capital market in the context of stocks and other traditional assets. These studies result in the

Table 4. Change of risk characteristics of an investment portfolio with different portions allocated to weather derivatives

Weather derivatives (%)	Expected return	Standard deviation (%)	Probability of return worse than:		
			0%	−10%	−20%
0	10	20	0.31	0.16	0.07
15	10	17	0.28	0.12	0.04
30	10	15	0.26	0.09	0.02

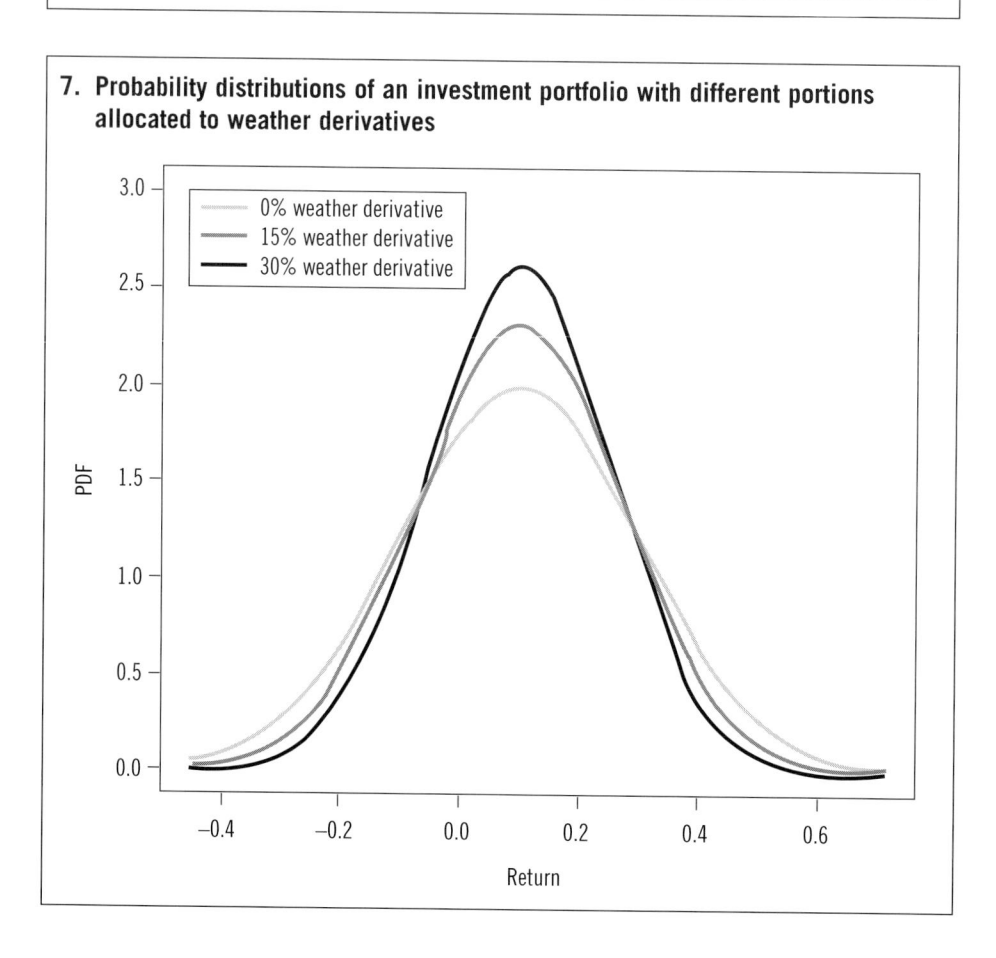

7. Probability distributions of an investment portfolio with different portions allocated to weather derivatives

modern portfolio theory (eg, Rudd and Clasing, 1988). Nevertheless, two important assumptions underneath the theory are not applicable to weather derivatives. First, modern portfolio theory assumes that the asset returns can be modelled by the lognormal distribution. Under this framework, the SD is almost exclusively used as the measure of risk and the covariance or correlation matrix is always used to quantify the relationship among individual assets. While these are reasonable assumptions for stocks and other traditional assets, their applicability to weather derivatives is questionable. As shown in the previous two sections, a better alternative is to model the joint probability distribution of weather indexes using a non-parametric simulation approach. Furthermore, we use the TVAR, a more appropriate measure of risk than standard deviation.

Another assumption under the modern portfolio theory is that assets are continuously traded. This allows an investor to build desired positions of different assets nearly simultaneously or within a relatively short period of time. As a result, an investor can maximise the diversification benefit by owning the market portfolio. Although this assumption is not absolutely true for the traditional capital market, it is shown to be an excellent approximation. However, the weather derivative market currently does not offer such liquidity. To the participants of the weather derivative market, this presents the most serious challenge as well as the most intriguing profitable opportunities.

Theoretically, an investor can use offsetting contracts at different geographical locations to reduce the portfolio risk. For example, an investor may demand that the return/risk ratio (defined in Equation (17)) of the portfolio be at least 18%. This can be achieved by a portfolio of a short March New York HDD swap at the strike of 720 degree-days and a long March Philadelphia HDD swap at the strike of 685 degree-days. The notional amount of both swaps is US$1,000 per degree-day (Table 5). It is shown that the return/risk ratios for the individual contracts (3% and 2%, respectively) do not need to be even close to the portfolio target (18%) for the portfolio to achieve this level of return/risk ratio because of the diversification benefit between the two contracts that exhibit offsetting profit/loss patterns (Figure 8).

In practice, the investor may be able to sell the New York swap on one day but unable to buy the Philadelphia swap or execute any other offsetting positions at the same time. In fact, the investor does not know whether an offsetting position can be found in the near future at all. In this case, the investor must demand that the return/risk ratio for the New York swap be substantially higher than 3% just in case an offsetting position never becomes available. This situation has two important implications:

1. The investor must require the strike to increase. In general, this implies that a seller must ask for more and the buyer must bid for less, leading to a less liquid market.
2. This also implies that the contracts executed in this environment usually carry a substantial "risk premium" (ie, the investor is compensated for the risk that the portfolio may never be diversified due to market illiquidity). This contrasts sharply

Table 5. Risk/return profile of a sample portfolio

Contract	Strike (degree-days)	Expected settlement (US$)	TVAR (US$)	r(%)
New York	720	8,000	219,000	3
Philadelphia	685	6,000	252,000	2
Portfolio	–	14,000	76,000	18

8. Simulation representations of the weather indexes for New York and Philadelphia (upper left), and the profits of the New York and Philadelphia contracts (upper right), the portfolio and the New York contract (lower left), and the portfolio and the Philadelphia contract (lower right)

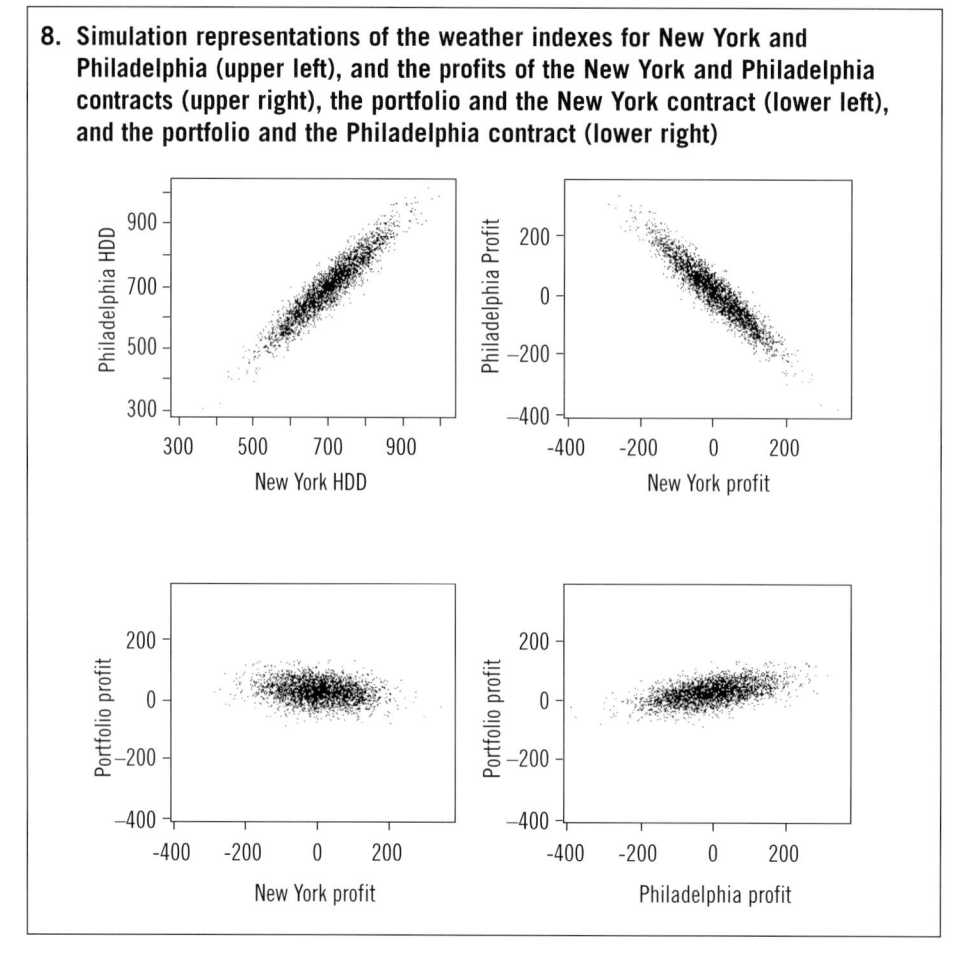

against the assumption in the modern portfolio theory that diversification is achievable almost instantaneously (hence investors are awarded no risk premium for asset-specific risk). Thus, if an investor is more effective at identifying and executing offsetting positions than most other investors, the investor can be the most competitive by, for example, quoting the best offer for the New York swap at the strike of 730. Assuming the investor manages to execute an offsetting contract a few days later that is the same as the Philadelphia swap in the example above, the portfolio return/risk ratio is dramatically improved (Table 6).

In the real world, offsetting contracts with such obvious arbitraging opportunities as the ones discussed above can be identified by most investors. This tends to be reflected in the prices and leave little opportunities for extraordinary profit. Thus, the key to substantial profitability is the ability to identify offsetting positions that are not obvious. For example, with an existing portfolio, the best offsetting position for an investor is usually not just another contract, but the combination of a selected

Table 6. Risk/return profile of a sample portfolio

Contract	Strike (degree-days)	Expected settlement (US$)	TVAR (US$)	r(%)
New York	730	18,000	209,000	8
Philadelphia	685	6,000	252,000	2
Portfolio	–	24,000	66,000	36

group of prospective contracts. Strategies for selecting contracts are discussed in the following subsection.

PORTFOLIO FORMATION THROUGH CONTRACT SELECTION

Since it is impossible to select desirable contracts simultaneously, the portfolio formation process is done iteratively as information related to new contracts emerges over time. At any given time, an investor is generally presented with the information related to two groups of contracts:

1. the existing contracts in the portfolio; and
2. the prospective contracts quoted on an exchange or from other sources (eg, over-the-counter brokers, direct counterparties, etc).

The investor must decide, within a relatively short period of time, how to respond to each of the prospective contracts in the second point above by executing at the ask/bid price, responding with a counter bid/ask price, or ignoring it altogether. For brevity, we do not discuss the issues related to creating new positions by the investor, although the principals are equally applicable.

Usually, the investment decisions are subject to certain constraints. It may be required that the portfolio TVAR should not exceed a critical value (eg, US$100 million) and the expected return should be no less than a threshold (eg, 15%). Presented with n prospective contracts, the investor must test each of the 2^n combinations to first eliminate the ones that violate the constraints and then select the combination that allows the final portfolio to achieve the greatest return/risk ratio. This approach is obviously not practical even for moderate n (eg, greater than 20), especially because the calculation of TVAR is highly non-linear.

We suggest two fundamentally similar alternatives developed in the (re)insurance community. One is based on a numerical approximation to the ideal approach above (Zeng, 2001). This approach reduces the number of evaluations from $O(2^n)$ to $O(n^2)$ and is shown to successfully approximate the ideal solution.

Another approach, which is less computationally demanding and essentially a simplified version of the above, is to calculate the marginal return on risk capital associated with this contract, defined as:

$$r_{mA} = \frac{\Delta EV_{1A}}{\Delta u_{\alpha A}} \qquad (28)$$

where ΔEV_{1A} and $\Delta u_{\alpha A}$ are the increases of the expected profit and TVAR, respectively, of the portfolio resulting from including contract A. $\Delta u_{\alpha A}$ is known as the "risk capital allocated to the contract". The contracts with the highest marginal returns on risk capital are selected to be included in the portfolio.

For example, an investor is presented with two prospective contracts (A and B) with the same expected profit (US$10,000) and TVAR (US$50,000). In this case, ΔEV_{1A} is equal to ΔEV_{1B}. However, the TVARs for the combined portfolio are different because the underlying weather indexes for A and B interact differently with the weather indexes underlying the existing portfolio (Table 7). Thus, A is allocated less

Contract	Expected profit (US$)	TVAR (US$)	Marginal TVAR (US$)	Marginal expected profit (US$)	Marginal r (%)
Existing portfolio	90,000	200,000	–	–	–
Portfolio + A	100,000	210,000	10,000	10,000	100
Portfolio + B	100,000	240,000	40,000	10,000	25

Table 7. Marginal risk/return of sample contracts

> **PANEL 1**
>
> **INVESTING IN WEATHER-LINKED INSTRUMENTS**
>
> The recent gyrations of the stock market have reminded investors that equity investments involve a significant element of risk and the stock market will not deliver guaranteed 20% returns year after year. Consequently, intelligent investors today are looking for investments that offer reasonable returns while being less susceptible to the wild swings of the economy or the fickle shifts of other investors' collective emotions.
>
> Weather-linked instruments represent an attractive complement to traditional equity and fixed-income investments. Since the underlying indexes are solely determined by weather conditions, the profits and losses associated with these instruments are little influenced by the state of the economy, interest rates, or the equity markets. Therefore, the introduction of weather-linked instruments into a portfolio of stocks and bonds can potentially result in a substantial decrease in risk and an increase in return.
>
> **Characteristics of weather-linked instruments**
>
> Weather-linked instruments are derivative contracts (futures, swaps, options, etc) based on weather indexes. The holders of these instruments typically include utilities, agri-businesses, reinsurance companies, commodity and energy traders and investment banks. There is rarely a balance between those who will suffer from extreme weather, and those who will profit from it. The desire to hedge risk is a consistent source of profit opportunity for the savvy investor.
>
> Weather-linked instruments cannot be priced using Black–Scholes or any other no-arbitrage-based option pricing model because the underlying indexes are not traded. Correct pricing must be built on an in-depth understanding of the underlying weather indexes and their correlations. This requires a seamless integration of expertise in meteorology, statistics and finance.

marginal risk capital and is a better candidate than B to be added to the existing portfolio.

Extreme care must be exercised when using this scheme for capital allocation. First, it is crucial that the capital allocation is done coherently to avoid misleading results (see, for example, Singh, 2002). Secondly, when prospective contracts are added to the existing portfolio, the marginal return/risk ratios for the remaining prospective contracts tend to change. The only way to alleviate this problem is by re-evaluating Equation (28) every time any single contract is selected. In this case, this capital allocation method essentially converges to the optimal risk selection approach introduced by Zeng (2001).

Conclusion

In this chapter, we discussed three issues that are crucial for an investor to successfully manage a portfolio of weather derivatives.

1. How to measure and quantify the risk/return profile of a portfolio. This provides the basis for differentiating portfolios and strategies with various risk/return characteristics.
2. Approaches for calculating the risk/return measures associated with a portfolio.
3. Strategies for building an optimal portfolio with a superior risk/return profile subject to constraints related to financial strength and risk tolerance of the investor.

We illustrated the unique challenges in the quantitative analyses and the optimisation of a portfolio of weather derivatives. These render inapplicable the methods developed for a portfolio of stocks and other traditional assets. Special approaches are introduced and discussed to address these challenges.

It is also demonstrated that the unique challenges and characteristics of weather derivatives make them an extremely attractive asset class. First, they offer returns that are truly independent from the state of economy and the fickle shifts of the collective emotions of capital market investors. Introducing these instruments to an investment portfolio can substantially reduce the portfolio risk without reducing the expected return. Secondly, but equally importantly, we believe that weather derivatives currently still carry a substantial risk premium due to market illiquidity. Successfully managing the risk through rigorous and efficient risk allocation can award the investors with significant profit potential and an extremely appealing risk/return profile.

Appendix 1: Calculating the joint probability distribution of multiple weather indexes

Which method to use for estimating the joint probability distribution and the difficulty of the task depends on the characteristics of the weather indexes and the amount of historical data available. Although it is impossible to include all possible methods, we discuss four that are commonly used.

THE COMPONENTS OF Y ARE INDEPENDENT

In this case, the joint PDF of Y is equal to the product of the marginal PDFs of the components.

$$F_\mathbf{Y}(\mathbf{y}) = \prod_{i=1}^{n} F_{Y_i}(y_i) \quad (A.1)$$

The marginal PDFs can be calculated using the approaches introduced in the subsection "A simple portfolio with cash and a single weather derivative contract". Equation (A.1) can be evaluated analytically or through simulations. For the latter, random samples are created based on the marginal PDF of the components and are grouped randomly to form the matrix of realisations seen in subsection "A portfolio with cash and multiple weather derivatives").

However, most weather indexes are not independent because weather conditions around the world are controlled by the same global atmosphere–ocean system. As a result, the above equation is often not applicable for a portfolio of weather derivatives. Nevertheless, it is useful to calculate the joint PDF of independent classes of assets (eg, as in subsection "A portfolio with cash, weather derivatives and traditional assets").

Y IS MULTIVARIATE NORMAL

Multivariate normal is one of the most commonly adopted models for multi-dimensional data. It is a natural extension of the univariate normal distribution. The formula for $f_\mathbf{Y}(\mathbf{y})$ is identical to Equation (24). The parameters, namely the mean for the individual weather indexes and their variance–covariance matrix, can be estimated based on historical data (eg, Figure 7, left panel). Detailed numerical methods can be found in many textbooks and references (eg, Genz, 1999). This concept also applies to other parametric probability distributions.

NON-PARAMETRIC ESTIMATE OF THE JOINT PROBABILITY DISTRIBUTION

Statistical tests or visual inspections often show that certain weather indexes cannot be modelled by the multivariate normal or any other parametric distribution. An alternative approach is to estimate the joint PDF in a non-parametric way. We illustrate this approach using the following example involving two weather indexes.

The historical observations (shown in Figure A.1) are first binned into grid boxes centred at $Y_1 = \{y_{11}, y_{12}, ..., y_{1n}\}$ and $Y_2 = \{y_{21}, y_{22}, ..., y_{2n}\}$. Let n_{ij} denote the number of observations that fall into the box centred at y_{1i} and y_{2j}. Then

$$P\left(Y_{1i} - \frac{\delta y_1}{2} \le y_1 < y_{1i} + \frac{\delta y_1}{2} \text{ and } y_{2j} - \frac{\delta y_2}{2} \le y_2 < y_{2j} + \frac{\delta y_2}{2}\right) \approx \frac{n_{ij}}{n_t} \quad (A.2)$$

where δy_1 and δy_2 are the sizes of the grid boxes along the directions of Y_1 and Y_2, respectively, and n_{ij} is the total number of observations.

Refined approaches based on the same principal do exist. One example is the kernel density estimation method (eg, Silverman, 1986). Moreover, this approach can be extended to PDF estimation beyond two dimensions, provided that the number of observations is large enough. Because the length of weather records usually does not exceed 100 years and the data are not completely independent due to autocorrelation (see Zeng, 2000), the actual information content (also known as "the degree of freedom") in the weather data is limited. Hence, we do not recommend using this approach for more than two weather indexes.

DYNAMIC MODELS

To produce a reliable joint PDF estimate, we must have at least one or both of the following:

1. a large amount of historical data; and/or
2. certain assumptions about the characteristics of the weather indexes that can reduce the free parameters that we must estimate.

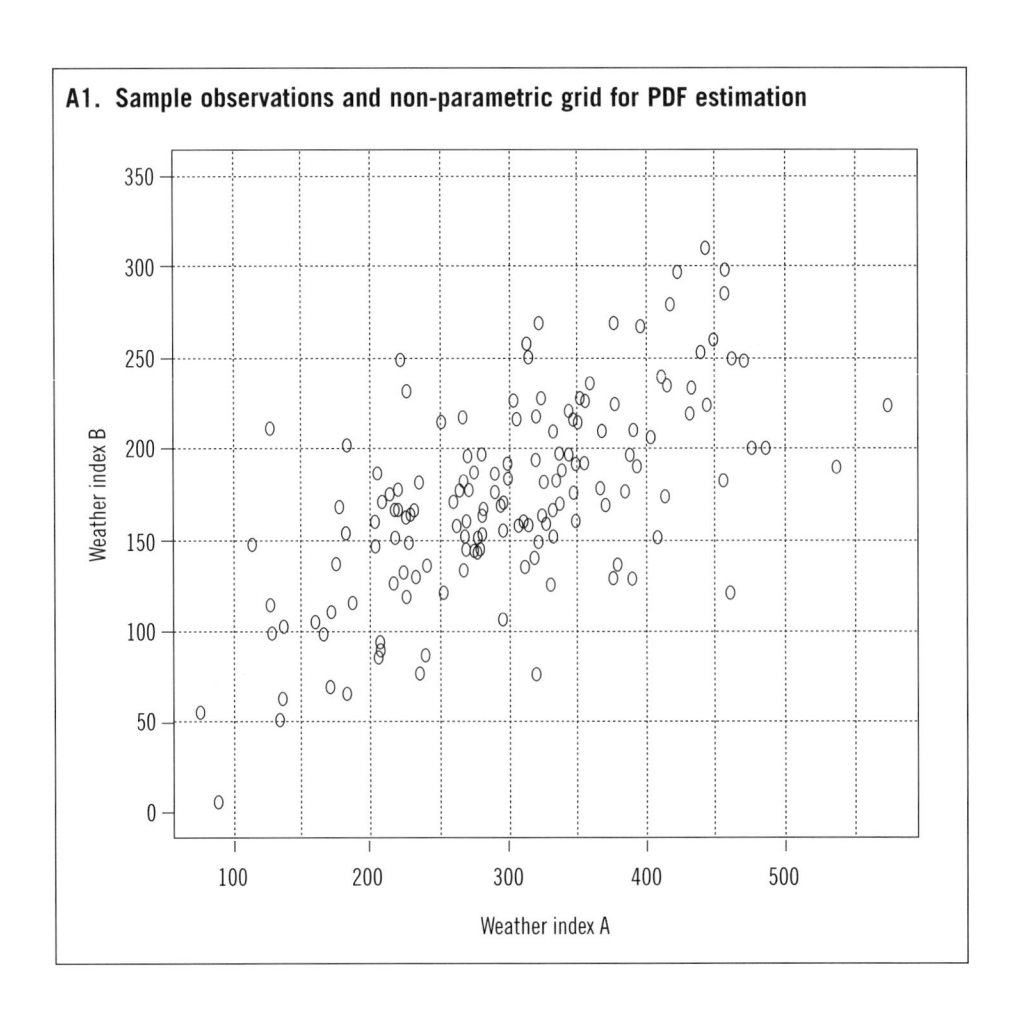

A1. Sample observations and non-parametric grid for PDF estimation

The approaches introduced in the previous subsection depend on the first point, but cannot be widely used because the volume of independent weather data is usually limited. This forces us to make certain assumptions about the weather indexes in order to produce probability density function estimates. Making very simple assumptions usually results in easy calculations (eg, the first two batches of approaches above) but the applicability is limited. Thus, the only truly viable option remaining is to use dynamical models to make sophisticated assumptions that account for the huge variability of weather indexes in general.

The most valid assumption about weather is the theory of fluid mechanics and thermal dynamics based on Newton's Laws. These theories lead to a set of dynamic and thermal dynamic equations that govern the motion of the atmosphere and ocean, linking various weather parameters at different times and geographical locations (see, for example, Gill, 1982). These equations cannot be directly used to make long-term climate predictions due to their non-linear nature and the resulting chaos. However, climate models (eg, Trenberth *et al.*, 1998) that have been developed based on the equations of atmosphere–ocean dynamics and thermodynamics can simulate the patterns of variation of different weather indexes. These simulations can form a set of simulated weather indexes that represents $f_Y(y)$ in the form of Equation (26). However, this type of model is currently not widely available because of its substantial demand of computing power and data processing capability.

There also exist dynamic models of weather indexes based purely on historical weather data. These models essentially assume that the weather process is governed by certain stochastic processes such as the Markov process. The parameters associated with these processes are then estimated based on historical weather data. Compared to the physics-based dynamic model, this type of method is easier to implement and computationally less expensive. However, the effectiveness of such models in reproducing realistic weather patterns is not adequately proven.

1 *The authors would like express our sincere thanks to Scott Mathews for extremely fruitful discussions and WeatherDecision LLC for providing the data and analytical tools that helped create many of the demonstrations used in this chapter.*
2 *For brevity, "portfolio" is used in place of "portfolio of weather derivatives" henceforth except when it is explicitly defined otherwise. Similarly, "contract" is used in place of "weather derivative contract".*
3 *In this chapter, we use the term "investor" broadly to include investors, traders, hedgers and speculators.*
4 *It is theoretically possible for x_α to be positive if α is small enough. In this case, v_α is negative and cannot be interpreted as a measure of capital at risk. However, in practice, α is often chosen to be large (eg, greater than 90%) because investors usually use VAR to quantify unlikely large losses. This usually leads to negative x_α for most realistic probability distributions of the portfolio return. Hence, we assume that v_α is always positive.*
5 *The fact that the value of a weather derivative cannot be determined precisely prior to the end of the contract also makes it necessary to describe the initial value of the portfolio (v_0) as a random variable for a portfolio that contains one or more weather derivatives at the beginning of the holding period. In this case, calculating the PDF of the portfolio return requires the joint PDF of the weather parameters at the beginning and at the end of the holding period. This can be done using the same approaches as those described in the subsection "A portfolio with cash and multiple weather derivatives". However, to avoid overcomplicating the formulas, we assume that the portfolio does not contain any weather derivatives at the beginning of the holding period, which guarantees that v_0 is always a constant. This is a fair assumption since most weather derivatives settle within a season (winter or summer). This is also true for multi-year contracts, as each year is settled separately.*

BIBLIOGRAPHY

Artzner, P., F. Delbaen, J.-M. Eber and D. Heath, 1999, "Coherent Measures of Risk", *Journal of Mathematical Finance*, 9(3), pp. 203–28.

D'Arcy, S., 1999, "Don't Focus on the Tail: Study the Whole Dog", in: *Risk Management and Insurance Review*, 2, pp. 4–14.

Genz, A, 1999, "Numerical Computation of Multivariate Normal Probabilities", *Journal of Computational Graphical Statistics* 1, pp. 141–9.

Gill, A., 1982, *Atmosphere-ocean dynamics*, (San Diego: Academic Press).

Nakada, P., H. Shah, H. U. Koyluoglu and O. Collignon, 1999, "P&C RAROC: A Catalyst for Improved Capital Management in the Property and Casualty Insurance Industry", *Journal of Risk Finance*, 1(1), pp. 52–70.

Rudd A., and H Clasing, 1988, *Modern Portfolio Theory: The Principal of Investment Management*, (Chicago: Probus Pub Co).

Singh, M., 2002, "Risk-Based Capital Allocation Using a Coherent Measure of Risk", *Journal of Risk Finance*, 3(2), pp. 34–45.

Silverman, B. W., 1986, *Density Estimation for Statistics and Data Analysis*, (Boca Raton, FL: CRC Press).

Trenberth, K. E., G. W. Branstator, D. Karoly, A. Kumar, N.-C. Lau and C. Ropelewski, 1998, "Progress during TOGA in Understanding and Modeling Global Teleconnections Associated with Tropical Sea Surface Temperatures", *Journal of Geophysical Research*, 103, pp. 14291–324.

Treynor, J., and F. Black, 1973, "How to Use Security Analysis to Improve Portfolio Selection", *Journal of Business*, January, pp. 66–86.

Zeng, L., 2001, "Pricing Weather Derivatives", *Journal of Risk Finance*, 1(3), pp. 72–8.

Zeng, L., 2000, "Using Cat Models for Optimal Risk Allocation of P&C Liability Portfolios", *Journal of Risk Finance*, 2(2), pp. 29–35.

Experience in Application

15

A Case Study of Heating Oil Partners' Weather Hedging Experience

Paul J. Forrest

Heating Oil Partners, LP

Heating Oil Partners, LP (HOP) was formed in 1995 and since then has acquired 33 heating oil distributors in the Northeast United States. Delivering 120 million residential gallons of heating oil a year, it is currently the second largest home heating oil distributor in the US.[1] HOP is involved in an annual programme of weather risk in order to prevent unnecessary revenue losses during warm winters that reduce oil sales.

Home heating oil, or #2 oil, is a by-product of the distillation of crude oil. It is produced by a distillation process by which crude oil is heated to the point that it vaporises, and then it is re-condensed. The result of this process is a distillate liquid product that is thicker than gasoline and has a much higher flash point. It is virtually impossible to burn liquid heating oil without first turning it into a vapour. For this reason, it has long been used as an energy source within the homes, and has many millions of users worldwide.

Heating oil is shipped from refiners to distribution terminals by pipeline, barge or truck. There is a labyrinth of residential heating oil distributors who pick the product up from the terminals and deliver it to individual homes. In the US, approximately 10 million households consume heating oil to heat their homes and/or to make hot water.[2]

The geography of the northeast United States

For the purposes of this chapter, the northeast United States is defined as the New England and Mid-Atlantic States or more particularly Maine, New Hampshire, Vermont, Massachusetts, Rhode Island, Connecticut, New York, New Jersey, Pennsylvania, Delaware, Maryland and the District of Columbia.

This area represents 4.6 % of the land mass and 22.9% of the population of the US. What makes this area rather unique is its use of #2 oil as the primary energy source for wintertime heating. Of the 10 billion gallons of heating oil sold to private residences in the US, approximately 75 % was consumed in the northeast. A little less than one in three households have chosen to use #2 oil over the alternative energy sources (natural gas, electricity, propane and to a lesser extent, coal, wood, solar and geothermal). There is no other area in the US that has such a high density of heating oil use.[3] There are reasons for the preference of heating oil in this region. The first is historical and relates to the relative ease of supplying heating oil from ocean or river barge or from pipelines built in the early 20th century. There were many homes that originally used coal as a heating source, but since there were not efficient systems to stoke furnaces with coal, homeowners sought alternative continuous flow sources.

A CASE STUDY OF HEATING OIL PARTNERS' WEATHER HEDGING EXPERIENCE

The second reason is related to the very rocky soil encountered throughout the northern parts of the northeast that causes the laying of natural gas pipelines to be extremely expensive. The third reason is linked to the relative cost efficiency of heating oil compared to other forms of energy; whereas natural gas and heating oil are somewhat similar in cost, electricity and liquid natural gas (propane) are much more expensive per btu produced.[4] New, strict environmental laws have all but eliminated coal and wood as viable sources of residential energy in the urban areas of the region.

> **PANEL 1**
>
> **TERMINOLOGY**
>
> Standards have been developed to help distributors and homeowners to better understand the outside air temperature and its effect on the need for warming or cooling one's residence. The terms used are "heating degree-day" (HDD) and "cooling degree-day" (CDD). They are both defined the same way as the difference between 65 degrees Fahrenheit (°F) and the average of the high and low temperature of the day. If the result is positive, it is a heating degree-day. If the result is negative, it is a cooling degree-day.
>
> As an example, if the highest temperature of the day was 45°F and the lowest temperature was 35°F, the average would be 40°F. Subtracting 40 from 65 gives 25 degree-days. The result is a positive number concluding that on this particular day, there were 25 heating degree-days. The more heating degree-days incurred, the colder the weather. Similarly, if the high for the day was 85°F and the low was 75°F, the average temperature would be 80°F. Subtracting 80 from 65 results in –15 degree-days. Since the result is a negative number, we would say that on this particular day, there were 15 cooling degree-days. The more cooling degree-days incurred, the warmer the weather.

The marketplace

The retail home heating industry in the northeast US is comprised of approximately 2,300 distributors. There are a handful of large companies delivering in excess of 75 million gallons a year; about one or two dozen companies delivering between 10 and 75 million gallons a year, and approximately 2,250 "Mom and Pop" delivery companies spread all across the area, most of whom deliver no more than 1–2 million gallons per year.[5] The Mom and Pop dealers are usually family-owned and cover a rather small geographic area. They may have one to five delivery trucks and fewer than 20 employees.

Being a large company has allowed HOP to develop a more sophisticated approach to the purchasing and marketing of its products than its smaller Mom and Pop competitors. Most of HOP's customers are looking for a full-service company to whom they will in effect, "outsource" their home heating requirements. HOP is effectively able to convince these customers to enter into automatic delivery plans; a delivery of heating oil is made to their homes when *the company* perceives their tanks to need refilling. This is a rather unique feature of the market; the buying decision by customers is a one-time event creating a number of future transactions over which they have very little control. Thus the initial sale of products and services to these customers during the summer months does not need to be repeated throughout the heating season.

HOP delivers approximately 74% of residential heating oil gallons during the prime winter heating season (November 1–March 31 in the northern part of the territory and December 1–February 28 in the southern extremes). There is an excellent correlation between HDDs and gallons of heating oil consumption during these months, which then forms the basis of the oil distributor's automatic delivery programme:

Expected HDDs during a Period ÷ K Factor = Gallons consumed

The K Factor is the common term in the heating oil industry to describe the relationship between the homeowner's consumption of heating oil and the HDDs that have occurred between deliveries. Each home has a different K Factor, dependent on the size and construction of the home, the number, age and habits of the inhabitants, and the efficiency of the heating plant. The lower the K Factor, the higher is the consumption of fuel. The distributors' computers track each homeowner's consumption of home heating oil and the number of HDDs that have occurred, and use this information to predict when next to refill the homeowner's tank.

Prior to the start of the heating season, HOP segregates the customers, highlighting those that have signed up for automatic delivery plans and displays their K Factors. HOP also calculates the normal HDDs during the winter period for their geographic area. Using the above formula, the company then calculates how much product these accounts will consume during a normal winter. This knowledge allows the company to place forward purchase contracts with wholesale distributors for the winter heating oil requirements for these customers.

Offering price stability
The oil market has been very volatile in the past few years causing many customers to demand price stabilisation programmes for the heating oil product they consume during the year. HOP offers both a fixed-price programme and a capped-price programme.

The fixed-price programme is simple: the customer is charged a fixed price for every gallon consumed. A "capped" price sets a maximum price the customer will pay for product, but resets the price at a lower level if the cost of product falls during the year. To cover the downside risk of the capped-price programme, HOP buys put options against the oil purchased forward. The put option will compensate HOP every penny that the heating oil product falls below the strike price at which the put was purchased.

As an example of the fixed-price programme, assume that the cost of a futures contract for January heating oil on the New York Mercantile Exchange is US$0.50 per gallon. If the aim is to make a US$0.40 gross margin on this product, the price would be set at US$0.90 for the customer. If, between now and January the market price of heating oil should rise, HOP is locked into a purchasing contract that allows the product to be obtained at US$0.50 a gallon, which can then be passed on to the customer at the fixed price of US$0.90. This satisfies both parties: the customer can budget how much will be spent on heating oil throughout the season and HOP will achieve the budgeted gross margins. However, only the price is fixed; the customer is still subject to his volume need.

In a similar example, HOP would buy the same futures contract for a capped-priced customer, but along with that contract, the company would also purchase a put with a strike price of US$0.50. Assuming that the put costs US$0.05, the cost of product in January would now be UD$0.55. With the target still at a gross margin of US$0.40, the company would offer a capped-price programme to the customer at US$0.95 per gallon. If the January market price of heating oil rises above US$0.50, HOP is still locked into the contract and the customer pays no more than the cap of US$0.95 for the product. If the January market price of heating oil should drop to US$0.40, the put option would be worth US$0.10 per gallon and that would be passed on to the customer, lowering their cost of product to US$0.85 per gallon. Since the product is capped if the price of heating oil goes up, and can be lowered through the profits realised from selling the puts if the price of heating oil falls, HOP maintains consistent margins whichever way the price moves.

In summary, HOP has taken the price risk out of the product sold to fixed- and capped-price customers. Each October, the new fiscal year begins, HOP is able to estimate fairly accurately what the gross margins will be for the preponderance of sales that year. This stability in gross margins is a very attractive selling point to investors. Yet, it does not remove all risk from the heating oil business.

The risks associated with the weather

There is another uncontrollable factor that must also be taken into consideration: the weather. As a reminder, the amount of heating oil that HOP sells each winter is very dependent on the number of HDDs that occur during the prime heating season.

As HOP uses HDDs to determine how much product to purchase on a forward basis, the company needs to understand the peculiarities of the HDD patterns for each of the cities in which it operates. Having its operations spread over a rather large geographical region, HOP has found there to be significant differences in temperature patterns among branch locations; the locations that are further north and the further inland (50 miles or so from the nearest shoreline) usually have colder winters (more HDDs) than those in southern locations or those residential areas that are tempered by neighbouring oceans. HOP uses eight separate weather stations located near its major branch locations to track the data used in forecasting HDDs and future consumption of customers.

Experience has shown it to be far better to use a span of years as a base for forecasting HDDs than to base it on a single year. Until 1998, HOP used the 30-year average HDDs for each reporting weather station. However, the three warm winters of 1997–8, 1998–9 and 1999–2000 caused the company to instead use a 10-year average for the forecasting of the winter of 2000–1 and 2001–2, as HOP wished to compensate for the apparent shift in the temperatures. In fact, the winter of 2000–1 was closer to the 30-year average than the 10-year average.[6] However, that merely resulted in more business than expected and the conservative forecast did no harm to HOP at all. HOP picks the measure of HDDs that errs on the conservative side. If the 10-year average were lower than the 30-year average – it would be a more conservative posture to use the 10-year average for HOP's budgeting and weather protection. If the HDDs fail to materialise during the prime heating season, HOP is unable to deliver the anticipated gallons and operating income is lost. In a results-oriented company such as HOP, a steady stream of predictable cashflow must be maintained to satisfy its investors.

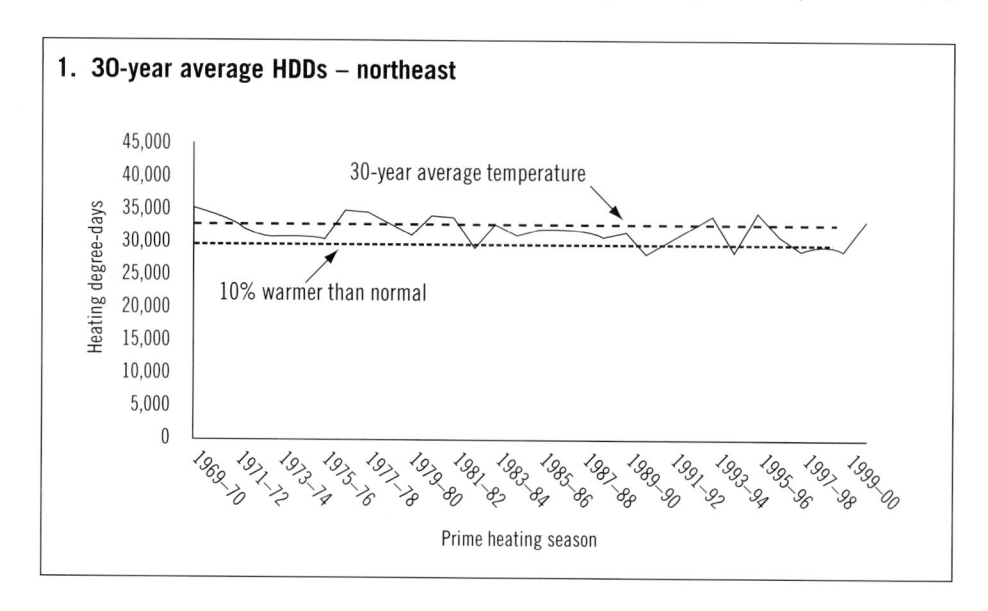

Figure 1 shows that fluctuations in HDDs from year to year rarely exceed 10%,

1. 30-year average HDDs – northeast

and in most cases stay ± 5–6% from a norm. However, for a company of HOP's size, this can mean a swing in earnings of 10–20%, a very meaningful number to the company's investors. For this reason, HOP decided it was essential to develop a risk strategy with reference to the weather and in particular to HDDs.

The beginnings of weather hedging at HOP

HOP's introduction to the world of weather risk hedging came in the winter of 1997–8. A local insurance broker brought a proposal from an insurance company to protect HOP against the possibility of a shortage of HDDs during a warmer than normal winter. Frankly, at this stage, HOP was a neophyte, relying on the broker to set up the programme. HOP picked a handful of weather stations near to branch locations, and set targets based on what the budget was based on, the published 30-year "normals" for those target weather stations. (Normally the option sellers would prefer to sell a shorter average, but HOP was able to get a deal set on 30-year normals.)

As it turned out, the winter of 1997–8 featured an "El Niño" and was one of the warmest on record for the northeast US, allowing HOP to collect approximately US$4.5 million from that policy that winter.[7] For HOP, it meant the difference between a decent year and a disaster. It also opened HOP's eyes to the possibilities afforded by weather risk management and HOP became instant believers in the product.

HOP's first step in developing a better understanding of the weather risk process was to gather as much historical weather information as possible for the areas in which the company is located. One problem encountered was finding dependable weather stations that published data on a continuous, reliable basis. The US National Weather Service (NWS), however, has equipped a number of their more important weather stations with the Automated Surface Observing System (ASOS). Such ASOS data sites have a data link to the Internet that allows accurate readings to be obtained with only a 24-hour delay and it was such sites that HOP chose to supply the data required for the measuring of HDDs.

Once the ASOS stations had been chosen, HOP further interrogated the Internet to find data history for each station. Enough US National Climatic Data Center (NCDC) data was available to provide HOP with a minimum of 50 years of historical temperature data for each station, two of which even had weather readings available back to the late 1800s. HOP constructed a database of these temperatures to allow the development of the company's own 10-, 20- or 30-year averages. Each month, HOP downloads the previous month's temperature data from the NCDC websites and updates its database accordingly.

A word of caution in dealing with weather statistics: the NCDC publishes what it calls its 30-year "normals" for each of its weather stations. These normals are updated once every 10 years in increments of a decade, and are widely used by newspapers, the television and the radio as a point of comparison for the day's weather; when the public is told that one day is two degrees colder than normal (for that time of year) this temperature comparison is being made by means of the NCDC's 30-year normal. As the interval between the 10-year updates of NCDC normals progresses, the data becomes more and more unrealistic to use as a basis of forecasting forward. For example, by 1998, the 10-year average temperatures in the HOP region were almost 5% warmer than the 1961–1990 NCDC published normals (possible reasons for this can be seen in Panel 2).[8]

In establishing the database for HOP's weather station cities, the company used historical high and low temperatures at each target city to develop its own un-rounded HDDs. Thus, when HOP develops a new year's programme, the same rounding techniques are used to set the strike point as will be used when actual performance is measured.

So far, this chapter has identified HOP as a business that is highly susceptible to changes in HDDs over a defined period of time. Reliable weather stations have been identified and a database of historical HDD data has been developed for each of

A CASE STUDY OF HEATING OIL PARTNERS' WEATHER HEDGING EXPERIENCE

> **PANEL 2**
>
> **ROUNDING TECHNIQUES**
>
> The NCDC reports both the daily high and low temperature rounded to the nearest whole degree, and the average of these rounded up to the next whole degree. The weather derivative industry calculates the average temperature for a 24-hour period from the rounded high and low temperature, but does not round up the result allowing for a difference between the two methods of up to a half degree.
>
> Over the course of a month, these rounding differences often cause an 8–10 HDD difference between the NCDC measurements and those used by the writers of weather derivatives. Over a five-month season, the difference has sometimes been 40–50 HDDs. When being compensated at US$12,000 per lost HDD (as HOP is), such rounding methodologies differences cause serious monetary consequences.

geographic target area. HOP is confident that, based on available historical records, it can develop reliable targets of HDDs for these weather stations and receive timely, accurate weather statistics to monitor company performance against its targets.

Measuring exposure

There is a high correlation between HDDs and gallons delivered during the five-month November–March period (the linear correlation coefficient is higher than 0.99). The tertiary heating season months of October and April have too few HDDs to use in the insurance coverage period and do not correlate as well with the number of gallons consumed as they do during the November–March period (the linear correlation coefficient for the months of October and April is approximately 0.80). As HOP delivers about 74% of its product during November–March, this is the period in which the company has its greatest exposure to reduced deliveries due to warmer than normal climatic conditions. Thus, we have defined the second variable needed in the programme: the measuring period.

Each year HOP develops a product-purchasing budget based on the number of residential customers purchasing oil in each of its major geographic markets (Boston (MA), Providence (RI), Hartford (CT), New Haven (CT), Newark (NJ), Allentown (PA), Philadelphia (PA) and Wilmington (DE)), their K-factors and a forecasted number of HDDs during the year. In the summer and autumn, as part of its price-risk management programme, HOP procures this product using forward contracts, futures or calls. Simultaneously, the company sets fixed and capped prices for customers, based on the predicted gross margins from sales. These forecasted gallons are based on a HDD target used as the basis for that year's budget. Thus, if the number of HDDs forecasted is not achieved, HOP will not realise the forecasted gross profits.

As with any form of insurance, 100% cover against any risk is very expensive. The cost of weather derivatives would be prohibitive for HOP if it was to seek recovery exactly at the 10-year average HDDs in each market. Experience has taught that the risk of the first 1–4% shortfall of HDDs has to be absorbed by the company before it can expect to initiate recovery from the weather derivative programme.

For HOP, the process of setting this recovery threshold required the patience of the market-makers who sell weather derivatives whilst the company developed a number of alternative HDD programmes and asked the derivative traders to price each one. Their offers were then entered into a matrix of cost versus recovery, similar to Table 1.

Using Table 1, HOP judged how much risk the company could afford to self-insure and how much risk the derivatives should cover. The choice here is somewhat

Table 1. Representative costs of alternative HDD thresholds

HDD deviation from normal (%)	Lost gross profits (US$)	Representative cost of programme (US$)
1	−115	420
2	−230	345
3	−345	275
4	−460	205
5	−575	145
6	−690	120

subjective and it depends on each company's "pain threshold" in terms of cost; a public reporting company that needs to make its forecasted profits might purchase coverage a little closer to the 10-year average, while a company that is only interested in catastrophic coverage might assume more risk by setting their programme on a higher strike point, thus not receiving any payback until temperatures are 4–5% warmer than normal.

In determining how much risk to absorb, it is important to study and understand the history of HDDs at each weather station. A station such as Boston, Massachusetts that is located on the Atlantic Ocean has a relatively stable history of HDDs over a prolonged period of time with a standard deviation over 50 years of ± 5.5%. Philadelphia, Pennsylvania, about 300 miles further south and further removed from the ocean, has a much less stable pattern of weather, with a 50-year standard deviation of ± 8.3%. As a result of this relative stability of past history, the cost of a derivative for Boston is less than the cost of a similar derivative for Philadelphia.

Studying the weather patterns will also help set the upper limits of the chosen protection. For example, the HDDs in Boston have only exceeded 10% warmer than normal once in the past 50 years and there are only two years in the past 50 that Boston's HDDs exceeded 7.5%. Philadelphia, on the other hand, exceeded 10% six times and 7.5% nine times over a similar period.[9]

Each city has a different historical pattern of winter temperatures, so once analysed, the upper limits of protection can be set city by city. In Boston, for example, the choice may be to insure from a strike point of 2% warmer than the 10-year average and protect to a maximum of 7.5% warmer. Accordingly, in Philadelphia, the strike target would be 2% warmer than the 10-year average and the limit of payback would be achieved at 12% warmer than normal.

The company's internal annual budgeting process allows the development of the potential gross profits (those that would be achieved if all the gallons that would be consumed if the HDDs reached the average temperatures for each city were delivered). Since one of the key budgeting assumptions is the number of HDDs expected for that city, by simple division, we can derive the gross profit per HDD for each city or the "tick size". The tick size is defined as the amount of payback needed from the company's derivative programme to compensate for the loss of one HDD in a particular city.

For example, assume that the gross margins expected at Boston are US$15,000,000 for normal temperatures for the November 1–March 31 period. If there are 5,000 HDDs in Boston during that period, by division we derive that HOP has an exposure of US$3,000 per HDD; a tick size of US$3,000 per HDD in Boston. We do a similar computation for each of the geographic areas chosen for weather derivatives protection.

At this point, knowing the strike point (the point at which the derivatives begin to compensate), the tick size (the gross profit recovery needed per HDD), and the upper limit of recovery (assumed above to be 7.5% warmer in Boston and 12% warmer in Philadelphia), there is just one missing ingredient: the upper monetary limits of protection. We simply take the upper percentage limit and multiply it by the HDDs for the city for the total period, resulting in the total number of HDDs that are needed to protect in that city. We then take the total number of protected HDD's and multiply it by the tick size to get the monetary amount of protection needed.

For example, if the 10-year average number of HDDs at Boston is 5,000, and we want to insure out to 7.5% warmer than normal, we need protection for 375 HDDs (5,000 x 7.5%). Multiplying these 375 HDDs by the tick size (US$3,000 per HDD), quantifies the maximum recovery of US$1,125,000 in Boston.

Once we have set these three elements, (strike point, tick size and upper monetary limit of protection), a weather derivative broker can price a corresponding put contract. If any of these elements were to change, the derivative traders would give a different price. Generally, a change in the strike point will have the largest effect on the cost of the derivative (altering the tick size would have only a slightly smaller effect). The monetary limit of coverage is not as important in pricing as the upper limit passes 1.5–2 standard deviations from the experienced average temperature for a city. The statistical significance of such a large deviation is remote and there is very little cost associated with increasing limits beyond that point. In fact, it is even possible to buy an open-ended contract with no upper limit.

Alternative pricing strategies

Although weather hedges could be bought for a single month at a time, to do so would be much more costly; the cost of five single monthly weather derivative contracts for November–March is about 30–35% higher than the cost of one programme covering all five months. The reason for this disparity is twofold. First, there is more risk assumed for a shorter period as the average over longer periods is more stable – that volatility is less in a longer period within a season.

PANEL 3

THE WEATHER DERIVATIVE

When purchasing a weather derivative from a seller, the buyer receives a standard International Swap Derivatives Association Inc (ISDA) contract. The transaction type is a weather floor derivative or a "put" contract. The contract specifies the calculation dates, weather stations, strike points, notional (tick size) amount, the maximum payoff amount and the premium charged. The contract specifies the weather reporting service (US National Weather Service) on which the calculation of HDDs will be based and the formula for rounding that will be employed. It also specifies that the final settle-up of the contract will be made on the fifth business day following the issuance of weather data by the weather reporting service. The contract is not restricted and may be traded freely once it is drawn.

It is very important to know the other party with whom the deal is made when entering a weather derivative contract; the buyer of the contract must pay the premium up front, and compensation (if any) from the counterparty will not be received until the end of the contract, some six or seven months later. It is critical to know that at the end of the contract, the other side will be present, willing and able to pay what is due.

The traders and dealers with whom HOP has dealt have all been credit-worthy institutions who trade these derivatives as part of a much larger portfolio. But remembering that Enron was one of the original weather derivative market-makers re-emphasises the importance of evaluating the counterparties with whom to trade.

Also, there is less trading liquidity in a monthly market compared to a five-month market. Weather derivative traders attempt to manage a portfolio of risks, balancing contracts that protect against risks from warmer than normal weather against contracts protecting against risks from colder than normal weather. The weather traders will protect themselves by purchasing contracts of risks opposite to those they are writing. In our example, if we wanted to protect against warmer than normal December HDDs, the traders need a market on the other side to balance their portfolio, and a one-month contract is relatively rare.

The cost of a three-month programme from December 1 to February 28 is about 90% of the cost of a five-month programme. However, by limiting coverage to the shorter three-month period, two months would remain without a hedge, and the cost per month to insure a HDD within the three-month window would be a lot higher per unit. Again, the same reasons apply to the three-month hedge as they do to a one-month hedge as to why they effectively cost more than the five-month hedge.

The weather forecast

For companies wishing to protect themselves with weather derivative contracts, there are a number of informative sources of weather forecasts available, detailing what is expected to happen in the relevant cities throughout the year. Each month the NCDC provides on its website a 30-day forecast and a rolling three-month forecast looking forward to the ensuing year.[10] The NCDC website supplies weather charts for both temperature and precipitation forecasts along with a technical (but understandable) prognostic discussion of the forecasts. The same site also provides 6- to 10-day and 8- to 14-day close-in forecasts.[11]

Each summer HOP's management spends much time reading and discussing these forecasts. As the company is firmly committed to covering itself against the risks associated with warmer winters, long-range weather forecasts tend to be disregarded and instead HOP's weather hedges are created without depending on what the forecasters predict. This is not a matter of trusting the ability of the weather forecasters, but that when trying to protect a company against the perils of warmer winters; it is more prudent to be careful than to wonder, in hindsight, why enough protection was not bought. In speaking with the traders from who the derivatives are purchased, weather forecasts do play a part in their calculations when pricing the programmes for HOP, but in the end, they are more likely to depend on historical data than forecasted weather. It is also true that an actual weather event like an El Niño will tend to influence the weather derivative traders more than a pure forecaster's prediction, but most of the derivative pricing is based on statistical probability associated with historical data rather than forecasts.

HOP's success with weather derivatives

As previously mentioned, HOP purchased its first weather protection for the 1997–8 winter, a year with an El Niño event. Temperatures in the northeast US hit a 103-year high point, and HOP collected about US$4.5 million from weather insurance that season.

The following year, with the underwriters sensitive to their losses the year before, HOP set up a more modest programme but still collected US$2.4 million when temperatures averaged 6% warmer than normal in the company's geographic region. The winter of 1999–2000 was again about 7% warmer than normal and the weather hedge programme repaid US$2.2 million for lost profits.[12]

Finally the 2000–1 season worked in favour of the weather derivative writers; the winter temperatures in the northeast US averaged about 5% colder than normal and HOP made no recovery from derivative contracts (though this was still acceptable as product sales during the 2000–1 season exceeded budgeted sales by approximately 5%). This was the first year in the last five that HOP experienced colder than normal temperatures throughout the region in which it operates.

A CASE STUDY OF HEATING OIL PARTNERS' WEATHER HEDGING EXPERIENCE

HOP's current year's programme (2001–2) set strike points in the selected cities approximately 1.2% warmer than the 10-year average; coverage originally extended out to 10% warmer than normal. Almost every long-range weather forecast read in the summer and the autumn of 2001 called for a colder than normal winter during 2001–2 and a weather pattern somewhat similar to the previous heating season which had resulted in 5% colder than normal temperatures. As a result of these consensus forecasts, and not having incurred losses the previous year, the weather derivatives market was a little softer: HDD puts were less expensive than the previous year, allowing HOP to pick up slightly more coverage for the same premium as the prior year.

Defying all forecasts, the November 1, 2001–March 31, 2002 period was the warmest on record for the 100+ years of history recorded by the NCDC.[13] As of the middle of December 2001, HOP had experienced 18.5% warmer than normal temperatures throughout the region and had accrued approximately US$2.7 million in value for the derivatives it had purchased. Remember, the period of coverage of HOP weather derivatives is from November 1–March 31, so there was no certainty that these accrued values would be recovered at the end of the five-month period. A colder than normal January and February could create additional HDDs and easily wipe out the shortfall of HDDs that had occurred during November and December's warm weather.

Since the weather forecasters were calling for a very cold January 2002 in the northeast US, HOP approached the market in mid-December and asked a number of derivative traders for a quote to buy (or cash-in) the full programme of five-month weather derivative contracts from HOP. HOP also asked traders for prices for a new programme beginning February 1, running until March 31, giving the same coverage during that period as under the original contract. Although numerous offers were made to buy the put contracts from HOP, most monetary amounts were discounted by about 10–25% from the 2.7 million already accrued under the programme. Their quotes for a new February–March contract revealed that the net cost of selling the existing programme and purchasing a replacement programme for February–March was equivalent to experiencing 10% colder temperatures than normal HDDs during January.

In other words, if HOP took its profits in cash in mid-December, went without protection in January and reinstituted a programme in February, it would be no different than if HOP let the existing contracts run and it was 10% colder than normal in January. If it was warmer than that, HOP was better off holding what it had. If HOP expected January to be more than 10% colder than normal, then it should take its cash profits and institute the substitute programme. After due consideration, HOP left the contracts in place and, in spite of what the forecasters had predicted, January was 17% *warmer* than normal, accruing an additional US$2 million in recovery from derivative contracts for that month alone.

As the end of January 2002 approached, the winter 2001–2 derivative contracts were already nearing the upper limits of their protection. HOP went back to the market and procured an additional US$3 million of upper-limit protection beyond that which had originally been projected to be adequate. The derivative traders were willing to write this additional coverage on the basis that attaining such high limits was a statistically unlikely event. Never in the 100+ years of recorded weather history, had the northeast US experienced a winter with a divergence from average that even approached one and a half standard deviations.[14] In retrospect, it would have been very inexpensive if HOP had undertaken higher limits at the beginning of the season rather than to add an additional upper layer of coverage mid-way through the season. Possibly, at the time which the original programme was set, HOP was somewhat persuaded by the previous summer's consensus from weather forecasters that the forthcoming year would be colder than normal, making the excessive coverage seem wasteful, and the added protection was much more than was needed.

As can be seen, there is a noticeably large amount of flexibility and depth in the weather derivative market. Hence, when HOP put its winter 2001–2 eight-city programme out to bid, it received responses from nine derivative traders, plus two unsolicited proposals from other brokers. There is a very active market for weather derivatives, and it is a continual market that does not stop once the original programmes are set. There is an active secondary market of derivatives trading throughout the year providing liquidity the need arise to turn an accrued profit into cash.

HOP's use of weather derivatives has been very well-placed. January 2002 was 17% warmer than the 10-year average; February was 14.3% warmer and March was 7.3% warmer. As a whole, the five-month period from November 1, 2001 to March 31, 2002 was 15.2% warmer than normal. HOP's total recovery for the 2001–2 winter season under its original and extended weather derivative contracts was approximately US$7.1 million dollars. Under the terms of HOP's derivative contracts, it submitted its claim on April 1, 2002 and was paid in full by the counterparty on April 5, 2002.

Conclusion

As such a staunch proponent of using insurance and derivative contracts to help defray the risks of business associated with the weather, HOP, along with others who have protected their companies against their particular weather risks, have been invited to speak at numerous risk seminars and gatherings of derivative traders to explain how these derivatives have helped to stabilise the company's profitability. At these conferences, representatives from a number of electric and natural gas utilities throughout the US spoke about how they purchase weather insurance against shortfalls in winter HDDs or overages in summer CDDs. For instance, an electric utility buys forward contracts from neighbouring suppliers to purchase expected excess electricity consumption during peak seasons. Should a summer season prove to be warmer than normal and air conditioning electricity consumption spikes above the amount forecasted by the utility, they would be forced to buy additional peak capacity at a very high spot market price. Thus, by purchasing excess CDD protection, the electric utility would be compensated from the insurance or

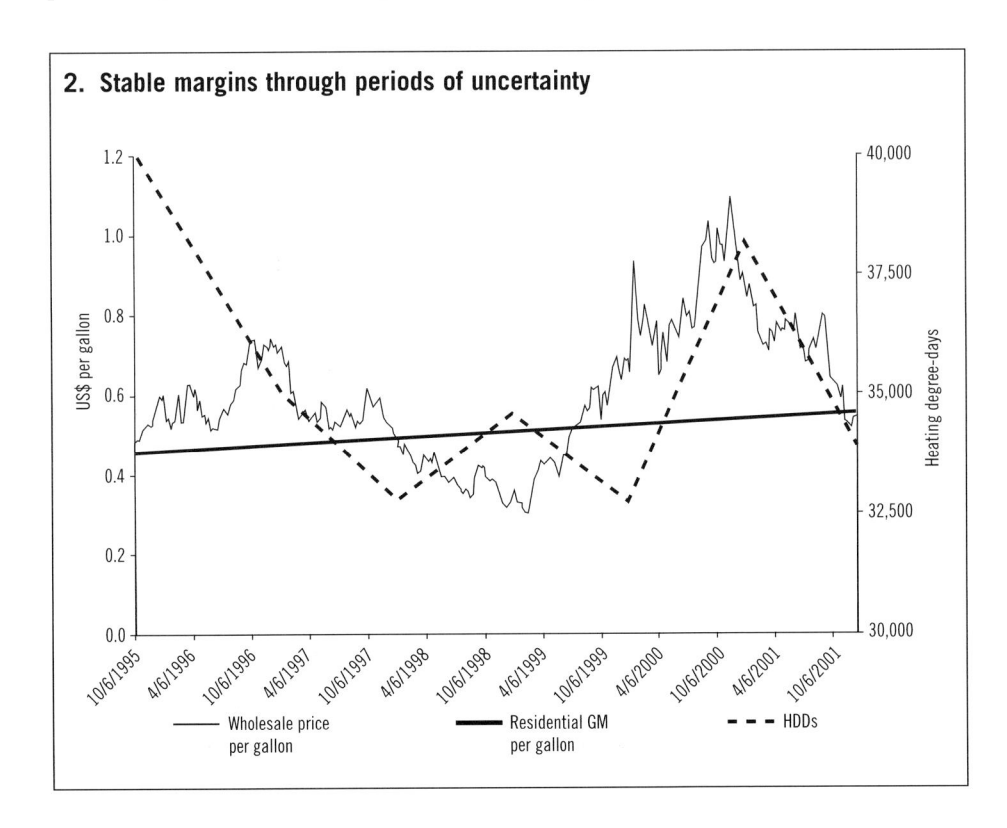

2. Stable margins through periods of uncertainty

A CASE STUDY OF HEATING OIL PARTNERS' WEATHER HEDGING EXPERIENCE

derivative contract to offset the costs of purchasing the high-priced extra peak capacity.

HOP recently undertook a programme to educate the large commercial heating oil accounts serviced by the company on the advantages of purchasing HDD derivatives to protect their budgets against colder than normal temperatures. This is the opposite side of the protection sought by HOP; if it is colder than normal HOP delivers more gallons, while the people to whom HOP deliver are forced to spend more. They could be protected in the same way as HOP.

Weather insurance can now be based on indexes other than temperature; at one conference a large retailer explained how important the spring season was to the business and how they purchased a combination of temperature and precipitation protection to ensure that bad weather did not crimp their profitability for the important spring selling season. Likewise, ski slope operators purchase HDD protection to insure against the possibility that there would not be ample snow.

This case study can be seen as a stimulus for others to consider how weather hedging can benefit their businesses. The following chart shows that while the experienced HDDs and the wholesale price of heating oil have varied dramatically over the past five years (the blue background and the yellow line), HOP's gross margins (the red line) have remained very stable. This chart is a dynamic representation of what a well thought-out risk protection programme can do for a company.

Two suggestions for those considering setting up a weather derivative programme would be:

1. Determine the protection you need, and to stick with it year in and year out regardless of what the meteorologists are saying.
2. Treat the weather hedge instruments much in the same way as you do business interruption insurance: set aside part of your budget each year to protect against the effect weather will have on your earnings streams. Buy the hedge every year and just hope you never have to collect on it.

1 There are very few US heating oil distributors that are publicly owned, thus data on competitors is limited. Among the Industry Trade Organizations, Heating Oil Partners is generally recognised as the second largest distributor in the US.
2 US Energy Information Administration. Household Energy Consumption and Expenditures, 1977. Table CE1-9c.
3 US Energy Information Administration. Household Energy Consumption and Expenditures, 1977. Table CE1-1c.
4 To compare heating oil with natural gas and other fuel sources, it is necessary to identify a common measurement. Since all of these fuel sources create heat, the measurement of comparison is the btu (British Thermal Unit). The cost and efficiency of producing one btu is different for each type of fuel.
5 Based on Industry Trade Organization data.
6 During the winter of 2000–1, the eight cities HOP measured experienced 38,203 HDDs vs the 10-year average of 34,876 and the 30-year average of 35,740 for these stations.
7 See Figure 1. The HDDs for the eight cities in our index were 8.8% warmer than the 30-year average, making it the fourth warmest winter in 50 years.
8 The US NCDC website and online store contain a wealth of historical data. See: http://ols.nndc.noaa.gov/plolstore/plsql/olstore.main?look=1.
9 Data for New England cities is available on the website: http://www.erh.noaa.gov/er/box/AveragesTotals.shtml. Historical HDD data for Pennsylvania is available on the website of the Pennsylvania State Climatologist: http://pasc.met.psu.edu/PA_climatologist/ids/index.html.

12 *During the winter of 1999–2000, the seven cities HOP measured that year experienced 29,240 HDDs vs the 10-year average of 31,323 and the 30-year average of 31,928 for these stations.*
13 *During the winter of 2001–2, the eight cities HOP measured experienced 30,043 HDDs vs the 10-year average of 35,408 and the 30-year average of 35,162 for these stations.*
14 *HDDs experienced during winter of 2001–2 exceeded the previous warmest winter (1990–1) by 6.1%.*

16

Weather Indexes for Developing Countries

Panos Varangis; Jerry R. Skees; Barry J. Barnett[1]
World Bank; the University of Kentucky; the University of Georgia

The challenge of coping with natural disasters has increased as populations have increased. These challenges have been particularly acute in developing countries where both the human and economic loss can be staggering and financial resources are limited. Reducing economic vulnerability to weather events in developing countries may very well be the most critical economic development challenge of the new millennium. As a proportion of GDP, natural disaster losses in developing countries are 20% greater than in industrial countries.[2] The economies of many developing countries rely heavily on agriculture and agricultural success is directly tied to weather.

In recent years, the international community has focused more attention on the relationship between weather disasters and poverty. Measures taken to reduce the economic impact of weather disasters can provide substantial advances in the fight against poverty. Although weather cannot be controlled, weather risk markets can be used to offset the financial impacts of adverse weather events, and possibly compensate for human suffering in developing countries. This chapter begins with a brief review of the degree to which weather events contribute to the human and economic suffering that accompanies weather events. The story is compelling – extremes in rainfall and temperature account for the vast majority of documented problems. These issues deserve attention from both the public and private sector.

Introduction
The questions raised in this chapter surround the possible linkages between the emerging weather markets and public and private solutions to the problems created by weather-based natural disasters. While this book deals with many of the conceptual issues surrounding weather risks, we also provide our conceptual base that is tied to how these risks can be segmented and layered. By segmenting and layering out weather risks, there are numerous opportunities for risk aggregators within a country to share risks globally. We examine three possible applications of weather insurance in developing countries: providing weather risk coverage to:

1. mutual insurance groups,
2. farmers and agri-businesses (directly); and
3. governments to protect their exposure when they offer catastrophic insurance and provide disaster aid.

Finally, as is emphasised throughout this book, we close by focusing on the needed infrastructure for measuring weather and making a strong argument that a supporting infrastructure has vast social benefits. These social benefits can only be recognised once the numerous ways that weather indexes can be used to cope with

WEATHER INDEXES FOR DEVELOPING COUNTRIES

and manage weather-based risks are understood. Private companies within developing countries, global markets that move the weather risks out of developing countries, governments of developing countries and the international donor community that is quick to respond when there are natural disasters in developing countries, can all play a role in using the same information and structure to address the problems brought on by extreme weather events.

Why is weather risk important in developing countries?

The Office of US Foreign Disaster Assistance (OFDA), in collaboration with the Centre for Research on the Epidemiology of Disasters (CRED) has compiled an emergency event database in an attempt to document both technological and natural disasters around the world.[3] While such data are difficult to obtain and certainly do not represent all natural disasters, the picture that emerges from this database is clear: weather events dominate the documented natural disaster problems. Table 1 clearly shows that either too much or too little rain accounts for the vast numbers of people affected by natural disasters. Events back to the early 1900s are documented in this database. The total number of people affected by weather events is well over four billion during this time period. The documented deaths due to flood and drought are ten times greater than those from earthquakes. Windstorms and extreme temperature events also account for significant problems.

The same natural disasters that contribute significantly to human suffering also create economic problems in developing countries. Beyond the direct economic losses, economic growth in many developing countries can be slowed due to adverse weather events. In part, this is because the economy of the majority of developing countries is highly dependent on agriculture. Severe weather events can have devastating impacts on farmers and on the wider population that relies on local agricultural production.

Table 1. Top natural disaster events ranked by total people affected and cost

	Total affected	Damage (US$)
Flood	2,349,000,000	307,800,000
Drought	1,673,900,000	51,042,445
Typhoon	220,550,000	48,419,742
Cyclone	149,600,000	19,377,285
Storm	88,177,056	54,054,337
Earthquake	73,191,744	320,830,000
Drought	26,714,267	–
Food shortage	26,330,940	93,449
Hurricane	16,043,903	85,838,545
Tropical storm	11,397,020	4,786,050
Crop failure	10,168,094	–
Winter storm	5,408,961	39,932,957
Heat wave	4,605,588	6,715,809
Volcano	3,754,080	3,668,779
Forest fire	3,636,236	25,312,899
Cold wave	2,500,063	15,544,150
Tornado	1,378,358	23,940,841

Source: OFDA/CRED International Disaster Database

Consider the example of Morocco. Table 2 shows the historic relationship between rainfall and economic growth. Between 1992 and 2000, real GDP growth was negative or very low in years with poor rainfall during the agricultural growing season. Conversely, high real growth in GDP occurred during years of average or better than average rainfall.

Weather risk markets and developing countries

Weather risk markets are among the newest and most innovative of markets for transferring financial risks. The financial instruments traded in weather risk markets are often called weather derivatives. Though there is some variability in the specific characteristics of weather derivatives, most are essentially index options. That is, they are financial options that settle based on the value of some underlying index measured by an objective third party (Hull, 2000). A common example would be an option on a stock index, such as the S&P500. The underlying index for weather derivatives is a measurement of some weather phenomenon, by an objective third party, at a given location over a specified period of time.

Participants in weather risk markets are drawn from a broad range of economic sectors such as energy, insurance, banking, agriculture, leisure, construction and entertainment. To date, the US energy sector has been responsible for most activity in weather risk markets but new participants from Europe, Asia, Australia, Latin America and Africa are now entering the market.[4] According to a survey conducted by PricewaterhouseCoopers, more than US$7.5 billion of weather risk has been transferred in weather risk markets since 1997.[5] As economic agents from different sectors gain a better understanding of the impacts of weather on their economic activities, weather markets are likely to expand further.

The agricultural sector has thus far made only limited use of weather risk markets. In the Canadian provinces of Ontario and Alberta, weather risk instruments have been used to hedge forage production risk. In late 2001, AGROASEMEX, the state agricultural re-insurance company in Mexico, used weather derivatives to reinsure part of its weather-related crop insurance risk. AGROASEMEX found that weather derivatives offered protection at lower layers of risk exposure and at lower prices than traditional retrocession reinsurance. There are current plans to launch weather risk contracts in Argentina for insuring dairy production against low rainfall and in Morocco for insuring sunflower and wheat production against low rainfall.

Yet many in the agricultural sector are unaware of emerging markets for weather derivatives. Further, in many developed countries highly subsidised agricultural

Table 2. Moroccan rainfall and GDP growth rates

Year	Rainfall (November of previous year to March of current year)	Real GDP growth (%)
1992	Poor	−4.0
1993	Poor	−1.0
1994	Heavy	10.4
1995	Very Poor	−6.6
1996	Exceptional	12.2
1997	Poor	−2.3
1998	Average to Good	6.5
1999	Poor	−0.4
2000	Poor	0.7

Source: Bank al-Maghrib, EIU

insurance schemes crowd out the demand for agricultural weather risk instruments. Developing countries, on the other hand, typically have no government-subsidised agricultural insurance. They are, however, highly susceptible to weather risks and often, highly dependent on agriculture. Thus, developing countries would seem to be a potential growth market for weather risk instruments.

Classifying risk – tying weather risk to realised losses

Users of weather derivatives must understand the complex relationships between weather events and losses. First, they must identify the specific weather event(s) that caused the losses, such as lack or excess of rainfall, high or low temperatures, high winds, or a combination of these. Second they must establish the period over which the occurrence of certain weather event(s) is critical. Third they must investigate the duration and intensity of the weather event that provoke losses. With frost, freeze, or high winds, losses can occur within a few hours. Drought losses generally occur over extended periods of time. Intense rainfall can cause some losses within a few hours while other losses would not occur unless the rainfall was sustained over several days.

Having identified the source of risk, the time period when losses are most likely to occur and the duration of the weather phenomenon that triggers losses, one must next determine the frequency, severity, and spatial-correlation of the weather phenomenon. It is important to note that this second set of risk characteristics are conditioned on the first set.

Extreme weather events can cause loss of life, loss of property and loss of subsistence and/or income-generating production. Sometimes, a single weather event can cause all of these types of losses. For example, in the autumn of 1998, Hurricane Mitch's torrential rainfall and high winds left a wake of death and destruction across the affected Central American countries.[6] Since the same weather events can cause various types of losses, the infrastructure to measure, classify and pay for extreme weather events can serve multiple purposes.

Advances in computer modelling have greatly improved the understanding of the relationships between weather events and losses and meteorological models continue to improve our understanding of weather phenomena. Models of hydrology and plant growth are among the many computer simulation applications that allow us to better understand the relationships between weather phenomena and potential losses. Geographic information systems map the spatial dimensions of affected areas. Many computer applications have been employed to improve our understanding of the relationships between weather events and losses of life, property and sustenance.

Once these important relationships are understood, many uses of weather derivatives are possible. To illustrate, consider the problems associated with extreme rainfall in a developing country. If unusually high rainfall occurs around harvest, crops can be ruined. Even more severe rainfall may cause flooding with resulting loss of life and property. Similarly, unusually cold temperatures in a developing country may damage crops but the extreme cold can be a threat to animal or even human life. Depending on the severity, drought can also cause losses ranging from crops to human or animal life.

A number of studies have examined the relationship between weather events and crop yields. Crop growth processes are complex and multiple weather events during the growing season influence the realised yield. Dischel (2001) describes the impact of both rain and temperature on California almond yields. Skees, *et al.* (2001) examines a portfolio of cereal crops for Morocco and shows that the same rainfall events influence the yields for three different crops. This of course opens the possibility of using the same weather index for managing yield risks on all three crops.

Skees and Zeuli (1999) demonstrates how rainfall derivatives can be used to aid in hedging against irrigation risk. Many developing countries depend heavily upon irrigation for agricultural production. Major conflicts over water are commonplace. Water markets are likely the best mechanism for addressing conflicts over water usage.

Many have argued that water markets can allocate water more efficiently than governments.[7] In developed countries, rainfall derivatives have been used to protect hydroelectric plants from the risk of low water levels when there is limited rainfall in the watershed that feeds the reservoir. Similarly, in developing countries, farmers who depend on reservoirs for irrigation could use rainfall derivatives to protect against insufficient rainfall in the watershed. Further, Skees and Zeuli (1999) argues that the water authority could purchase rainfall derivatives and then sell water rights with an embedded warranty to deliver either water or an indemnity payment to users of water. Such an innovation might make water markets more palatable in many places in the world.

BASIS RISK

Demand for weather index instruments may be affected by the presence of basis risk. Spatial variability in weather is one source of basis risk. For example, consider a situation where a farmer purchases a weather derivative that will pay an indemnity if rainfall measured at a nearby weather station is below some trigger level. It is possible that rainfall measured at the weather station may be above the trigger level while the actual rainfall at the farm is well below the trigger level. In this case, though the farmer has likely experienced losses, no payment will be received from the derivative. Basis risk can also occur due to less than perfect correlation between the weather event and losses.[8]

It may be possible to offset some basis risk by triggering payments using average measurements over a longer period of time (monthly or even quarterly) and/or using average measurements over more than one weather station. Also, it is important to note that basis risk due to spatial variability in weather is more of a problem for individual buyers of weather contracts (eg, a farmer) than for producer associations, financial institutions and agro-industries, whose exposure to loss risk is spread over much larger geographic regions. Whether basis risk overwhelms the significant benefits of weather index instruments is an empirical question, the answer to which depends largely on the spatial correlation of weather events and the correlation between weather events and actual losses.[9]

LAYERING

A private insurance provider may be interested in designing a weather index insurance that protects farmers and others at risk from weather events. There are

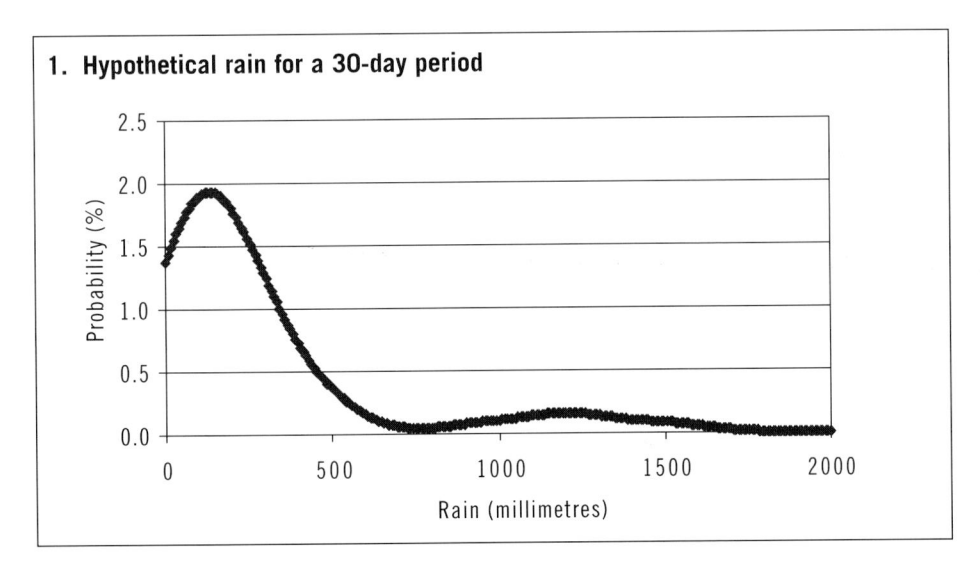

1. Hypothetical rain for a 30-day period

many possible ways to design weather index insurance and policies could be sold through various market channels. However, for extreme weather events, institutional arrangements may be required that go beyond what insurance markets in developing countries can typically provide.

Figure 1 presents a hypothetical probability distribution for rainfall over a 30-day period.[10] The high frequency of low rainfall likely affects crop yields. Rainfall above 500 millimetres (mm) may also damage crop quality. Very extreme rainfall events likely cause floods with resulting loss of life and property. The extreme rainfall events beyond 1,000mm may be created by hurricanes. Thus, those wishing to hedge against hurricane risk may also desire weather insurance based on wind speed.

Consider a rainfall derivative based on the probability distribution in Figure 1. Suppose the policy pays US$100 for each millimetre of realised rainfall between 500 and 1,000mm. If realised rainfall was equal to 1,000mm, the payoff on the derivative would be US$50,000 ((1,000mm − 500mm) × US$100 per mm). No additional payments would be made for rainfall realisations higher than 1,000mm. In the terminology of weather derivatives, this contract would have a strike of 500mm, a tick rate of US$100 per millimetre and a limit of US$50,000.

Without the limit, the contract would be very expensive. Insuring against the risk in the upper tail of the probability distribution or the "catastrophic" risk is extremely expensive. There are few historical observations in the tail of the distribution so it is difficult to accurately assess the likelihood of these extreme events. Because of this ambiguity, weather risk market participants are often reluctant to trade derivatives with very much exposure to catastrophic risk (sometimes called "tail risk"). Derivatives that do cover tail risk are usually very expensive since sellers will load the price to account for the ambiguity associated with the tail of the distribution.

Within the range of 500–700mm, trading of weather derivatives is more likely (as the distribution is smooth and "well-behaved" up to roughly 700mm). This layer of the rainfall probability distribution may offer adequate protection for many types of risks. Risk protection is often sold in layers. Once the probability distribution has been estimated, it is relatively straightforward to identify different layers of risk.

Different institutional structures seem to work better for transferring different layers; weather risk markets are more efficient at trading layers that are closer to the mean of the probability distribution, whereas standard insurance and reinsurance markets have typically been more efficient at providing protection for tail risk. As compared to weather markets, standard (re)insurance markets generally have much more liquidity for layers in the tail of the distribution.[11] In weather risk markets there is generally more liquidity for layers that are closer to the mean of the probability distribution. This liquidity improves pricing efficiency. For this reason, some users of weather derivatives prefer to protect against tail risk by scaling up the liability that they purchase in these layers (by buying more protection than they need in this layer) rather than by trying to purchase layers in the tail of the distribution. The additional liability purchased for the layer closer to the mean is then used to compensate for any deeper losses should a weather event in the tail of the distribution occur. Such tactics are common among those who use exchange-traded instruments such as futures and options contracts to hedge risk.

Evolving catastrophe (cat) bond markets may one day facilitate the bundling of weather derivatives for catastrophic layers, thus allowing for more efficient pricing. cat bonds offer contingent funding for well-specified natural disasters. Those exposed to losses from the specified natural disaster sell cat bonds to investors. If the specified natural disaster does not occur, investors recoup their initial investment and earn a highly favourable rate of interest. If the specified natural disaster does occur, investors lose their interest payment and some or all of the initial investment, depending on the structure of the contract. Since natural disasters are not correlated with other equity markets, cat bonds provide an excellent opportunity to diversify a portfolio of investments.[12]

Possible uses of weather indexes in developing countries
Various institutional arrangements could be used to transfer weather-related risks in developing countries. Once an index is developed for a particular weather phenomenon, there are many potential applications for the index. For example, consider an index for rainfall measured at a particular weather station over a given period of time. The index could be used to:

1. allow collective mutual assistance groups to transfer their systemic exposure to rainfall risk to outside investors;
2. provide private-sector weather-based insurance directly to economic agents at risk from too much or too little rainfall; or,
3. establish weather-based parametric triggers for the provision of government disaster assistance.[13]

WEATHER INDEX INSURANCE AND COLLECTIVE MUTUAL ASSISTANCE GROUPS
Economic agents in developing countries often use informal risk management arrangements. For example, in a village or among a group of farmers, there may exist an informal understanding that the group will assist individuals who suffer losses. These arrangements, which are not unlike mutual insurance companies, provide a very important mechanism for dealing with idiosyncratic risks. However these arrangements can break down when regional weather events generate simultaneous losses for all members of the collective assistance group. Examples of such weather events would include severe droughts, floods, hurricanes or freezes. By providing opportunities to transfer systemic, spatially-correlated, risks, weather risk markets can complement such informal arrangements that are better suited to sharing losses from idiosyncratic risks. Several researchers have proposed similar arrangements in the US for farmer-owned processing or marketing cooperatives.[14] The cooperatives could become mutual insurers of crop yield risk that purchase weather risk instruments to offset their exposure to regional weather events such as drought.

WEATHER INDEX INSURANCE SOLD DIRECTLY TO PRODUCERS AND AGRI-BUSINESSES
In many developing countries, markets for insurance and re-insurance are either underdeveloped or nonexistent. This is particularly true for agricultural insurance. Traditional crop insurance is based on individual yield losses and thus, requires extensive field inspections and human judgments. The administrative costs are quite high, particularly in developing countries where arable land is often divided into very small plots. Crop insurance also has the usual problems of adverse selection and

---

PANEL 1

FONDOS DE ASEGURAMIENTO

A good example of how weather insurance could enhance mutual insurance arrangements amongst farmers is the Mexican *fondos de aseguramiento*, or mutual funds. These are non-profit, civil associations and they operate in such a way that the collected premiums create reserves to pay indemnities and cover operational costs. However, in the event of severe regional (or large-scale) events such as drought, excess humidity and frosts, the collected premiums and reserves are not sufficient to cover losses because the weather events affect all farmers at the same time. Obtaining insurance for these weather perils is critical for the financial viability of the mutual insurance funds in Mexico.

WEATHER INDEXES FOR DEVELOPING COUNTRIES

moral hazard.[15] These problems are aggravated in developing countries where information on individual farm yields is scarce.

Weather insurance, based on the occurrence of weather events rather than on actual losses such as crop failure, could be used to cross-hedge agricultural production risk. The underlying assumption is that certain weather events (such as rainfall or temperature) are highly correlated with crop losses. For example, a weather insurance contract could be written that protects against severe rainfall deficiencies over a given period. The contract would specify a rainfall trigger and an indemnity that would be paid for each millimetre that realised rainfall fell below the specified trigger. The contract would settle based on actual rainfall measured at an official weather station.

A number of studies have emphasised the advantages of weather-based insurance over traditional crop insurance in developing countries.[16] The traditional crop insurance problems of moral hazard and adverse selection occur because farmers have better information than the insurer about the probability of crop losses and/or the potential magnitude of losses. But this is unlikely to be the case with weather phenomena measured at official weather stations. Since no field inspections are required, weather-based insurance should also have significantly lower administrative costs than crop insurance. Weather insurance is also more transparent and easier for farmers to understand than traditional crop insurance. Weather insurance could be marketed to farmers through a variety of institutions in developing countries. Among these are banks, farm cooperatives, input suppliers and micro-finance organisations.

Farmers are not the only ones who might be interested in purchasing weather insurance. Agricultural traders, agro-processors, even shopkeepers or landless labourers, anyone whose income stream can be negatively affected by weather events, might benefit from purchasing weather insurance. Banks and rural finance institutions could purchase such insurance to protect their portfolios against defaults caused by severe weather events.[17] In Argentina agricultural input suppliers often provide inputs to farmers on credit. Some suppliers have considered requiring farmer-creditors to purchase a weather insurance policy. The weather insurance policy would help insure that farmers would be able to repay the input suppliers should crops be lost due to extreme weather events. If banks, input suppliers and other creditors can use weather insurance to offset some of their exposure to credit risk, they may be willing to provide more credit and at better terms. This is an important issue for many developing countries as credit availability to agriculture is typically constrained, in part because of exposure to systemic weather risks.

Sellers of weather insurance could use international weather risk markets to transfer their systemic weather risk exposure to investors outside the country. Some developing countries have formal insurance sectors that provide standard insurance products like property and casualty cover. The loss risk from many of these insurance products can be highly correlated with weather risks. If spatially correlated severe weather events occur, all or nearly all of those covered by insurance policies must be compensated at the same time. This can pose an intolerable level of risk exposure for local insurance providers. For this reason, local insurers use mechanisms to spread these financial risks internationally. Until recently, the only viable way for national insurance companies to transfer spatially correlated risks was to purchase reinsurance from various international providers. Reinsurance markets remain an extremely important mechanism for spreading spatially correlated risks, but these markets are subject to cyclical pricing, high transactions costs and large premium loads.[18] Further, even major international reinsurance companies have limited capacity to absorb large amounts of covariate risks. Weather risk markets provide alternative mechanisms for transferring these covariate risks to entities outside of the region or the country.

INDEX INSURANCE AND GOVERNMENT DISASTER ASSISTANCE

It is not uncommon for governments in both developed and developing countries to provide disaster assistance following major natural disasters. In developing countries, government assistance is often supplemented by international donor organisations. One of the problems with such intervention is that it becomes highly political. Without clear rules for how funds will be distributed when there is a disaster, serious inequities are common.[19] The question of how best to distribute disaster aid is beyond the scope of this paper. However, it can be said that it is important to leverage both government funds and foreign aid to obtain the best possible solutions.

Disaster assistance can also become self-perpetuating. If individuals expect assistance following a natural disaster, economic decisions will be made without the full social costs of those decisions being taken into consideration. In effect, while individuals receive the benefits of their private economic decisions, the risks inherent in those decisions are socialised. Resources that might generate more social benefits if used elsewhere, are instead used to shield risk-takers from the full consequences of their private economic decisions. Thus, private decision-makers are likely to engage in more risky activities setting off yet another cycle of losses and disaster payments.[20]

Various indexes of natural hazard risks are now being used to construct financial instruments. For example, the Richter scale is used to construct financial instruments based on earthquakes; wind speeds are used to construct financial instruments that protect against hurricanes; rainfall is used to protect against drought or flood. Since these indexes can be represented by a probability distribution, parametric estimates can be made of the expected payoff (the pure premium) on the instruments.

Relative to current methods, there are many potential advantages to providing disaster assistance based on weather indexes. The first is transparency. If the criteria for triggering disaster assistance have been well defined *ex ante*, politicians may find it more difficult to use disaster assistance as a means for transferring funds to favoured regions of the country.

The second potential advantage is that with sufficient years of accurate data, the parametric triggers can be set so as to normalise the expected value of disaster assistance across various regions of the country. This would reduce the perverse incentives inherent in disaster assistance programmes that continually compensate those whose private decisions expose them to more risk. Furthermore, the information about relative differences in triggers across regions could be made publicly available. This would provide valuable information to both public and private decision-makers about the relative risk exposure to different extreme weather events in different regions of the country.

National governments might consider providing regions, provinces, communities, or businesses, with either catastrophic weather insurance or disaster assistance based on clear triggering mechanisms for extreme weather events. In effect, the national government would be pooling catastrophic risks across the country. In large countries or climatically diverse countries, such pooling would allow diversification to reduce the aggregate exposure to catastrophic risk.

In small or climatically homogeneous countries, pooling may offer few diversification benefits. In these situations, cat bonds have been promoted as a means for spreading low probability-high consequence catastrophic risks to risk takers around the globe.

Still, the cat-bond market has not matured as expected. There are a number of reasons for this. Since cat bonds are relatively new instruments, many still do not understand how they can be used. Furthermore, there can be significant development costs for cat-bond contracts. Parametric cat bonds, which trigger payments based on an objectively measured index of a natural event (eg, Richter scale measurements, flood levels, wind speed, etc), are likely to have the lowest development costs.

WEATHER INDEXES FOR DEVELOPING COUNTRIES

Disaster assistance itself can also be too expensive for many developing countries. This is particularly true following a major disaster. To the extent that natural disasters disrupt GDP growth, financing disaster assistance internally (which, in essence, is a form of self-insurance) is even more problematic.

However, societies must provide immediate emergency response to victims of natural disasters. In addition, societies also need markets that allow individuals to protect against financial losses when natural disasters occur. Ironically, parametric cat bonds can facilitate both of these needs. Furthermore, the same instrument and structure could be used for both. This, of course, means that transaction costs could be reduced significantly.

A number of international relief agencies are involved in providing disaster assistance in developing countries. These agencies need ready access to cash when a natural disaster occurs. Parametric cat bonds could also provide this contingent funding. A relief agency could identify donors who are concerned about natural disasters in a specific region of the world. The agencies would ask those donors to provide the funding needed to purchase cat bonds that provide funding contingent on specific natural hazards occurring in that region. If the natural disaster occurred, the cat bond would provide funding needed to provide disaster assistance. The same donors could be encouraged to sell the very cat bonds they are helping to purchase. Then if the event did not occur, the donors would profit from the earnings on the cat bond. If the event did occur, donors would know that the payments they are required to make on the cat bonds are, in turn, being used for disaster assistance in their regions of concern. In essence, this is a sophisticated and efficient way to provide contingent funds for relief agencies and spread catastrophic risk around the world. Over time, the expected cost to the donor would equal the expected payoff on the cat bond. Currently, this is not the case as significant premium loads are added to cat bonds to account for ambiguity and lack of understanding. If donors become familiar with cat bonds through philanthropic activities, they might be more likely to include such instruments in their investment portfolios.

A ROLE FOR INTERNATIONAL CAPITAL MARKETS
These are just some examples of how contingent claims instruments such as weather indexes and parametric cat bonds could be used in a developing country context. What is common across these examples is the need to transfer out of the country or region systemic weather risk that is un-diversifiable in a local context. Collective mutual assistance groups, cooperatives, insurance companies, banks or other creditors, government disaster relief agencies, international relief agencies – all are exposed to potentially devastating natural disaster risks. Many of these risks are caused by systemic weather events. Weather risk markets could be an important mechanism for facilitating the transfer of these risks outside of the local area. With reduced exposure to covariate weather risks, economic agents should be willing and able to increase their investments in local economic opportunities.

Weather risk instruments and other instruments for transferring covariate risks offer a unique opportunity to link world financial markets and developing countries in a partnership that is mutually beneficial. Investors and portfolio managers should find these instruments attractive since the returns would be largely uncorrelated with those of most other investments. Systemic weather risk that was undiversifiable at a local level would provide significant diversification benefits in highly aggregated portfolios of financial instruments.

Information systems

Several important issues must be addressed before weather risk management instruments can be used effectively in developing countries. These issues are not necessarily exclusive to developing countries, but they may be more problematic in a developing country context.

> **PANEL 2**
>
> ## POSSIBLE GOVERNMENT STRATEGIES IN WEATHER AND CATASTROPHIC RISK SHARING
>
> Governments in developing countries can use many different strategies, that are not mutually exclusive, to facilitate risk-sharing market activities.
>
> 1. The government may assume weather risks for its own account, and explore means to spread that risk outside the country. For example, the Turkish government assumes some of the risk of earthquakes destroying homes. The US government, through the national flood insurance programme, assumes some of the risk of flood damage to homes. In both cases, the governments clearly limit the amount of loss risk they are willing to assume for each home. Further, in most countries the government "owns" the risk of weather events affecting the poorest of the population. These loss risks are socialised across all taxpayers.
> 2. The government may act as a facilitator that bundles and transfers risk, without assuming the risk itself. The California Earthquake Authority (CEA) is an example. The CEA provides a mechanism to accumulate, and subsequently spread the risk to insurance and capital market intermediaries. It does not assume the risk for its own account. The government provides the technical assistance (with limited, defined, credit support) to facilitate the functioning of the risk transfer markets. Over time, the government intention should be to make the assisted programmes self-supporting.
> 3. The government may actively work to remove regulatory and structural barriers that limit the operations of private risk-sharing markets. One such example is that insurance regulations in many countries do not envisage the use of weather derivatives as insurance or reinsurance instruments. This is particularly true in developing countries. For example, a local insurance company may wish to use weather derivatives to hedge its portfolio of weather-related insurance policies. However, if insurance regulators do not recognise weather derivatives as an effective mechanism for retroceding risks, they may require the company to keep in reserves the full notional amount of outstanding insured risks. If the potential benefits of new financial instruments, such as weather derivatives, are to be fully realised, regulatory change will be needed so that these instruments are recognised as legitimate mechanisms for obtaining retrocessional risk protection. Some countries, such as Mexico, are already working on enacting these regulatory changes.
> 4. With regard to weather risk markets, governments can provide access to meteorological data and make the necessary investments in weather stations to ensure accurate and tamper-proof measurements. Governments can also invest in identifying and modelling weather risks that are particularly important to the overall economy or to specific economic sectors. Governments in developing countries should view the timely availability of quality weather data as an important public good that would greatly enhance the transfer of weather risks through both public and private means.

It is critical that reliable and verifiable weather measurement systems be in place. More specifically, there must be:

1. accurate, complete, and available historical weather data; and
2. mechanisms in place to insure the security of future weather measurements.

WEATHER INDEXES FOR DEVELOPING COUNTRIES

ACCURATE, COMPLETE, AND AVAILABLE HISTORICAL WEATHER DATA
For most weather derivative applications, a minimum of 30–40 years of historic daily weather data are necessary. It is important to note that data requirements depend, in large part, on where the trigger is set relative to the expected value of the underlying weather measure. The availability, quality and cost of such data can vary greatly across countries. Yet, many developing countries do have historical weather data and, compared to many other types of official government data, it is hard to imagine why anyone would have had an incentive to falsify historic weather data.

A more troubling problem can occur if a region is undergoing long-term climatic change and/or if the region is exposed to climatic cycles of varying length.[21] In these situations, historic data may not provide an accurate measure of the current expected frequency and magnitude of extreme weather events. Thus, it becomes very difficult to price weather derivatives.

In developing countries, problems with historical weather data are often related to missing observations and inconsistencies in measurements.[22] In the US, the UK, Germany and Japan, less than 1% of the observations will typically be missing in a sample of daily weather data. In many developing countries this percentage is significantly higher. In Mexico, for example, it is around 4%.[23] It is extremely important to try and assess whether these missing observations are random occurrences or whether they are in some way related to realisations of the weather phenomenon being measured. For example, serious data deficiencies will exist if missing observations reflect situations where measurement devices were overwhelmed by extreme events. Due to the possibility of recording errors, it is important to test data for consistency across time (at the same weather station) and across space (compared to nearby weather stations).

Weather data can be quite expensive in many countries. The Weather Risk Management Association, a non-profit organisation for weather derivatives professionals, has recently started lobbying the EU and European governments to lower their prices for weather data. In developing countries, that do not have centralised databases of historic weather data, the costs of collecting weather data and transferring it into an electronic format can be prohibitive. Economic development agencies could provide an important service to developing countries by financing efforts to build readily accessible databases of existing historic weather data.

For a variety of reasons, officials in some developing countries are reluctant to make weather data accessible. In some cases, government leaders are suspicious of how the data will be used – particularly if foreign organisations are requesting the data. In these situations it is important to recognise that weather data are an important national resource. Government leaders are unlikely to make this resource widely available on a timely basis unless they can be convinced that doing so will generate sufficient economic benefits. Thus, it is important that government leaders be educated about the potential economic benefits of weather derivatives.

In other cases, poorly funded local weather service offices recognise that they are a monopoly supplier of a data resource that is now suddenly in demand. Simply giving away, or even selling, their historic weather data greatly reduces the potential to extract further economic rents from the use of that data. Thus, these offices may be willing to provide only limited historic data (specific weather stations and/or specific years) and then only after negotiating an arrangement where they sell their services in the form of "value-added" data rather than raw weather measurements.

SECURITY OF FUTURE WEATHER MEASUREMENTS
Markets for weather derivatives will not develop unless tamper-proof weather stations are in place to ensure reliable readings of insured events. In most developing countries, the infrastructure for measuring and recording weather phenomena is very basic and lacks security. Without adequate security, weather observations could

be altered so as to trigger, or not trigger, payments on weather derivative contracts. Thus, creating a market for weather derivatives may require that automated weather stations be put in place with upgraded weather measuring systems and improved security. New hardware systems, such as optical precipitation sensors, can eliminate any direct human involvement in the recording process. Readings can also be verified by comparing with adjacent stations or with remotely sensed data from satellite imagery. Comparing measurements between several weather stations within a region can indicate inconsistencies in readings, signalling a problem. Satellite instruments can provide estimates of rainfall and perhaps, in the longer run, become the primary means of providing rainfall measurements. At present, calibration is an important issue with satellite measurements of rainfall. For any given location, there is generally not a sufficient time-series of satellite imagery to compare to conventional measurements.

Overall, despite the problems and issues highlighted with weather data from developing countries, weather data may still be the most reliable and verifiable data that can be used for risk management purposes in many of these countries. Further, the utility of these data could be greatly improved if international donors would invest in providing advanced weather stations and in building electronic archives of historic weather data.

Conclusion

Weather risk markets provide new opportunities for developing countries to transfer weather-related loss-risks. Policy-makers can use weather risk markets to develop and reinsure effective disaster assistance programmes that are activated by specific catastrophic weather events that have been defined *ex ante*. Further, the assistance should not unduly distort economic incentives. In particular, care should be taken to assure that aid does not spur unsustainable new economic activity in areas that are highly vulnerable to natural disaster risks. Failure to do this will only result in more losses and suffering when the next disaster strikes.

Government disaster assistance should also be structured so as not to crowd-out private risk management initiatives. While infrequent catastrophic events may require government assistance, events that are more frequent but still cause serious losses are more appropriately left to formal insurance markets or informal collective mutual assistance groups. Solutions should involve segmenting and layering risks so that the most catastrophic risk is handled with government aid and the less catastrophic risk is left to private market mechanisms. International weather markets can support the development of private insurance and reinsurance markets within developing countries by providing opportunities for transferring locally undiversifiable, systemic weather risk to financial intermediaries outside of the country.

International weather trading companies stand to gain diversification benefits from including developing country weather risk in their portfolios. This may contribute to reducing the overall risk of their weather portfolio and reducing the overall costs of covering weather risks. Weather trading companies and international organisations could try to develop better understanding of weather risks in developing countries and engage in a dialogue with governments and the private sector on issues related to data availability and quality, necessary weather measuring infrastructure, and educational activities to raise awareness.

1 *The views expressed in this paper are entirely those of the authors.*
2 *World Bank (2000).*
3 *EM-DAT: The OFDA/CRED International Disaster Database, Université Catholique de Louvain, Brussels, Belgium, http://www.cred.be/emdat/intro.html.*
4 *Dischel (2001).*

5 Lancaster (August 2001).

6 See the LANIC newsroom website: http://lanic.utexas.edu/info/newsroom/mitch.html.

7 Anderson (1983), Gardner and Fullerton (1968), Randall (1981), Wahl (1989) and Weinberg, Kling and Wilen (1993).

8 This basis risk is analogous to that faced by one using exchange-traded financial instruments such as futures contracts. For exchange-traded commodities, basis is the difference between the futures market price and local price of the commodity. Variability in the basis over time, or basis risk, introduces a random source of error for those who attempt to hedge price risk using commodity futures contracts.

9 For example, in areas with micro-climates, the spatial correlation in weather events may be quite low. The resulting basis risk may be so high that weather index instruments are not a reliable mechanism for transferring local weather risk.

10 This distribution is only slightly different to the distribution of May rainfall at a selected weather station in Mexico.

11 Cole and Chiarenza (1999), Doherty (1997), Lamm (1997), Skees and Barnett (1999) and Skees (1999).

12 Skees, Varangis and Larson (2001).

13 Black, Barnett, and Hu (1999), Skees, (1999) and Zeuli (1999).

14 Skees and Reed (1986), Quiggin, Karagiannis and Stanton (1994), Just, Calvin and Quiggin (1999).

15 Gautum, Hazell and Alderman (1994), Sakurai and Reardon (1997), Skees, Hazell, and Miranda (1999) and Skees, (2000).

16 For example, see the impact of natural disasters on micro-finance institutions (Nagarajan, 1998).

17 Froot (1999).

18 Barnett (1999), Rossi, Wright and Weber-Burdin (1982) and Freeman and Kunreuther (1997).

19 Kaplow (1991).

20 An example would be if rainfall levels have a negative trend and/or droughts occur with more frequency over time.

21 Inconsistencies in measurements usually result from: 1) relocation of the weather station; 2) human error in measuring or recording the weather observation(s); or 3) equipment failure.

22 These figures are estimates based on the authors' experience working with these data.

BIBLIOGRAPHY

Anderson, T. L., (ed), 1983, *Water Rights: Scarce Resource Allocation, Bureaucracy, and the Environment*, (Cambridge, MA: Ballinger Publishing Co).

Barnett, B. J., 1999, "US Government Natural Disaster Assistance: Historical Analysis and a Proposal for the Future", *Disasters*, 23, pp. 139–55.

Black, J. R., B. J. Barnett and Y. Hu., 1999, "Cooperatives and Capital Markets: The Case of Minnesota-Dakota Sugar Cooperatives", *American Journal of Agricultural Economics*, 81, pp. 1240–6.

Cole, J. B., and A. Chiarenza, 1999, "Insurance Risk-Securitization: The Best of Both Worlds", *Risk Magazine, Insurance Risk Supplement*, July.

Dischel, R., 1998, "The Fledgling Weather Market Takes Off", *Applied Derivatives Trading*, November, URL: www.erivativesreview.com

Dischel, R., 2001, "Double Trouble: Hedging Rainfall and Temperature", *Weather Risk Special Report, Risk Magazine and Energy and Power Risk Management*, Risk Waters Group, August, pp. 24–6.

Doherty, N. A., 1997, "Financial Innovation in the Management of Catastrophe Risk", Fifth Alexander Howden Conference on Disaster Insurance, Gold Coast, Australia, August.

Freeman, P. K., and H. Kunreuther, 1997, *Managing Environmental Risk Through Insurance*, (Boston: Kluwer Academic Publishers).

Froot, K. A., (ed), 1999, *The Financing of Catastrophe Risk*, (University of Chicago Press).

Gardner, B. D., and Fullerton, H. H., 1968, "Transfer Restrictions and Misallocations of Irrigation Water", *American Journal of Agricultural Economics*, 50, pp. 556–71.

Gautum, M., P. Hazell and H. Alderman, 1994, "Rural Demand for Drought Insurance", The World Bank, Policy Research Working Paper, 1383.

Hull, J. C., 2000, *Options, Futures, & Other Derivatives*, Fourth Edition, (Upper Saddle River, NJ: Prentice Hall).

Just, R. E., L. Calvin and J. Quiggin, 1999, "Adverse Selection in Crop Insurance: Actuarial and Asymmetric Information Incentives", *American Journal of Agricultural Economics*, 81, pp. 834–49.

Kaplow, L., 1991, "Incentives and Government Relief for Risk", *Journal of Risk and Uncertainty*, 4, pp. 167–75.

Lamm, R. M. Jr, 1997, "The Catastrophe Reinsurance Market: Gyrations and Innovations amid Major Structural Transformation", *Bankers Trust Research*, (New York: Bankers Trust Company), pp, 1–13.

Lancaster, R., August 2001, "Educating End-users." *Weather Risk, an Energy Power Risk Management and Risk Special Report*, Risk Waters Group, London.

Nagarajan, G., 1998, "Microfinance in the Wake of Natural Disasters: Challenges and Opportunities", Microenterprise Best Practices (MBP) Project, Bethesda, Maryland, Development Alternatives, Inc.

Quiggin, J., G. Karagiannis and J. Stanton, 1994, "Crop Insurance and Crop Production: An Empirical Study of Moral Hazard and Adverse Selection", in: *Economics of Agricultural Crop Insurance*, D. L. Hueth and W. H. Furtan, (eds), (Norwell MA: Kluwer Academic Publishers), pp. 253–72.

Randall, A., 1981, "Property Entitlements and Pricing Policies for a Maturing Water Economy", *Australian Journal of Agricultural Economics*, 25, pp. 195–220.

Rossi, P., J. Wright and E. Weber-Burdin, 1982, *Natural Hazards and Public Choice: The State and Local Politics of Hazard Mitigation*, (New York: Academic Press).

Sakurai, T., and T. Reardon, 1997, "Potential Demand for Drought Insurance in Burkina Faso and its Determinants", *American Journal of Agricultural Economics*, 79, pp. 1193–207.

Skees, J. R., 1999, "Opportunities for Improved Efficiency in Risk-Sharing Using Capital Markets", *American Journal of Agricultural Economics*, 81, pp. 1228–33.

Skees, J. R, 2000, "A Role for Capital Markets in Natural Disasters: A Piece of the Food Security Puzzle", *Food Policy*, 25, pp. 365–78.

Skees, J. R., and B. J. Barnett, 1999, "Conceptual and Practical Considerations for Sharing Catastrophic/Systemic Risks", *Review of Agricultural Economics*, 21, pp. 424–41.

Skees, J. R., S. Gober, P. Varangis, R. Lester and V. Kalavakonda, 2001, "Developing Rainfall-Based Index Insurance in Morocco", World Bank Policy Research Working Paper 2577, April.

Skees, J., P. Hazell and M. Miranda, 1999, "New Approaches to Crop Insurance in Developing Countries", International Food Policy Research Institute, EPTD Discussion Paper No. 55, November, Washington, DC.

Skees, J. R., and M. R. Reed, 1986, "Rate-Making and Farm-Level Crop Insurance: Implications for Adverse Selection", *American Journal of Agricultural Economics*, 68, pp. 653–59.

Skees, J. R., P. Varangis and D. Larson, 2001, "Can Financial Markets be Tapped to Help Poor People Cope with Weather Risks?", Paper presented at the UNU/WIDER Conference on Insurance against Poverty, Helsinki, Finland, June.

Skees, J. R. and K. A. Zeuli, 1999. "Using Capital Markets to Increase Water Market Efficiency" Paper presented at the International Symposium on Society and Resource Management, Brisbane, Australia, July 8.

Wahl, R. W., 1989, *Markets for Federal Water*, (Washington DC: Resources for the Future).

Weinberg, M., C. Kling and J. Wilen, 1993, "Water Markets and Water Quality", *American Journal of Agricultural Economics*, 75, pp. 278–91.

World Bank, 2000, "Managing Economic Crises and Natural Disasters", in: World Development Report 2000/01: Attacking Poverty, Washington DC.

Zeuli, K., 1999, "New Risk Management Strategies for Agricultural Cooperatives", *American Journal of Agricultural Economics*, 81, pp. 1234–9.

17

Weather Risk Management for Agriculture and Agri-Business in Developing Countries

Ulrich Hess; Kaspar Richter; Andrea Stoppa[1]
IFC; World Bank; Procom Agr, Rome

Risk is a pervasive characteristic of life in developing countries, especially in rural areas. The economies depend heavily on weather conditions, and experience frequent weather hazards, such as drought, floods and windstorms. These factors typically affect most households and companies in the same area at the same time. Furthermore, as households and companies typically have a low asset base and little access to well developed insurance and credit markets, they are financially ill-equipped to deal with weather shocks. As a result, their weather risk management (WRM) is inefficient, resulting in negative implications for economic and social development. New WRM insurance instruments, like area-based weather indexes, provide a viable alternative to traditional insurance instruments, and offer real advantages to households, companies and governments in developing countries.

Introduction
This chapter has three components. The first part argues that weather risk causes substantial inefficiencies in developing countries; agri-businesses, faced with underdeveloped formal financial markets, have to rely on traditional WRM that is associated with underinvestment and overdiversification. We discuss how new WRM can overcome the pitfalls of traditional WRM and have a large development impact. In the second section, we discuss the range of potential uses of new WRM. The final part turns to the operational aspects of a new WRM, studying in detail the case of WRM for cereals in Morocco.

WRM and rural development
VULNERABILITY TO WEATHER-SHOCKS
Agriculture and agri-business are the prime source of income for most families and businesses in developing countries; in 1999, 69% of the population in low-income countries lived in rural areas, compared to 50% in middle-income countries and 23% in high-income countries.[2] Agriculture accounted for 27% of GDP in low-income countries, compared to 10% in middle-income countries and only 2% in high-income countries (World Bank, 2001). These numbers understate the importance of agriculture for economy growth, which is magnified by multiplier effects (through linkages from agriculture to other economic sectors), the role of agricultural exports

WEATHER RISK MANAGEMENT FOR AGRICULTURE AND AGRI-BUSINESS IN DEVELOPING COUNTRIES

as a foreign exchange earner, and the overriding importance of subsistence farming for the livelihood of the bulk of the population.

Agriculture is inherently dependent on the vagaries of weather, such as the variation in rainfall. This leads to production (or yield) risk, and affects the farmers' ability to repay debt, to meet land rents and to cover essential living costs for their families. But the effects of weather events also matter for rural lending institutions and agri-businesses, as they determine the risk exposure of borrowers and input providers. With weather conditions affecting a large share of business activity, many developing countries in Sub-Saharan Africa and other parts of the world display a high sensitivity of both agricultural and GDP to fluctuations in rainfall (Benson and Clay, 1998 and Guillaumont, Guillaumont, Jeanneney and Brun, 1999).[3] Ultimately, the precariousness of farmers and producers translates into macroeconomic vulnerability.

Developing countries are not just more dependent on weather conditions but also suffer the brunt of natural disasters (due to the hazardous environmental conditions), many of which are caused by weather hazards. According to World Bank (2001), between 1988 and 1997 natural disasters claimed an estimated 50,000 lives a year and caused direct damage valued at more than US$60 billion a year. Developing countries incurred the vast majority of these costs: 94% of the world's 568 major disasters between 1990 and 1998 took place in developing countries. In Asia, which experiences 70% of the world's floods, the average annual cost of floods over the 1990s was estimated at US$15 billion. On the basis of current trends, these numbers are likely to rise in the future. The incidence of El Niño events, associated with anomalous floods, droughts and storms, has increased over the last 10 years (Freeman, 1999).

TRADITIONAL WRM IS COSTLY AND INEFFECTIVE

Farmers in developing countries have always been exposed to weather risks, and for a long time have developed ways of reducing, mitigating and coping with these risks (Besley, 1995 and Dercon, 2002). Traditional risk management covers actions taken both before (*ex ante*) and after (*ex post*) the risky event occurs (Siegel and Alwang, 1999). Examples of *ex ante* strategies include the accumulation of buffer stocks as precautionary savings and the diversification of income-generating activities through changing labour allocation (working in farm and non-farm small businesses, and seasonal migration) or varying cropping practices (planting different crops, like drought-resistant variants, planting in different fields and staggered over time, inter-cropping, and relying on low-risk inputs). Similarly, companies may self-insure through high capitalisation and diversification of business activities. Communities collectively mitigate weather risks with irrigation projects and conservation tillage that protects soil and moisture. Examples for *ex post* strategies range from farmers seeking off-farm employment, to distress sales of livestock and other farm assets, to withdrawal of children from school for farm labour, and to borrowing funds from family, friends and neighbours (Jacoby and Skoufias, 1998).

While such risk management has assisted developing countries in coping with weather risks (Hazell, Pomareda and Valdes, 1986), it has important shortfalls. These strategies are costly, as they often lower vulnerability in the short term at the expense of higher vulnerability over the longer term. For example, when a farmer diversifies he gives up higher income due to specialisation in return for a lower variability of income. Equally, a farmer who sells productive assets, like draught oxen, to make ends meet lowers his future income stream. Similarly, a company misses out on profitable business opportunities if it decides to draw credit below its optimal level in order to keep a credit reserve in case of a weather shock.

Additionally, some of the informal risk management strategies are ineffective to deal with weather risks. Weather-related events constitute covariate risk, as they typically affect many households in a community or region at the same time. Yet, in times of great stress, like crop failure due to drought, informal arrangements tend to break down, as the members of the community, or "risk pool", are jointly affected. The income of the village as a whole is reduced, triggering a collapse of community-based informal insurance arrangements (Morduch, 1998). As in the above example of the farmer who attempts to sell livestock to make ends meet after a drought, livestock prices will fall as supply outstrips demand. Similarly, when farmers seek off-farm employment in response to a natural disaster, the sudden rise in the labour supply will drive down market wages.

FORMAL FINANCIAL MARKETS ARE UNDERDEVELOPED
While traditional WRM mechanisms provide at best partial coverage, formal financial markets are insufficiently developed to fill the gaps. Private insurance markets are impeded as a result of information asymmetries, the covariance of weather risks, lack of acceptable forms of collateral and government programmes. These factors all lead to high unit transactions costs, limited spread of institutions and less access for the poor.

The classical problem of asymmetric information limits the scope for crop insurance schemes; farmers will always be more knowledgeable about their production risk than credit institutions. From the insurer's point of view, this makes it difficult to separate farmers accurately into low- and high-risk groups, raising the possibility that only the high-risk farmers will take up insurance. Additionally, once insured, farmers may reduce their efforts to control production risks, leading to higher losses for the insurance company. In order to control such adverse selection and moral hazard problems, insurance companies have to raise rates, invest in monitoring mechanisms, and require marketable collateral as precondition for borrowing, which increases premiums and reduces demand for insurance.

In addition, weather risks are correlated within a region. This spatial covariance makes it difficult for local insurers with limited regional diversification to pool risks and offer affordable insurance coverage. While in principal primary insurers could pass on risks to an international reinsurance market, there is little transfer of such risk from the emerging markets for a number of reasons. The size of weather risk readily available for underwriting is limited, and transaction costs are high due to lack of standardisation and asymmetric information between insurer and reinsurer (Skees, 2000).

Finally, government risk management programmes may crowd out private sector risk management. In many countries, government have stepped in with a range of interventions for farmers. Governments mitigate risk for example through price stabilisation, subsidised crop yield insurance and drought relief. Most programmes, especially the multiple-peril crop insurance, have absorbed large sums of public resources, yet there is little evidence that these interventions had positive effects on agricultural lending or production. Instead, they have led to excessive risk taking of farmers and a growing dependency from public disaster relief (Skees, Hazell and Miranda, 1999).

EFFECTIVE WRM – A "CATCH 22" SITUATION?
The discussion so far raises serious questions about the possibility to effectively insure farmers and agri-businesses against weather risks, be it through formal or informal insurance. The failure of formal and informal risk management mechanisms preserves high production risks and disadvantages farmers in dealing with the numerous other risk sources deriving from markets, policies and institutions (Siegel and Alwang, 1999). For example, weather risks can be linked to price fluctuations, especially when the natural hazard has a broad spatial spread. Equally, health, institutional and political risks can trigger and exacerbate production risks.

These uncertainties not only hurt the livelihood of farmers, but also impede the development of a financial market. This leads to a 'Catch 22' situation (Skees, 2000):

credit institutions realise that income of farmers and agri-businesses are subject to large risks, and either ration credit or charge higher interest rates to cover these risks; without access to affordable credit, farmers delay the adoption of new technologies and the introduction of new farming systems, and keep relying on ineffective informal risk management strategies.

Hence, successful weather risk-sharing arrangements in developing countries would offer potentially huge benefits not just to farmers, but also to agri-business and financial markets (Skees, 1999). With the advent of effective WRM, finance institutions would be able to collateralise rural credit more efficiently and extend loans to groups of weather-exposed farmers that otherwise would not be bankable. Other sectors of the economy would benefit as well; for example, companies in the energy sector are also exposed to weather risk through the impact on energy demand, and could improve their insurance coverage through effective WRM.

PANEL 1

A MARKET FOR NEW WRM

One important example of new WRM is "weather index insurance" (Turvey, 2001). The key innovation of such contracts is that insurance is linked to the underlying systemic risk (ie, low rainfall), defined as an index (measuring rainfall) and recorded at a regional level (local weather station), rather than the extent of the loss (the resulting reduction in crop yields). In other words, the economic incentive of a farmer to work hard for a good harvest is unaffected by the weather-based insurance, avoiding moral hazard. Adverse selection is minimised as premiums are fixed without taking into account the composition of the risk pool of farmers in the insurance scheme. At the same time, as long as weather parameters correlate sufficiently with yields, it will result in a substantial reduction in a farmer's risk exposure.

Weather index insurance has a number of advantages (Skees, Hazell and Miranda, 1999). It is inexpensive to administer, as it allows for standardisation, avoiding the need to draw up and monitor individual contracts. It can be supplied by the private sector with little or no government subsidy, as it avoids the incentive problems of crop insurance programmes related to asymmetrical information. It is affordable for poor and rich farmers alike, and accessible to agri-business and other sectors. Furthermore, by eliminating the systematic production risk component linked to the weather index, it improves the risk profile of farmers and companies. As a result, insurance companies can offer "wrap-up" contracts for the remaining independent risk. An example of such an insurance programme is the Nicaragua Risk Management Project, funded by the World Bank. This pilot project started in 1999, and provides farmers and agri-business – or anybody else who may want to protect themselves against weather risk – with the option of purchasing "rain lottery tickets". They are sold in small denominations and entitle the holder to a payoff whenever rainfall in a given area drops below a specified level.

Another example of new WRM is the use of weather derivatives to lower reinsurance or retrocession cost. In 2001, Agroasemex, the Mexican state-owned agricultural reinsurance company, transacted a first weather derivative with a leading weather derivative market maker. Following extensive development work of the World Bank, Agroasemex and the provider developed indexes that track the performance of the company's portfolio for the period of autumn and winter 2001–2 based on 10 weather stations: four for low temperature, five for excess humidity and one for drought. The crops and regions covered were: tobacco in Nayarit (low temperature), tobacco in Nayarit (excess humidity), beans in Sinaloa (low temperature), beans in Sinaloa (excess humidity), maize in Sinaloa and Sonora (low temperature), maize in Sinaloa

> and Sonora (excess humidity), garbanzo beans in Sinaloa (excess humidity) and sorghum in Tamaulipas (drought).
>
> The covered exposure was portfolio risk of an agricultural reinsurer. The use of weather derivatives lowered the reinsurance cost of Agroasemex. As a result of this transaction, Agroasemex's retrocession costs dropped by up to two thirds. This supports Agroasemex's efforts to offer competitive products to Mexican agriculture.
>
> With improved risk insurance, agri-business will accelerate the adoption of new technologies, specialise in the most competitive activities, and become more adaptive and flexible. In practice, rural finance institutions could be the primary customers as they hedge their exposure to weather events, and require borrowers to take out insurance as liquid collateral to mitigate part of the default risk. Small and subsistence farmers could become customers through associations and cooperatives, while large farmers could buy weather insurance directly as a hedging tool.
>
> The demand for new WRM is difficult to measure, but it is clear that the inefficiencies in current risk management techniques indicate a substantial market potential. It will be attractive for farmers and agri-businesses as long as the "basis risk" (ie, the probability of incurring a loss that is not covered by the insurance) is not too high. World Bank research concluded that there is demand in developing countries for weather index-based insurance in rain-fed agriculture. Worldwide, weather derivative markets have reached a cumulative transaction size of more than US$8 billion from 1997 to 2001. While a large majority of deals are still based in the US, the market is expanding in Europe and Asia in very diverse sectors ranging from tourism to agriculture and power (World Bank, 2001). The market potential for weather hedging instruments in emerging markets is large.

WRM for agriculture and agri-business in developing countries

This section looks at the key success factors of new WRM in emerging markets. Five factors are highlighted:

1. good weather data in key locations;
2. the client profile as financial intermediary;
3. facilitation by development organisations;
4. a benign regulatory framework; and
5. risk transfer mechanism into international weather markets.

WEATHER DATA
A pre-condition for the emergence of a competitive weather risk market is comprehensive and accurate weather data for the past and future. Historical data, usually at least 30 years of daily information on key parameters, need to be accessible and reasonably priced. Operational weather stations have to be identified and basis weather variables defined, also with regard to cleaning procedures. All countries run weather stations that report SYNOP data to the World Meteorological Organization (WMO).[4]

Many countries provide long series of adequate weather data. Examples include advanced economies such as Argentina, Chile, South Africa, Turkey, Morocco, Tunisia and Mexico, but also poorer economies like Nicaragua. Nevertheless, in some countries, the availability and quality of rainfall measures is compromised for different reasons. Many developing countries do not provide easy or affordable access to weather data, although the quality and comprehensiveness of the databases are often good and mostly cleanable and usable.[5] In addition, moral hazard issues related to weather data collection are accentuated in developing economies with weaker institutional frameworks. Finally, just as in OECD (Organization for Economic Co-operation and Development) countries, the lack of universally accepted quality control procedures as well as different characteristics of non-SYNOP data – varying definitions of daily average or maximum temperature data, for example – pose a problem to the weather markets.[6]

Various remedies exist to provide secure and reliable rainfall data:

❑ Incentive structures have to be geared towards accurate data measurements.[7]
❑ Matching weather station series with information from third-party sites, together with comparing historical raw data and cleaned data for pricing purposes, can help to provide an understanding of the cleaning methodology used by weather stations.[8]
❑ Meteorological services may have close ties with major established weather service authorities in OECD countries, as does the Moroccan weather service with France Météo. These partnerships can be useful in ensuring international quality standards.
❑ The risk of tampering with the data can be effectively addressed through fallback stations of the weather risk provider as well as crosschecks with nearby stations. Weather data sensors placed directly on premises of the end-user clients – for example, multiple moisture or tiny temperature gauges placed on farm land – can be very effective in deterring data manipulation. In a project with the World Bank, RADARSAT, the Canadian Earth observation satellite, investigates the feasibility of using of satellite data for back-up and verification purposes. Contract design can also help to prevent data tampering. For example, proportional contracts provide fewer incentives for manipulation than digital contracts, where payoffs are fixed on an "all-or-nothing" basis.

END-USERS: FINANCIAL INTERMEDIARIES AND THEIR CLIENTS
Most demand for new WRM instruments comes from insurance and reinsurance companies as well as other financial intermediaries who either seek to hedge their exposures, or to intermediate or retail weather risk protection. Very few markets allow for direct sale of weather derivatives by international derivative dealers. The profile of end-users and end-user deals of weather markets in developing countries is different from OECD markets in various ways, including the importance of agriculture for most developing countries and the lack of risk management instruments (as were discussed in the previous section), and the link to credit, higher (perceived) credit and regulatory risk (as will be discussed in the following paragraphs).

Link to credit
As debated in the previous section, access to formal credit is often limited, and burdensome collateral requirements ultimately make very few rural businesses bankable. For example, traditionally only state-owned agricultural banks would lend to rain-fed agriculture. The exposure to high credit risks and covariate weather risk led to low loan repayment rates and periodic recapitalisations of these banks by the state. The rescheduling of loans not only represented a burden on state finance, but also provided farmers, anticipating the next government bail out, with incentives to default. Recently, a number of countries adopted new regulations to impose market discipline on state banks.[9]

These changes lead to market opportunities for weather insurance. Banks require more liquid collateral from their clients as a pre-condition for crop financing, and demand that customers buy weather insurance from insurance companies. Alternatively, banks become end-users of wholesale weather contracts that protect their systemic weather exposures. Potential clients for weather insurance products therefore include micro-finance institutions, input suppliers, contract farming companies and other intermediaries that lend to agriculture and agri-business.

Demand assessments for weather index insurance or straight weather derivatives do not exist yet for other developing countries, as the products are not marketed. Studies of insurance demand, however, suggest that farmers are willing to pay between 7% and 10% of their input costs. Currently, Moroccan farmers pay a 9% premium of the maximum indemnity for traditional crop insurance. Almost all farmers in the eligible areas take up the insurance. This insurance is mostly sold through the main agricultural bank along with seasonal credit.

Credit risk
Credit risk arises with all over-the-counter contracts as both parties have promised to pay the other in the future, depending on the final value of an index, and must be trusted to live up to the promise. This can be contrasted with exchange-traded securities where the exchange assures final payment. Credit risk or the risk of default of the counterparty in emerging markets is compounded by currency transfer risk. In other words it does not matter to the weather risk provider whether the default is triggered by a macro problem (the Peso crisis, for example) or counterparty default – the risk rating will be equal or lower to the country risk rating.

This risk can be mitigated by dealing with subsidiaries of OECD companies that are guaranteed by their parent companies. These cases can be structured as OECD jurisdiction contracts. One example are OECD manufacturing companies who outsourced production to the Caribbean and Central America and have difficulty finding business interruption insurance covering production shortfalls due to hurricane disruption.[10] Weather risk providers would enter into hedging contracts with the US parent companies, who in turn would pass on the protection to their affiliates.

A second option for risk mitigation is the purchase of political or credit risk insurance. International organisations such as the Multilateral Investment Guarantee Agency (MIGA) (political risk insurance) or International Finance Corporation (IFC) (partial credit guarantees) are also providers of these instruments for emerging market clients.

FACILITATORS
The introduction of weather risk management into emerging market economies requires development work that sometimes cannot be recuperated in trading margins of contracts. Development institutions, such as the World Bank Group, promote pilot cases to generate demonstration effects to raise awareness. The IFC, as part of the World Bank Group, entered into a partnership with a leading weather risk market maker in order to promote WRM in emerging markets. IFC also obtained board approval for an investment in a weather insurance company to be established in Morocco (Haggerty, 2002).

REGULATORY FRAMEWORK
Credit risk and regulatory risk will result in WRM in the form of insurance contracts. Up-front premium payments protect the weather risk providers by reducing credit risk to transaction risk. Documentation of the deal as an insurance or reinsurance contract is often the only way to introduce WRM in emerging markets. Derivatives are seldom accepted by regulators and are generally associated with gambling or negative cases of over-hedging, such as the so-called "Ashanti Goldfields" problems.[11]

Insurance contracts usually require an "insurable interest" by the insured, which may be viewed as incompatible with a weather contract settled on the basis of third-party data as opposed to losses suffered by the insured. However, the emerging experience in several countries shows that this departure from the traditional insurance concept is not a major obstacle. Mexican, Moroccan and Turkish regulatory authorities indicated that weather index-based insurance policies could comply with existing insurance regulations.

Weather risk transfer into international markets will mostly follow the insurance and reinsurance route, before the risk becomes transformed into a weather derivative. Derivative providers are usually not licensed to engage in reinsurance business in developing economies. Highly rated reinsurers step in and write a reinsurance treaty for a local insurer that represents the business. The reinsurer then passes on the risk to a weather risk market maker and thereby effectively transforms the risk into a derivative.

POTENTIAL APPLICATIONS
New WRM has a range of applications.[12] Precipitation contracts are preponderant due to the dominance of the agriculture sector in the work of the authors, but the first wave of major deals in emerging markets will probably be covering power sector operators, in particular hydropower generators. These applications are either weather hedging structures or substitutes for traditional insurance.

Weather hedging
Income smoothing through the financial mitigation of weather shocks is relevant for few large sophisticated operators. Most of the deals will be closer to the catastrophe zone of insurance, or the one in seven- or even one in ten-year event range. Extreme events such as the 100-year hurricane are *de facto* covered by emergency and government handouts, so weather risk management in emerging markets comes in for the mezzanine range of risk between hedging and cat insurance. Another class of applications are operators facing penalties in case of failed deliveries, such as high-value agricultural exporters, eg, broccoli farmers from Mexico face heavy penalties in case of non-delivery to the large supermarket chains in the US.

Substitute for traditional insurance
Insurance and reinsurance markets have generally hardened during the last years and in particular after September 11, 2001.[13] Higher premiums and shrinking capacity spur demand for alternative risk transfer, such as cat bonds, but also WRM instruments. The main application of new WRM would be as an alternative to crop insurance, but also as business interruption insurance, which is almost unavailable for most businesses in emerging markets. The risks are often weather-related, as illustrated by the case of Venezuela where the cause of recent business interruption indemnity payments was flooding and soil erosion resulting from the combination of high wind speed and excessive precipitation. Similarly, hurricane exposure of property and casualty risk in the Caribbean could effectively be (re)insured through weather contracts.

The case of rainfall index insurance in Morocco
OVERVIEW
Efforts to develop insurance programmes related to weather events are not new to Morocco. Drought is recognised to be one of the main risks for Moroccan agriculture (Skees, 2001), if not the single most important cause of crop failure, and the attention of both the public sector and the insurance industry has been long focused on developing appropriate safety nets to protect farmers from its dangerous effects.

In 1995 the Moroccan government, in partnership with the insurance industry, activated the "Programme Sécheresse" (drought plan). Despite the clear reference to drought, the scheme, revised and improved in 1999, is in fact a yield insurance programme, the only connection to the weather event being the ministerial declaration that officially declares the existence of a drought period and allows the insurance company to activate the indemnification procedure.

In order to evaluate the possibility of developing an insurance programme directly related to weather events, in 2001 the World Bank helped the Moroccan government to launch an on-field international research project. The research team, after an accurate analysis of the productive environment in agriculture as well as rainfall patterns and agricultural yields, concluded that Moroccan agriculture could significantly benefit from a rainfall insurance programme and recommended the adoption of a pilot area-based rainfall insurance scheme (Skees, 2001). The programme recommended by the World Bank study is a rainfall insurance programme for crops (cereals and sunflower in particular) that indemnifies producers if rainfall levels fall below a specified threshold. Rainfall would be measured at the synoptic stations of the National Meteorological Service (Direction de la Météorologie Nationale, DMN), and should be accessible in real time to all parties involved in the transaction.

To help the local insurance industry design the practical details of such a programme and facilitate access to international weather risk management markets, the IFC, assisted by the Italian Government, sponsored a project to help structure the weather contracts and set up a company that would launch and manage such products.

CONTRACT DESIGN

The structure of the rainfall insurance programme recommended by the World Bank's study was developed in analogy to a European put option, where the option price is the cost of the coverage and the strike is the rainfall threshold below which an indemnity is triggered.

The idea underlying such types of contracts is that, once the existence of a sufficient degree of correlation between rainfall and yield is established, an agricultural producer can hedge production risk by entering into a contract under which payments would be made if rainfall levels fall below the selected strike. In order to structure the contract, the issues to evaluate are therefore how to determine the strike and at what level to set it.

In the case of cereal and sunflower production in Morocco, the adopted procedure for developing rainfall insurance contracts was:

Step 1. Production and rainfall data were collected and organised.
Step 2. The most appropriate rainfall period was selected estimating correlations between yields and different rainfall periods.
Step 3. Specific rainfall indexes were constructed assigning "weights" to different rainfall periods in order to maximise correlation between yields and rainfall.
Step 4. Different payment schemes were analysed and evaluated.

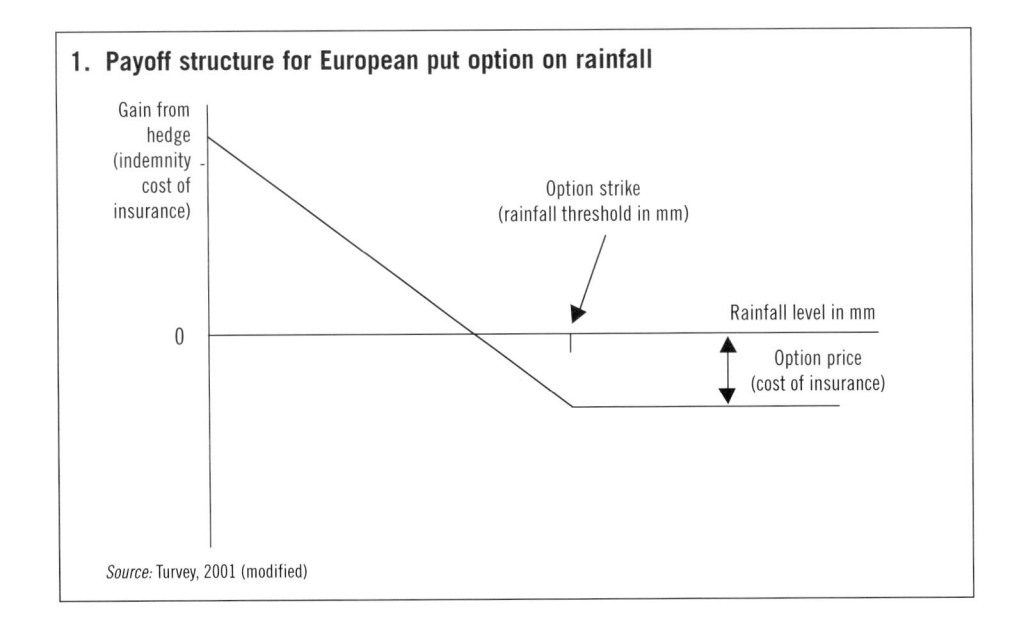

1. Payoff structure for European put option on rainfall

Source: Turvey, 2001 (modified)

After having collected and validated the data, the first choice to make in designing the contract is to define the rainfall time period, which should be considered for coverage purposes. Such a choice, of course, is mainly dependent on climate and plant physiology, but marketing issues have their relevance: in order to avoid the possibility of producers making an informed decision on whether to enter into the contract or not, it is clearly not advisable to include rainfall periods that precede contract signing time.

Once the appropriate period has been selected, the issue becomes structuring the rainfall index. In this respect, the general concept is that despite the high level of yield-rainfall correlations measured for crop production in Morocco (close to 0.8 in the case of wheat), it is nevertheless an advantage to incorporate agronomic information in the contract structure that enhances the measurement of the yield-rainfall relationship. In fact, precipitation in different stages contributes in different measures to plant growth and, in addition, an excess of rain may be of no use for production. Hence, it is useful to develop a weighting system that allows one to differentiate the importance of rainfall in different growth periods and to shape the model so as to take into account the fact that excess rain may be wasted without contributing to plant growth.

In order to structure the index, trends in yield and rainfall series were examined, rainfall for each synoptic station aggregated in 10-day periods and weights assigned through a mathematical programming procedure that maximises correlation between yields and the rainfall index. The vector of weights is then adjusted through an ad hoc procedure that slightly modifies the optimised vector in order to make it consistent with logic and agronomic intuition. This last step may somewhat reduce correlation between the two series, but allows homogenous rainfall periods to be established, that help to make the contract more understandable and more marketable. An example of one the indexes developed for wheat in Morocco is given in Table 1:

Table 1. Structure of rainfall-index insurance for wheat in Morocco

Month	November			December			January			February			March		
10-day period	1	2	3	1	2	3	1	2	3	1	2	3	1	2	3
Weight	2.0	2.0	2.0	0.5	1.0	1.0	1.0	0.5	0.5	1.0	1.0	1.5	1.0	0.5	1.0

Source: IFC, Morocco Weather Index Insurance Project

The final value of the index (the value which, when compared with the threshold, indicates whether or not the insured should be granted an indemnity) is calculated by summing the values obtained by multiplying rainfall levels in each period by the specific weight assigned to the period.

Customers participating in the rainfall-index programme receive a payment if the level of the index falls below a predetermined threshold. The payment is equivalent to the percentage of rainfall-index shortage multiplied by the level of coverage selected.

In applying the programme, the customer should be allowed to select different levels of coverage in order to have the opportunity to insure different levels of potential revenue.

Figure 2 provides a graphical description of the performance of the rainfall-index insurance in the case of wheat production for a specific synoptic station in Morocco. The figure represents the different level of wheat revenue with or without rainfall insurance. It should be noted that the insurance programme prevents revenues from falling below a threshold of approximately Dh3,000 (approximately US$300).

One other useful way to describe the performance of the rainfall index insurance

2. Wheat revenues with and without rainfall insurance (index threshold 275 mm)

Source: IFC, Morocco Weather Index Insurance Project

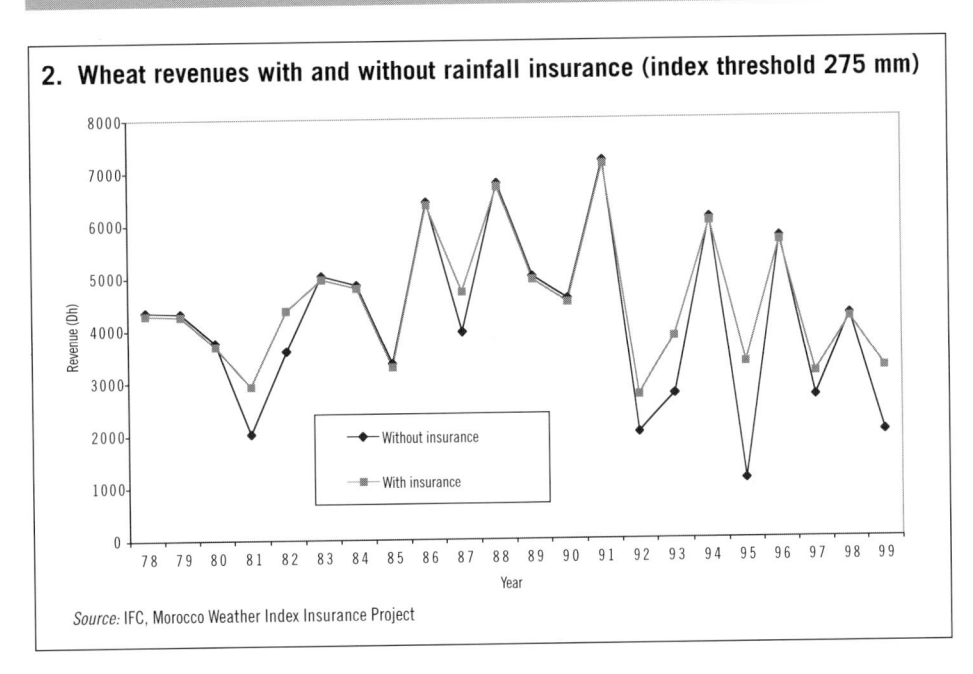

is to analyse the dynamics of revenue loss and the payments triggered by the programme in each of the crop years.[14]

Figure 3 shows that the programme triggers a payment in each of the years for which a revenue loss is recorded and also that it does not generate "false positives", ie, payments in case of no revenue loss. Figure 3 also helps illustrate the issue of the selection of the threshold level. Quite logically, the higher the threshold set for the contract, the better the coverage provided, but, by way of a trade-off, a higher threshold results in a higher cost of the insurance coverage. In Figure 3, for example (coverage level 375 mm), indemnities for years 1982 and 1987 are probably too high, wasting resources that are accounted for in the premium of the policy. Figure 4 shows a case for a different threshold level (275 mm) for which the coverage is probably not as good as in the preceding case, but for which the actuarial premium is more than halved.

The proposed proportional rainfall-index payment scheme is obviously only one of the possible solutions for structuring a weather-related crop insurance programme. Several different alternatives, all aiming at making the coverage as extended and as comprehensive as possible, were evaluated by the IFC research team. From the payment structure point of view, non-proportional contracts (ie,

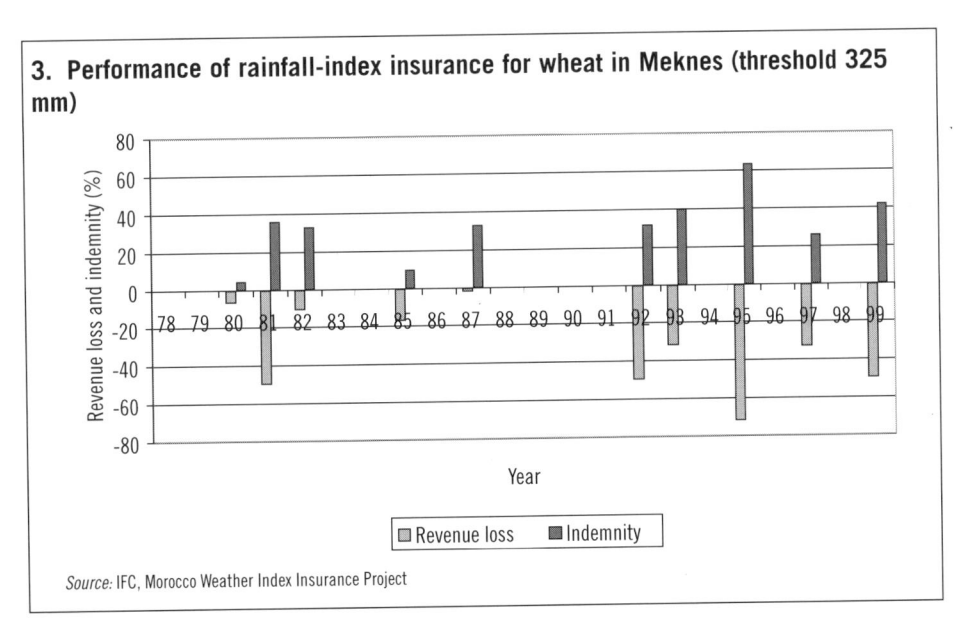

3. Performance of rainfall-index insurance for wheat in Meknes (threshold 325 mm)

Source: IFC, Morocco Weather Index Insurance Project

increases in unit payments as rain shortfall increases) were tested and other weather variables like temperature were added to the structure of the contract. Overall, however, the simplicity of the rainfall index and the comparatively lower cost of the coverage led to the selection of the simple proportional rainfall index as the preferred model for implementation.

An interesting marketing feature of the rainfall insurance programme that should be launched for the 2002–3 crop year is that, following the successful experience of the drought programme, the contacts will be most probably marketed by linking the insurance policies to farmers' credit requests. Agricultural producers need resources for anticipating cultivating costs and part of the loan granted to the farmer can be devoted by the credit institution to financing the insurance coverage. This marketing procedure will certainly help the development of the programme in its infant stage, at the same time granting revenue coverage to the producer and reducing default risk for credit institutions.

KEY SUCCESS DRIVERS IN MOROCCO

The proposed rainfall-index insurance programme for crop production in Morocco has several interesting features. As mentioned in the preceding sections, rainfall indexes are free of most moral hazard and adverse selection problems.[15] In addition, significant cost savings are achieved by eliminating the need for costly on-field damage assessment activity.

For the programme to be successful, however, certain conditions have to be met. First, of course, rainfall-index insurance should provide adequate coverage of farmers' revenue. The high levels of yield-rainfall correlation – before and after optimisation, see Skees (2001) – seem to satisfy this prerequisite, but, the programme being an area-based insurance scheme, good levels of yield-rainfall correlation at the area level are not sufficient. Like all area-based insurance programmes, the issue of different risk patterns among producers in the same area (ie, basis risk) is extremely relevant. In particular, if rainfall is not homogeneously distributed in the area, a good level of rainfall recorded at the regional level might not correspond to sufficient rainfall at farm level. In this situation the insurance payment would not be triggered and, despite having purchased an insurance contract, the producer incurs a revenue loss. Consequently, a crucial issue to investigate is the homogeneity of agronomic conditions and of rainfall distribution within a given base area.

In the insurance scheme proposed for Morocco, rainfall will be preliminarily measured at the DMN synoptic stations but the programme designers are evaluating

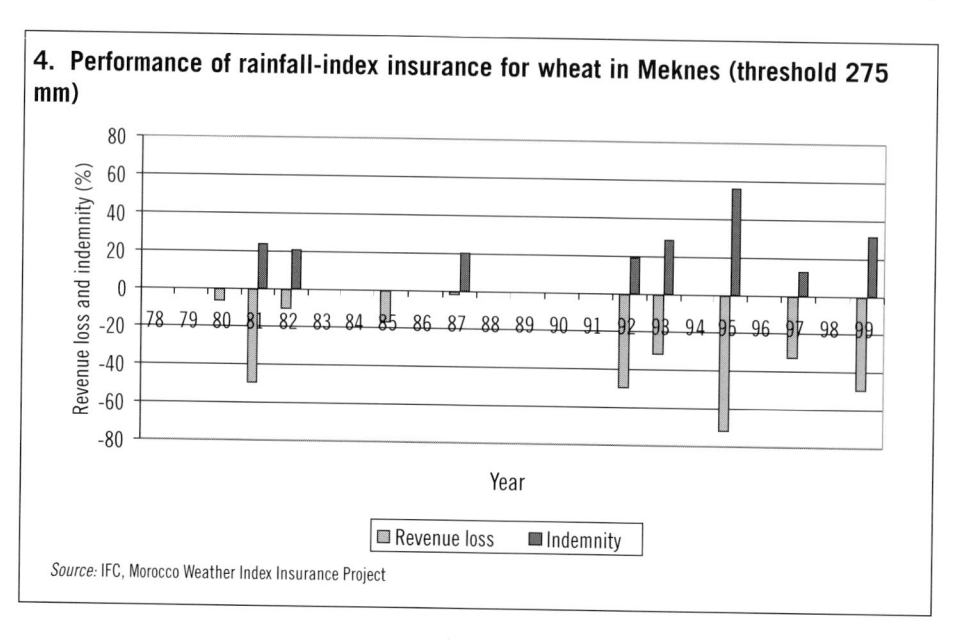

4. Performance of rainfall-index insurance for wheat in Meknes (threshold 275 mm)

Source: IFC, Morocco Weather Index Insurance Project

the opportunity of relying on secondary weather stations to reduce the size of the base areas. Secondary stations are not as reliable as the automatic synoptic ones but, in this respect, the progress in rainfall measurement technology grants low price and high standard measurement devices that could be used for fallback verification purposes.

The ultimate condition for the success of the programme is the price at which the coverage can be provided. This will be decided by the market, but certainly careful programme design, reliable data measurement and access to international risk management players give the markets opportunities to best express themselves. Weather insurance certainly cannot modify weather conditions, but it can help manage weather risks in a more efficient way.

Conclusion

This chapter has demonstrated the enormous potential for weather risk management in the agri-business sector in developing countries. Theory and practical examples point to higher entry barriers, but also higher margins for weather risk market makers in these countries. Major barriers include data problems and credit risk concerns. Data quality varies but sophisticated verification mechanisms such as satellites or temperature gauges allow for weather insurance to be offered almost anywhere. A key factor in determining demand for weather risk hedges is credit – farmers do not buy insurance, they are required to collateralise credit with insurance. In current regulatory environments, weather hedges will generally be sold in the form of insurance. End-users will often be intermediaries such as agricultural banks or insurance companies, or input suppliers and agro-processing companies exposed to throughput risk. The weather risk market is able to substitute some of the traditional reinsurance covers and can efficiently offer yield protection to farmers where crop insurance fails due to high expense ratios. The Moroccan example demonstrates how a few modifications to the basic cumulative rainfall contract can minimise basis risk for a particular crop and at the same time provide income protection to farmers.

Ultimately the emergence of a vibrant weather market will be driven by knowledge transfers to local "champions" who grasp the opportunity, as well as the demonstration effects of a few emerging market transactions and the willingness of global weather risk market makers to shoulder some up-front costs in order to reap the benefits of a globally diversified weather market.

Table A1. Applications by sector and weather parameter

Weather parameter	Agriculture/agri-business Crop production	Livestock	Fisheries/aquaculture
Precipitation	Sugar (Fiji) Cereals (Mexico, Romania, Morocco, Tunisia, Turkey)	Argentina, Uruguay: Milk production Cattle and sheep (Morocco)	
Precipitation: Monsoon risk	AG and AGB (South Asia)		
Hurricane (wind speed and flooding)	Nicaragua: RI of national emergency fund		Shrimp farm (Honduras)
Temperature: Freeze	Coffee (Brazil)	Meat production (Mongolia, Argentina)	
Temperature: Heat	RI of area-yield index (Argentina)	Goats: cashmere (Mongolia)	Shrimp disease risk (Belize)
Precipitation and Temperature	Various crops – agricultural products fund (Argentina) Input supplier insurance (Argentina) RI of AG INS (Argentina) Citrus Business Interruption INS (Caribbean) RI for national emergency find for farmers (Mexico + Nicaragua)		
Sea temperatures	Cotton (Peru)		Hake (Namibia)
Niño risk			Sardine (Peru)

Note: These applications are based on proposals and expressions of interest by end-users and intermediaries.
RI Reinsurance   AGB Agribusiness
AG Agriculture   INS Insurance

1 *The views expressed in this article are those of the authors alone and do not in any way reflect or engage World Bank or IFC policies. The authors would like to thank Luc Christiaensen for helpful comments and Rachid Guessous (MAMDA, Morocco) for valuable input.*

2 *The World Bank groups countries into low-, middle-, and high-income categories on the basis of their gross national income per capita (gni-pc). In 2000, the cutoff levels for a low-income country were no more than US$755, for a middle-income country at least US$756 and no more than US$9,265, and for a high-income country at least US$9,266.*

3 *Sub-Saharan Africa includes all of Africa except the five nations bordering the Mediterranean.*

4 *SYNOP data is recorded at a specified time across the world. Formats are standardised by the WMO.*

5 *Usually there is no bias in the data, as incentives were not skewed towards over- or understating measurements.*

6 *The OECD provides governments a forum in which to discuss and develop economic and social policy. Members are: Australia, Austria, Belgium, Canada, Czech Republic, Denmark, Finland, France, Germany, Greece, Hungary, Iceland, Ireland, Italy, Japan, Korea, Luxembourg, Mexico, Netherlands, New Zealand, Norway, Poland, Portugal, Slovak Republic, Spain, Sweden, Switzerland, Turkey, United Kingdom and the United States.*

7 *For example, weather stations in Saudi Arabia have an incentive to underreport temperatures once the measurements reach the vicinity of 50°C, which is the threshold that allows employees to stay home.*

8 *The International Rice Research Institute, for example, provides many developing countries with excellent data.*

9 *Both Morocco's and Mexico's major agricultural banks tightened their lending policies in 2001–2. Insured or highly collateralised farmers (possibly with movable assets) will receive input credits.*

10 *The weather risk market should be able to substitute weather contracts for business interruption insurance on more competitive terms, as they can price weather risk more precisely on standardised and transparent terms. More importantly, the weather market is hungry for diversification out of US temperature contracts and therefore quotes on a portfolio-adjusted basis that drives down premiums.*

11 *Like other mining companies, Ashanti Goldfields (AGF) had hedged against falls in the gold price by contracting forward sales (or options) that locked-in the current price. In a falling market this strategy would protect revenue and profits, but in a rising market the hedge book became a liability. AGF's hedge book was unusually large, representing about 10 million ounces of gold. The counterparties in these derivatives transactions were 17 banks that were entitled to call in margin deposits once the negative value of the hedge book – or its 'replacement cost' at the current market price – exceeded US$300 million.*

Following the gold price hike, the replacement cost of AGF's hedge book reached about US$570 million at a gold price of US$325 per ounce, requiring deposits of up to US$270 million which the company was unable to find. AGF later reached an agreement with its counterparties.

12 *As an example, Table A1 lists a number of proposals and expressions of interest by end-users and intermediary related to IFC and World Bank work in various countries.*

13 *After steady erosion throughout most of the 1990s, reinsurance rates had increased in almost all sectors at 2000 year-end renewals. Since January 1, 2001, the pace of reinsurance rate increases has accelerated.*

14 *Revenue is calculated setting a fixed price of milling wheat at Dh250 per ton for the entire period.*

15 *For an interesting discussion of moral hazard and adverse selection in weather insurance see Turvey (2001).*

BIBLIOGRAPHY

Benson, C., and E. Clay, 1998, "The Impact of Drought on Sub-Saharan African Economies: A Preliminary Examination", World Bank Technical Paper 401, Washington, DC.

Besley, T., 1995, "Saving, Credit, and Insurance" in: J. Behrman and T. N. Srinivasan, *Handbook of Development Economics* 3(A), pp. 2123–207, (Amsterdam: North Holland).

Dercon, S., 2002, "Income Risk, Coping Strategies and Safety Nets", World Institute for Development Economics Research Discussion Paper 2002/22, Oxford.

Freeman, P., 1999, "The Indivisible Face of Disaster" in: World Bank, Investing in Prevention: A Special Report on Disaster Risk Management, Washington, DC.

Guillaumont, P., S. G. Jeanneney, and J.-F. Brun, 1999, "How Instability Lowers African Growth", *Journal of African Economies*, 8(1) pp. 87–107.

Haggerty, J., 2001, "Weather Derivatives: IFC Moves to Assist Developing Economies", *International Financing Review*, 1414, December 15.

Jacoby, H. G., and E. Skoufias, 1998, "Testing Theories of Consumption Behavior Using Information on Aggregate Shocks: Income Seasonality and Rainfall in Rural India" *American Journal of Agricultural Economics*, 80(1) pp. 1–14.

Hazell, P., C. Pomareda, and A. Valdes, 1986, *Crop Insurance for Agricultural Development: Issues and Experience*, (Baltimore: The John Hopkins University Press).

Morduch, J., 1998, "Between the Market and State: Can Informal Insurance Patch the Safety Net?" Harvard Institute for International Development Discussion Paper 621.

Siegel, P., and J. Alwang, 1999, "An Asset-Based Approach to Social Risk Management: A Conceptual Framework", World Bank Social Protection Discussion Paper 9926, Washington, DC.

Skees, J., S. Gober, P. Varangis, R. Lester and V. Kalavakonda, 2001, "Developing Rainfall-Based Index Insurance in Morocco", World Bank Policy Research Working Paper 2577.

Skees, J., 2000, "A Role for Capital Markets in Natural Disasters: A Piece of the Food Security Puzzle", *Food Policy*, 25(3), pp. 365–78.

Skees, J., 1999, "Opportunities for Improved Efficiency in Risk Sharing Using Capital Markets", *American Journal of Agricultural Economics*, 81(5), 1228–33.

Skees, J., P. Hazell and M. Miranda, 1999, *New Approaches to Public/Private Crop Yield Insurance*, (Washington, DC: World Bank Mimeo).

Turvey, C. G., 2001, "Weather Derivatives for Specific Event Risks in Agriculture", *Review of Agricultural Economics*, 23, pp. 333–51.

World Bank, 2001, World Development Report 2000/2001: Attacking Poverty, (New York: Oxford University Press).

End Piece

Speculations on the Future of the Weather Market

Ulrich Hess, Kaspar Richter and Andrea Stoppa

Satellite-based insurance indexes[1]

Satellite imagery has long been used to assess the scope of natural disasters. More recently, researchers have examined how satellites can help crop insurers handle claims. For example, Bentley, Mote and Thebpanya (2002), report how they used Landsat data to estimate the cost of crop damage from hail and windstorms. Using satellite imagery, the researchers found damage that went undetected by traditional approaches that rely on site inspection. On average, thunderstorms cause between US$1–3 billion in property and crop losses in the US and the researchers expect that satellites will be increasingly used to supplement traditional insurance. In a similar fashion, the Saskatchewan Crop Insurance Corporation is looking to see if satellite images can help speed claim response time on tens of thousands of registered claims in Canada.

But what of less traditional forms of insurance? Index insurance – most often associated with weather insurance – pays out when an agreed upon trigger is hit, for example, when the number of consecutive days without rain reaches an agreed upon level. Index insurance is much cheaper to administer than traditional forms of insurance since there is no need to assess damage. The World Bank Group currently offers reinsurance for weather index insurance and researchers looking for ways to extend the benefits of weather insurance markets to developing countries. But insurers may not be willing to cover risks in areas where weather stations are remote and potentially subject to tampering. Can satellites help? Researchers at the World Bank think that the answer may be yes and have teamed up with Canadian satellite company, RADARSAT International, to see if satellite data can be used to supplement other data to spot falsified claims.

Can insurance triggers be built directly upon satellite data? Technology limits some applications of satellites because they only periodically pass over sites that ground stations monitor continuously: this means that satellite measurements would be less frequent than ground-based measurements. Even so, satellites may be well suited for contracts based on area yields or based on grazing conditions. Moreover, satellites are becoming increasingly sophisticated and satellite-based indexes may be a reality in the not too distant future.

Panos Varangis, Jerry R. Skees and Barry J. Barnett

Based on his experience developing innovative financial instruments, Richard Sandor concludes that markets for these instruments generally take a full generation to mature and be widely accepted.[2] By this gauge, weather risk markets are somewhere in their early adolescence. Nurturing and patience are certainly in order. While these markets have progressed and hold significant promise, their future is still uncertain.

SPECULATIONS ON THE FUTURE OF THE WEATHER MARKET

The collapse of Enron and revelations since provide a sobering reality regarding the potential dangers facing participants in largely unregulated over-the-counter markets. Further, in recent months, other firms in the US energy sector have experienced financial problems. Many of these problems may be the result of competition and an increasingly efficient market. However, these events underscore the importance of identifying major issues that will ultimately determine if weather risk markets are just a passing fad or if they can provide the risk-sharing services needed by a vast array of potential users.

The authors believe there are three major issues that will determine the ultimate success or failure of weather risk markets:

1. pricing transparency;
2. new users beyond the energy sector; and
3. effective use of emerging technologies.

While some steps have been taken to make more price information available on websites, there is still relatively little information available about actual weather risk market settlement prices. In this regard, weather risk markets still operate much like insurance markets.

Information on initial asking prices, however, is widely available (for example, see *http://guaranteedweather.com/*). Initial asking prices are generally established using standard reinsurance procedures. Actual settlement prices, however, are based on negotiation between buyers and sellers and depend, in part, on the correlation between the loss risk in the deal being negotiated and the sellers overall book of business. Information on actual settlement prices is generally not widely available.

Further, buyers often have little knowledge of the risk-pricing algorithms employed by sellers. Of course, buyers can slowly learn through the process of negotiation but the transaction costs of doing so can be quite steep. Perhaps, since the products are not homogeneous, weather risk markets will always be plagued by limited pricing transparency and associated inefficiencies. However, market success will likely require creative efforts to increase transparency.

The risk pooling benefits of weather risk markets will not be fully realised unless the set of end-users becomes more diverse. A more diverse set of end-users implies more heterogeneous weather risks being traded in the market and hence, greater opportunities for risk pooling. The authors believe that production agriculture and agri-business could be an important source of new users from around the globe. In many developed countries, highly subsidised crop insurance crowds out the potential demand from production agriculture for weather risk products. In some cases, opportunities may exist to partially reinsure crop insurance portfolios using weather risk products. Agri-businesses are another potential source of new end-users. Many are subject to weather-related throughput risk but have no access to government-subsidised insurance.

Opportunities for attracting new agricultural end-users may be greatest in developing countries. However, one must be patient and cautious in these markets. Significant investments are required to assess risk exposure and design optimal products for various potential end-users. Much of this will also involve effective use of emerging technologies, including satellite images and Doppler radar. As indicated in Chapter 16, developing countries need cost-effective mechanisms for transferring weather risks. Let's hope that weather risk markets are up to the challenge.

J. Scott Mathews

The ultimate forecast – whither the weather market?

An educated guess about where the weather market is going must be based on a look-back at its beginnings. The dual nature of the weather market – "insurance vs derivative" – became standard fare at weather conference slideshows that contrasted weather risk contracts from these two perspectives. Both sides of the risk transfer aisle, however, seem to see the same truth: that assigning a financial price to a contract means that weather must be taken as either an event or a commodity.

Insurance benefits based on accumulated rainfall had been written into Lloyds of London's "pluvius" insurance in the mid-19th century: pluvius coverage may have been the first weather derivative in underwriter's clothing (see Chapter 12). The contemporary insurance perspective clearly boxes weather into event definitions – heat wave, freeze, tornado, blizzard or torrential downpour. The derivative point of view considers the tangible manifestations of the event. "You see a torrential downpour," a derivative professional might say to his insurance rival at a conference cocktail reception, "but we see five and a half inches of water in the rain gauge." Water is, unquestionably, a commodity.

As the debate moves over to the cheese board, the insurer might announce – "Weather isn't a commodity because you can't deliver it – no delivery, no commodity!" The commodity protagonist lays it on the line – "Of course weather is deliverable. It is just that we don't do the delivering. We take delivery, from Mother Nature!"

The future direction of the weather market into the 21st century is clear. The insurance perspective will remain, but it is the commoditisation of weather that shall lead the market on its path to maturity.

Commoditisation of items not normally considered to be commodities began when Wall Street discovered financial engineering. It sounded wrong at the time as engineers deal with tangibles, but the Street soon found that anything could be commoditised, securitised, monetised or otherwise derivitised.

Richard L. Sandor, PhD, often dubbed the "Father of Financial Futures", was the proponent of non-tangible futures contracts and soon US Treasury Bond contracts were listed at the Chicago Board of Trade. It didn't take long for other financial "instruments" to be engineered: stock index contracts were next, and soon there were Eurodollar contracts.

Traders of agricultural commodities have always kept one eye on the weather since weather has a fundamental impact on of everything that sprouts; and traders of energy commodities are also weather sensitive, as the demand side of fuel use ebbs and flows with the temperature. In fact the "weather trade" in certain commodity markets had been around for quite awhile. Expecting rain after a Midwest dry spell in the summer? Sell corn. Prolonged cold snap forecast in Florida? Buy orange juice futures. Warmer than normal winter in the Northeast? Sell heating oil short.

It was only a matter of time before Mother Nature, herself, would be commoditised: Dr Sandor, and others, engineered the CBOT's catastrophe insurance futures contracts in 1997, and the stage was set for the weather asset to emerge.

To point out one of the wobbly wheels dragging the weather market in its fifth year, it can be seen as a football match with only one team on the field; the ball gets kicked around, but there is no real excitement. The other team in this analogy is believed, by most weather hedge marketers, to be potential end-users. In the power and natural gas markets, end-users are the companies at the other end of the wires and pipes. They tend to be risk-avoiders who hedge their commodity price risk with derivatives. However, all mature instruments in the capital markets have their balance between the natural buyers and sellers: eg, producers/consumers, lenders/borrowers, exporters/importers, etc. But who "uses" weather? So, the end-user is on the same team, just as is nearly everyone who has been playing in the weather trading in the first five years.

Who then, really, is the missing team? The investor is the key to a livelier match. Tomorrow's weather investors are those who currently provide essential liquidity in equities, interest rates, foreign exchange and commodities markets. They are hedge

SPECULATIONS ON THE FUTURE OF THE WEATHER MARKET

fund managers, commodity pool operators and other similar fiduciary trading decision makers.

Without the investors who manage billions of dollars, liquidity is absent and so the weather market will migrate its activity onto the regulated futures exchanges where millions of contracts trade every day. It will do this specifically to attract the investor searching for risk diversification: Balance means positions spread across many asset classes, and weather could be one of them. An investment in a weather risk contract is, unequivocally, a diversification decision.

An investor who avoids the weather derivative slice in the asset class pie may have a more accepting view of the weather market if it comes in a different package – a weather portfolio mix or a weather bond (see Chapter 13). Value-seeking investors will begin to see the weather asset as a "natural resource play" instead of the "probability play."

Certainly, climate is a natural resource and is of no lesser importance than other natural attributes such as navigable rivers, deep harbours, forests, arable land and mineral reserves. Market analysts observe climatological effects on the economies of cities, regions and nations, fostering new studies in economics, agronomy and atmospheric sciences. In addition, the climate-change debate could contribute much to "bullish and bearish" decision-making. For fundamental reasons, investors will be guided, respectively, in taking and holding weather assets grounded in one end, or the other, of this controversial spectrum.

Every particle of risk will soon be disaggregated from its asset. In the weather risk transfer game of the future, the economic nuclei of a rainy day or a scorching heat wave, will be stripped out of businesses and offered to the markets as distinct products. In the wake of the zero-coupon bond, could zero-weather stock be far behind?

Copyright © 2002 J. Scott Mathews, New York. All rights reserved.

Stephen Jewson

Arbitrage pricing for weather options

As the weather market develops it is possible that arbitrage pricing could become more important for some contracts. This section develops a simple arbitrage pricing model for weather options. Since the model involves understanding the variations in the price of a weather contract during the contract period, a crucial part will be to understand the effects of changing weather forecasts.[3]

THE SWAP MARKET

The basis for this arbitrage model is the idea that if the swap market becomes liquid, then linear swap contracts could be used to dynamically hedge the risk in options. In the current swap market this would be prohibitively expensive. When liquidly traded swaps are used to hedge options, the arbitrage price of the option is then given in one of three (equivalent) ways which come from standard arbitrage pricing theory. The price is:

1. the discounted expected payoff of the option plus the discounted expected loss on the swap;
2. the discounted expected payoff of the option under a measure in which the swap price process is a martingale;[4] or
3. the solution of an appropriate partial differential equation (PDE).

Methods 2 and 3 depend on being able to derive a stochastic process for the swap price.[5] The form of this process would depend on the particular dynamics of the swap market. We assume that the swap market is "balanced", in that there is an equal amount of supply and demand for each contract, and that there are no trading costs.

This would result in the swap price settling at the discounted mean payoff level, with no risk loading in either direction. The players in the market will estimate this discounted mean payoff by using historical data and forecasts, although different players will use different numbers of years of data, different detrending assumptions and different forecasts. We neglect these differences and assume that a market-driven consensus emerges.

We can now jump straight to the price of an option using the concept of market price of risk. The market price of risk associated with the swap index is zero. Assuming no arbitrage, the market price of risk on all options settling on the same index will also be zero, and hence all such options will also be priced at the discounted mean payoff. Thus, the no arbitrage price for a weather option is just the discounted mean payoff. It is tempting to think that we could calculate this mean payoff using actuarial methods. However, this could lead to a price that was inconsistent with the current market price for the swap and thus create an arbitrage opportunity. The current swap price level must be incorporated into the calculation of the mean payoff in order to avoid this.

DERIVING A PRICE PROCESS FOR THE SWAP

What causes day to day changes in the swap price during the contract period? We assume that players in the market calculate the swap price using historical data and forecasts, and update prices on a daily basis. We make the additional assumption that the swap is linear (without caps or limits) and, for simplicity, that it is based on cumulative temperature with a tick of one.[6] The mean swap payoff is then simply the mean cumulative temperature minus the strike, and the mean cumulative temperature is the sum of the mean daily temperatures during the contract period. These mean daily temperatures can be evaluated using historical data up to the prior day, forecasts from the current day out for the forecast period, and historical statistics beyond that. Since the swap is linear and we are only interested in the mean, we can use a single damped forecast or the mean of an ensemble forecast: this is the one case where ensembles do not add any useful information.

When this price is re-evaluated on the following day, it will change because of:

1. the difference between the newly available historical data and the old forecast it replaces;
2. the difference between the newly available forecasts and the old forecasts they replace;
3. the difference between the last day of the newly available forecast (which extends one more day into the future) and the historical statistics it replaces; and
4. appreciation at the risk-free rate.

The dynamics of the first three terms are affected by the forecast errors for the zero-day forecast, the differences between successive forecasts, and the extra day in the new forecast. We invoke the efficient forecast assumption to understand and model these dynamics. The efficient forecast assumption states that forecast errors and forecast changes are unpredictable. This is justified as follows: if forecast errors were predictable, that information would already be included in the forecast. The efficient forecast assumption is a good approximation of reality.

Assuming the forecast error and forecast changes are random, and because we can reasonably assume that both the temperature and the forecast are normally distributed, we can model the shocks caused by factors 1, 2 and 3 above using a normally distributed random variable. These shocks are highly correlated with each other, but are not correlated in time. The total random shock, which drives the random change in the discounted swap price from day to day is thus also normally distributed and uncorrelated in time.

This total shock is not dependent on the current level of the swap price, but only on the statistics of the forecast errors. These forecast errors are likely to vary in time, partly deterministically due to seasonal effects and predictable flow-dependent effects, and partly stochastically due to unpredictable flow-dependent effects.[7]

The final expression for changes in the swap price is thus:

$$dS = \sigma(t)dB + rSdt$$

where dS is an increment in the swap price S, $\sigma(t)$ is the volatility which depends on the sizes of the forecast errors and varies with time, dB is an increment of Brownian motion, and r is the risk-free interest rate.

CALCULATING THE OPTION PRICE

We have now derived a stochastic process for the swap price. The discounted price is a martingale, and so by applying standard theory we can see, once again, that the option price is simply the discounted expectation of the option payoff. There are a number of ways of calculating the option price consistently with the market swap price. They all depend on estimates of the volatility, $\sigma(t)$. This can be derived from the statistics of past forecast errors, from the spread of ensemble forecasts, or from historical variability of the swap price.

Julian Roberts

A successful future for the weather risk management market is surely predicated on overcoming a number of constraints that currently seem to hinder a more widespread take-up, especially by end-users outside the energy sector. There are distinct roles for the varying participants in the weather market to play in facilitating access to cost-effective and operationally practicable solutions. However, it is clear that outside certain sectors, the 'one size fits all' approach simply does not work and the industry has to work hard to attract end-users with products that suit their needs rather than vice versa; trying to shoe-horn disparate and heterogeneous requirements into the trammels of two or three boilerplate products ill serves both buyer and seller alike.

A broader understanding of the impact of weather upon commercial activity alongside the realisation that viable hedging strategies exist will lead inexorably to a change in the way that companies are compared and valued. In a more formal sense, we should expect this to influence the ways in which weather-sensitive corporates are rated for investment purposes according to the actions that they may (or may not) have adopted to manage their weather risk. Evidence of this is already available, for example, in the enthusiastic response by equity analysts to the recent weather hedge purchased by Atmos Energy Corporation. Their decision to implement an HDD put led to an increase in expected earnings with a consequential increase in stock price.

The weather risk management market itself needs to adopt a structure that best suits the challenge it faces if it is to mature and sustain long-term growth. An effective primary market is needed to originate clients, warehouse (where necessary) and assume basis risk. This will allow products to be tailor-made to suit specific requirements. In parallel to this, an effective secondary market, based upon volume and more standardised contracts, is needed to support the primary activity. This must bring greater transparency to pricing and competition between capital providers.

In the process of an orderly and efficient market for weather risk management products structuring itself we may expect to see distinct roles emerging for all existing participants according to their respective capacity and expertise. In this the providers of both insurance and derivative products will play their respective parts in a truly alternative risk marketplace.

Robert S. Dischel

Value-based weather trades

The weather market will not fade away as long it continues to serve end-users with valuable hedging tools, and as long as weather traders can profit. Yet, if it does not grow from its current size, weather asset originators will not be able to justify continuing in the market at current profit levels: they must grow their book of trades or they will withdraw from the market. This market is not likely to survive more withdrawals, so it must flourish if it is not to perish. There must be a substantial increase in liquidity for the market to grow.

The market could grow in its service to the energy sector. Climate volatility will continue and it is likely that, given time, more energy end-users from more nations will recognise the opportunity the market offers them to neutralise weather-related return volatility. But each of these end-users requires individualised weather products – these over-the-counter specialised structures are essential – but even the growth of specialised over-the-counter structures would not bring liquidity as secondary trades on specialised products are limited.

The market could grow into other, as yet untapped, sectors. It might find a way to profitably service more of the non-energy weather-exposed enterprises, including governments. It would have to provide structures that protect from weather events, not just on a single variable like temperature or rainfall. Before the originators expand on a large scale into non-temperature-indexed products they would require liquidity – then they could take risk and experiment with new structures. However, it might seem a doomed situation if liquidity must precede expansion.

The responsibility for enlarging the market falls squarely on the weather assets originators. They could allow it to languish, as they more or less have, by continuing to keep it a price-driven market, meaning less focus on the value to end-users than on trading. The tendency has been to pick low hanging fruit from the orchard of opportunity. In focusing on price, and trading among themselves and not on value for buyers, the originators have left rich fruit to wither on the trees; it is time to stretch to the higher branches.

We even see suggestions for using swaps in the absence of an underlying commodity as the basis for pricing, leaving each dealer to better outguess Mother Nature than does his trading counterparty. While arbitrage pricing works well in other markets where there is an underlying commodity, it cannot here – unless it is a layer of nuance on top of value-based pricing – the layer that incorporates supply, demand and forecasts on top of estimates of outcomes.

The fact remains that liquidity must come from capital market investors who, as a practice, often hold assets to maturity. Differences in prices make for good trading, but investors are willing to take risk only if they see measurable value. Few will invest where pricing is opaque and the products are obscure, as weather derivatives are to some. If this market is to thrive, it must attract investors seeking familiar, transparent and comprehensible products. One such product is the weather-linked bond. The time is upon us for securitising weather derivatives into less specialised products, just as the mortgage market securitised mortgages into collateralised mortgage obligations, with cashflow tranches designed to interest a broad range of investors.

In the early efforts at securitising weather assets, a portfolio of derivatives backed the weather-linked bond (see Chapter 13). At first glance, it might appear that before securitising weather assets, the originators must have a ready supply of derivatives that can be securitised, but this is not so – and it is not how the mortgage market works.

Investors look to be paid by the asset originator. A credit-worthy originator with a significant commitment of capital could line up buyers of tranches simply on the promise to pay; there is no requirement for specific derivatives to back the weather-linked bond. The weather-linked structure need only specify a publicly available

SPECULATIONS ON THE FUTURE OF THE WEATHER MARKET

weather index or indexes and offer attractive tranches with an assortment of principal or interest guarantees, and payments that float with the weather. The originator could issue the bond and simultaneously, or sequentially, trade in the futures and over-the-counter markets to "balance" the portfolio of derivatives and the bond: the portfolio could be built over time.

The tranche holders could also use the futures markets. They could capture earned, but unrealised return, in any futures-related tranches (those built around the weather futures market) or temporarily adjust their weather portfolios with futures trades. Activity in the weather futures alone would stimulate liquidity.

With the sales of weather-linked bond tranches, weather risk is transferred to the investor whose capital is returned to the weather market. Originators could recycle the capital – they could issue more weather assets, either bonds or derivatives, and liquidity would be enhanced. With the tranche buyer's focus on value, value-based pricing must follow. The over-the-counter end-user who also seeks value will be better served, and in a more liquid market.

– – –

1 *With thanks to Don Larson of the World Bank for his contribution to this section.*
2 *Personal communications with J. Skees, October, 2000.*
3 *The model described was first outlined in an article by Jewson et al. (2002).*
4 *A martingale is a stochastic process for which the conditional expectation at t+1, conditioned on time t, is equal to the value at time t. Loosely speaking, the process does not, on average, drift up or down with time.*
5 *Note that we assume that swaps have a varying price and a fixed strike. In practice it is more common to have a fixed price (zero) and a moving strike. The two views are interchangeable since the swaps here are priced at the mean, but we use the first because the parallels with arbitrage models for financial derivatives are clearer.*
6 *A tick of one means that a change of one in the settlement index gives a change of one in the payoff.*
7 *The partly stochastic nature of the forecast errors means that the shock is not actually previsible, which is a small complication. We ignore this small effect and assume that forecast error statistics are previsible.*

BIBLIOGRAPHY

Jewson, S., A. Brix and C. Ziehmann, 2002, "Getting the Price Right", Weather Risk Report, Global Reinsurance Review, pp. 10–15.

Wilmott P., S. Howison and J. Dewynne, 1995, *The Mathematics of Financial Derivatives: A Student Introduction*, (Cambridge University Press).

INDEX

A
Acock, M.C. and Y.A. Pachepsky (2000) 81
agriculture and agri-business, weather risk management in developing countries 295–310
Agriculture Financial Services Corp (AFSC) 51
AGROASEMEX 281
Alaton, P. et al. (2002) 139
Alexandersson, H. (1986) 88
Alexandersson, H. and A. Moberg (1997) 85
alternative derivative strategies 40
alternative pricing strategies 272–3
Alternative Risk Transfer (ART) 28, 215
alternative risk transfer market 215–30
  historical perspective 215–16
  nature of demand 216–19
  nature of supply 223–5
Ambaum, M.H.P. et al. (2001) 61, 179
ambient temperature, and demand for power 4–6
AMEDAS (Automated Meteorological Data Acquisition System) 76
American Meteorological Society 153
  (1998) 152
Anderson–Darling test 138
anomaly modelling 141–2
anomalies 172
AON Re Canada 51
arbitrage pricing 147–8
Arctic Oscillation (AO) 66, 106, 179

ARMA models 139, 142–3
artificial neural network (ANN) techniques 81
Artzner, P. et al. (1999) 246, 247, 248
atmosphere
  general circulation 107–8
  observation 118
atmospheric flow 187–8
Atmospheric General Circulation Models (AGCMs) 171
atmospheric phenomena, classification and scales 152
atmospheric prediction and predictability 194–200
Australia
  instrumentation changes/relocations 84
  meteorological networks 75–6
autocorrelation function (ACF) 142, 143
Automated Surface Observing Systems (ASOS) 56, 82, 269
average revenue flows 206–8
average temperature 9, 30, 32, 57
  call 35
  collar 33
  collars indexed to 32–3
  put 35
  swap 33, 34

B
banks 224–5
Barnes, S.L. (1964) 80
Barry, R.G. and R.J. Chorley (1998) 57, 62
basic weather derivatives 30–35

basis risk 283
Battisti, D.S. and A.C. Hirst (1989) 63, 64
Benson, C. and E. Clay (1998) 296
Bergeron, T. (1960) 57
Bergeron–Findeisen theory, rainfall 57
Besley, T. (1995) 296
binary derivatives 29
Bois, P. (1970) 88
Bowman, A. and A. Azzalini (1997) 133
Brier, G.W. (1950) 123
Brody, D.C. et al. (2002) 139
Brown, P.E. et al. (2001) 90
Bruce, J.P. (1994) 185
Burn analysis 135, 176
"butterfly effect" 119

C
Caballero, R. et al. (2002) 139, 143, 170
call, average temperature 35
call options 29, 34–5
Cao, M. and J. Wei (2000) 139
cap contracts 48
capital market investors, and market liquidity 21–2
capital markets, and weather assets 28–9
caps 30, 31
Casella, G. and R.L. Berger (2002) 136
catastrophic weather 59–62
catastrophe (cat) bonds 28, 60, 239, 284, 287–8, 302
catastrophes 220–21
Caussinus, H. and O. Mestre (1996) 88
Central England Temperature (CET) series 20

INDEX

Centre for Research on the Epidemiology of Disasters (CRED) 280
chaos 119–20, 152
Chapman, R.E. (1992) 188
Charney, J.G. (1975) 158
chi-square test 138
Clemmons, L. and P. VanderMarck (2000) 82
Cleveland, W. and S. Devlin (1988) 133
climate 62–8
   differentiated from weather 7, 55
   in Europe 3
   in the US 3
   and weather 98–9
   and the weather risk market 7–8
climate data 6, 73
   versus synoptic data 76–7
Climate Prediction Center (CPC) 158–9, 165
climate risk, and temperature risk 8–9
climate uncertainty 97–114
climate variability 97–106
   managing 106–12
climatology 12, 155
coalescence theory, rainfall 57
Coles, S. (2001) 136
collars 40, 48
   indexed to average temperatures 32–3
collateralised mortgage obligations (CMOs) 21, 317
collective mutual assistance groups, and weather index insurance 285
Columbia River 51
Community Climate Model, National Center for Atmospheric Research (NCAR) 154
computer models, grid point versus spectral 117
Conrad, V. and C. Pollak (1950) 82
contract form 27–8
convolution process 161–2
cooling degree-day (CDD) indexes 131
cooling degree-days (CDDs) 8, 44, 57, 266, 275

Coordinated Universal Time (UTC) 74
copulas 145
Coriolis force 58, 60
credit ratings 27
Cressman, G.P. (1959) 80
critical-day contract 50
critical-day derivatives 29
critical-event contracts 35–6
Crutzen, P.J. (1970) 63
cumulative distribution function (CDF) 128, 242
cumulative rainfall contracts 7
cyclogenesis 61

D

daily detrending 133–4
daily models
   and Niño3 forecasts 180
   and site-specific seasonal forecasts 180–81
   and weather forecasts 177–8
daily temperature models 138–9
   validation 143–4
D'Arcy, S. (1999) 245
data assimilation 118–19
Davis, M. (2001) 139
Davison, A. and D. Hinkley (1997) 142
DeGaetano, A.T. and R.J. Allen (1999) 85, 88
degree-days 7, 8–9, 30, 57
demand for power, and ambient temperature 4–6
Dercon, S. (2002) 296
deregulation 3, 6, 219
derivative prices, and weather probabilities 12
derivative products 7
derivative structures 29–36
derivatives
   advantages 226–7
   disadvantages 227
   vs insurance 225–7
developing countries
   use of weather indexes 285–8
   weather indexes for 279–94
   weather risk management for agriculture and agribusiness 295–310
   and weather risk markets 281–2
Diebold, F.X. and S.D. Campbell (2001) 139

differential heating 62
digital derivatives 29
digital payoff contracts 50
Dischel, R.
   (1998) 139
   (2001) 282
Dool, H.M. van den (1994) 116
Downing, T.W., A.J. Olsthoorn, and R.S.J. Tol (1999) 186
Dutton, J.A.
   (1995) 188, 194
   (2002) 188
dynamic climate models 153–5
dynamical models 198, 260–61
D'Zurko, D. and D. Robinson, (1999) 56

E

Easterling, D.R. et al.
   (1997) 129
   (1999) 76
Easterling, D.R. and T.C. Peterson, (1995) 88
ECMWF (European Centre for Medium Range Weather Forecasts) 172
ECMWF global model 117, 118
ECOMET 78–9
El Niño 5, 8
El Niño–Southern Oscillation (ENSO) 55, 63–6, 100–3, 121, 156, 170–71, 179
   analogs 198
   discovery 115
   dynamics 100–2
   effects 102–3
   observations 100
   predictability 103
   and tropical cyclones 65
electrical thermometers, types 56
Elektrizitatswerk Dahlenburg 52
Embrechts, P. et al. (2000) 145
end-to-end forecasts 124
end-user strategies 31
end-users 29–30
energy companies/traders 224
ENIAC (Electronic Numerical Integrator and Computer) 116, 192
Enron 8, 27, 28, 233, 235
ensemble forecasts 120–22, 154

INDEX

numerical (AGCM) ensembles 175
statistical ensembles 175
ENSO-CLIPER model 65
Entergy-Koch Trading 27, 51
envelopes 137
errors, in forecasts 177
ETC Lothar 61
ETC Martin 61
Europe
  10-day forecasts 170
  climate in 3
  meteorological networks 75
European Centre for Medium Range Weather Forecasting (ECMWF) 117, 120
evapotranspiration 157
Exchanges 225
extra-tropical cyclones (ETCs) 61
extreme value theory (EVT) 148

F
far from the mean severe events 220
Farman, J.C. et al. (1985) 67
Federal Aviation Administration (FAA) 56
Feynmann, R.P., R.B. Leighton and M.L. Sands (1989) 56
financial weather contracts 3, 8, 26–9
First Law of thermodynamics 62
floor contracts 48
fondos de aseguramiento (Mexican mutual funds) 285
forecast experiment 199
forecast skill 198–200, 206–10
forecast uncertainty 122, 124
forecasting, history of 115–17
forecasts 73
  benefits of 208–9
  errors in 177
  evaluation 123
  short-term and long-term 14–17
Fraedrich, K. (1990) 170
Freeman, P. (1999) 296
future weather measurements 290–91

G
gas demand, and temperature 10
Geman, H. (1999) 147
general circulation of the atmosphere 107–8
general circulation models (GCMs) 121
Genz, A. (1999) 259
Geographic Information System (GIS) 80
geostrophic balance 58
Germany, winter temperatures 13–14, 18
Gill, A. (1982) 261
Glahn, H.R. and D.A. Lowry (1972) 122
Glantz, M.H. (1996) 121
global climate change 67
global warming 15–16
globalisation 241
Goldenberg, S.B. et al. (2001) 60, 61
government disaster assistance 287–8
Green, P.M. et al. (1997) 63
grid-point models 117, 118
Guillaumont, P. et al. (1999) 296
Gullet, D.W. et al. (1992) 81

H
Haggerty, J. (2002) 301
Hansen, J. et al. (1996) 129, 133
Harries, J.E. et al. (2001) 121
Hazell, P. et al. 296
Heat index 59
heating degree-days (HDDs) 5, 8, 33, 120, 266, 267, 268, 270–73, 274
  cleaned and enhanced annual series 82
Heating Oil Partners LP (HOP)
  measuring exposure 270
  price stability 267–8
  and weather derivatives 273–5
  weather hedging 265–77
hedging 206–10
  precipitation risk 43–55
hedging alternatives 46–8
  for non-standard precipitation risks 48, 50
Heidke skill scores 199
hierarchical polynomial regression techniques 81
Hildebrandsson, H.H. (1897) 63
Hirst, A.C. (1988) 63, 64
historical weather data 108–10, 290
Hot Air Gas Company (HAGC) 10–12, 36–41
Houghton, J.T.
  (1986) 58
  (2001) 187
Houghton, J.T. et al. (2001) 67
Huber, P.J. (1981) 132
Hull, J.C. (2000) 281
Hurrell, J.W.
  (1995) 66
  (2001) 66
Hurricane Andrew 59, 60, 65

I
ice storms 62
index detrending 132–3
index models
  and Niño3 forecasts 179–80
  and site-specific seasonal forecasts 180
index-based pricing, and weather forecasts 176–7
Industry loss warrants 60
information, availability 203
information handicap 6
information systems 288–9
insurance
  advantages 226
  disadvantages 226
  vs derivatives 225–7
Insurance Services Office 59
insurance/reinsurance companies 223–4
international capital markets 288
international metadata 78
International Swaps and Derivatives Association (ISDA) 226, 272
interpretation, and post-processing model output 122
IPCC (Intergovernmental Panel on Climate Change)
  (1996) 156
  (2001) 67, 121, 129

J
Jacoby, H.G. and Skoufias, E. (1998) 296

INDEX

Japan, meteorological networks 75–6
Jeffreys, H. (1925) 187
Jewson, S. (2000) 178
Joint Center for Satellite Data Assimilation 193
Jones, K.F. and N.D. Mulherin (1998) 62

K

Kalnay, E., S.J., Lord and R.D. McPherson (1998) 197
Karl, J.T. et al. (1993) 156
Karl, T.R. and C.N. Williams (1987) 88
Katz, R.W. and A.H. Murphy (1997) 123, 189
Kelvin notes 233
"kernel smoothing" 135
Kimberlain, T.B. and C.W. Landsea (2001) 65
Knaff, J.A. and C.W. Landsea (1997) 65
Knappenberger, P. et al. (2001) 129
Koch 27, 28
Kohler, M.A. (1949) 87
Kolmogorov–Smirnov test 138
Kreps, R. (1999) 147
Kriging method 80
Kuligowski, R.J. and A.P. Barros (1999) 122
Kunkel, K.E., R.A., Pielke and S.A. Changnon (1999) 186

L

La Niña 8, 101, 102
Landsea, C.W. and J.A. Knaff (2000) 65
layering 283–4
Lempfert, R.G.K. (1932) 115
limited expected value 128
Linear Error in Probability Space (LEPS) score 110
linear regression 11
liquidity 227–8
Livezey, R.E. (2002) 189
Local Climatological Data (LCD) reports 75
London International Financial Futures Exchange (LIFFE) 9
long-term trends 155–6
long-term weather or climate changes 221–2
longer-range prediction 197

Lorenz, E.N.
  (1963) 152
  (1982) 119

M

McCullagh, P. and J.A. Nelder (1989) 136
McKee, T.B. et al. (2000) 83
Madden Julian Oscillation (MJO) 106
market liquidity, and capital market investors 21–2
Mason, S.J. and L. Goddard (2001) 65
Massachusetts Institute of Technology (MIT) 119
maximum or minimum temperature 57, 74
measurement risk 45
medium-term events 222
Mekis, E. and W.D. Hogg (1999) 81, 85
metadata 6, 74, 90
  international 78
  US 78
  and weather station history 77–8
meteorological indexes 179
meteorological networks 74–6
  Australia 75–6
  Europe 75
  Japan 75–6
  special networks 76
  US 75
meteorological records, simulation of 204–6
'misery' indexes 59
model output statistics (MOS) 122
modelling and valuation, weather derivatives 127–50
models 152
modern numerical forecasting 117–22
Morduch (1998) 297
Moreno, M. (2000) 139
Morocco, rainfall index insurance 302–7
mortgage-backed securities (MBS) 21
Morton, O. (1991) 193
Multilateral Investment Guarantee Agency (MIGA) 301
multiple linear regression approaches 81
multiple weather indexes, joint probability distribution 259–60
multi-variable contracts 9
Multivarite ENSO Index (MEI) 100
Murphy, A. (1993) 181

N

Nakada, P. et al. (1999) 147, 246
National Center for Atmospheric Research (NCAR), Community Climate Model 154
National Climatic Data Centre 75
national meteorological service (NMS) 73, 74
National Oceanic and Atmospheric Administration (NOAA) 7, 75
National Research Council (NRC) (1998) 188
National Weather Service (NWS) 49, 75
natural disasters 280
near the mean events 219–20
Nebeker, F. (1995) 115, 116
"nested modelling" 122
Niño3 forecasts
  and daily models 180
  and index models 179–80
non-parametric time-series modelling 142
North Atlantic Oscillation (NAO) 55, 66–7, 105–6, 121, 179
numerical weather prediction (NWP) 116, 194–7

O

OECD 299, 301
Office of US Foreign Disaster Assistance (OFDA) 280
operational seasonal weather forecasts, historic performance 158–60
option contracts 206
orbital dynamics, and the sun 62
Orrell, D. et al. (2001) 177
over-the-counter (OTC) weather market 231–2, 234

P

Pacific Decadal Oscillation (PDO) 55, 67–8, 103–5, 156–7, 179
Pacific North American pattern (PNA) 106
Pacific Ocean 8, 63, 100, 101
Palmer, T.N.
  (2001) 120
  (2002) 124
parametric time-series models 142–3
Penland, C. and T. Magorian (1993) 64, 121
Penland, C. and P.D. Sardeshmukh (1995) 64
persistence 156
Peterson, T.C. *et al.* 84, 85
Pettitt, A.N. (1979) 88
Philander, S.G. (1990) 121
"Pineapple Express" storms 61
pluvius insurance 217
portfolio optimisation 254
Potter, K.W. (1981) 88
power markets, US 3
precipitation 201
  measurement 49–50
  see also rainfall; snowfall
precipitation contract applications 50–52
precipitation contracts 9, 302
  future 52
precipitation risk
  definition of the hedge 44–6
  hedging 43–54
  identification and quantification 43–4
  and weather stations 44–5
Press, W.H. *et al.* (1988) 199
pricing models
  incorporation of seasonal forecasts 178–9
  incorporation of weather forecasts 175–6
probabilities 32
probability distribution function (PDF) 241–2
  calculation in weather derivative portfolios 249–54
put, average temperature 35
put options 29, 34–5

Q

Quasi-Biennial Oscillation (QBO) 55–6, 68

R

rainfall 44
  Bergeron-Findeisen theory 57
  coalescence theory 57
  measurement 57–8
  and wheat yield 45–6
rainfall contracts 9
rainfall index insurance, Morocco 302–7
rank correlation 145–6
RAROC (risk-adjusted return on capital) 147, 246
Raymond (1994) 63
relative humidity 59
residential gas use, and outdoor temperature 10–11
resistance temperature devices 56
return on allocated capital (RAC) 147
Rhoades, D.A. and M.J. Salinger (1993) 85, 88
Richardson, D.S. (2000) 123, 189
Richardson, L.F. (1922) 116, 119
risk loading 146–7
risk premium 255
Roulston, M.S. *et al.* (2002) 122
Roulston, M.S. and L.A. Smith (2002) 123
rounding techniques 270
Rudd, A., and H. Clasing (1988) 255

S

Sacramento Municipal Utility District (SMUD) 50–51
Saffir–Simpson scale 60
Santer, B.D. *et al.* (1993) 156
Schrumpf, A.D. and T.B. McKee (1996) 83
Scottish Hydro 8
sea-level air pressure (SLP) 100
sea surface temperature (SST) 100, 101, 102, 121, 153, 156–7
seasonal forecasts 17, 169–71
  incorporation into pricing models 178–9
  incorporation into traditional pricing models 163–5
  sources 173
seasonal outlooks 198
seasonal weather forecasting techniques 153–8
seasonal weather forecasts 151–69
  in weather risk management 160–63
securitisation market 231–3
  current activity 234
  first securitisations 233–5
series-based forecasts 170
severe thunderstorms 61–2
Shapiro–Wilks test 138
Sharpe ratio 245
shifts 108
short-range forecasts 117
short-term climate forecasting 110–12
short-term weather influences 222–3
Shorter, J. *et al.* (2002) 180
Siegel, P. and J. Alwang (1999) 296, 297
"Sigma" reports, Swiss Re 61
Silverman, B.W. (1986) 260
Singh, M. (2002) 258
single forecasts 174–5
site-specific forecasts 173–4
site-specific seasonal forecasts
  and daily models 180–81
  and index models 180
Skees, J.R.
  (1999) 298
  (2000) 297
  (2001) 302, 303, 306
Skees, J.R. *et al*
  (1999) 297, 298
  (2001) 282
Skees, J.R. and K.A. Zeuli (1999) 282, 283
skill of forecasts, measuring 181–2
Smith, L.A. *et al.* (2001) 124
Smith, S.E. (1996) 63
Snell, S.E. *et al.* (2000) 81
Sneyers, R. (1990) 85, 88
snow cover 157
snowfall, measurement 49, 58
snowfall contracts 9
soil moisture 198
  and vegetation state 157–8
Solow, A. (1987) 88

INDEX

sonic detection and ranging (SODAR) 59
Southern Oscillation Index (SOI) 63, 100, 179
spatial interpolation, weather data 80–81
spectral models 117, 118
speculators 22, 30, 40
Spencer, R.W., and J.R. Christy (1990) 156
standard deviation 245–6
statistical approaches, to weather and climate risk management 200–206
statistical models, in seasonal forecasting 155–8
statistical tools, for forecasting 157–8
Stensrud, D.J. et al. (2000) 120
streamflow, measuring 59
streamflow contracts 9, 30
Suarez, M.J. and P.S. Schopf (1988) 63, 64
sun, and orbital dynamics 62
swap, average temperature 33, 34
swap contracts 162
swap levels 34, 40
swaps 32, 48
Swiss Re, "Sigma" reports 61
SYNOP data 76, 299
synoptic data 73
  vs climate data 76–7
SYNOPTIC format 80
Szentimrey, T. (2001) 88

T

tail value-at-risk (TVAR) 247–9, 257
temperature 7, 200
  and gas demand 10
  measurement 56–7
  modelling daily temperatures 138–9
temperature risk, and climate risk 8–9
temperature-indexed contracts 9
temperatures 1650–2000, UK 20
temperatures
  capturing seasonality 140–1
  trends in 204
temporal interpolation, weather data cleaning 81
thermistors 56
thermocouples 56
thermometers, types 56

thunderstorms 61–2
"tick rate" 30
tornadoes 62
Torro, H. et al. (2001) 139
Trade Winds 100–1, 107
traditional pricing models 160–63
  incorporating seasonal forecasts into 163–5
Trenberth, K.E. et al. (1998) 261
trend analysis 198
trends 110, 129–35, 172
  in temperatures 204
Treynor, J. and F. Black (1973) 245
tropical cyclones 60–61
  and El Niño–Southern Oscillation (ENSO) 65
Tuomenvirta, H. (1998) 94
Turvey, C.G. (2001) 298

U

UK
  temperatures 1650–2000 20
  windstorm 87J 61
UK Met Office Hadley Centre 20
UK Meteorological Office (UKMO) 74
"urban heat island effect" 13
US
  climate in 3
  demand for electricity 4–6
  instrumentation changes/relocations 82–4
  metadata 78
  meteorological networks 75
  power markets 3–4
  winter temperatures 1890–2000 15
US Historical Climatology Network (HCN) 76
US National Center for Environmental Prediction (NCEP) 120, 193, 198
US National Climatic Data Center (NCDC) 269
US National Weather Service (NWS) 269
US NCEP MRF (US National Centre for Environmental Prediction Medium Range Forecast) 171–2

V

value-at-risk (VAR) 246–7
variational assimilation 119
Vincent, L.A. (1998) 88
Vincent, L.A., and D.W. Gullet (1999) 85
Visbeck, M. et al. (2001) 66
volatility 209–10

W

Walker, G.T. and E.W. Bliss (1932) 115
Walker, G.T.
  (1923, 1924, 1937) 115
  (1928) 63, 115
water, and local climate 4
weather
  and climate 98–9
  differentiated from climate 55
weather assets, and capital markets 28–9
weather and catastrophic risk sharing, government strategies 289
weather and climate contingencies 189–92
  observing 192–3
weather and climate risk
  categories and strategies 189
  decades to centuries 186
  hours to days timescale 185–6
  identifying 188–92
  seasons 186
  sensitivity of economy to 188–9
weather and climate risk management, statistical approaches 200–206
weather data 26–7, 73–94, 299–300
  availability and cost 78–9
  dates of discontinuities 85
  detecting discontinuities 85–9
  methods for treating discontinuities 84–5
  quantification of discontinuities 88
weather data cleaning 73, 79–82
  complications 81–4
  instrumentation changes/relocations 82–4
  spatial interpolation 80–81
  temporal interpolation 81

RISK BOOKS

INDEX

weather data reports, errors 79–80
weather derivative market 4
weather derivative portfolios
  calculation of probability distribution function (PDF) 249–54
  management 241–63
  portfolio optimisation 254–8
  quantification of risk 245–9
  risk/return profile 242–5
weather derivative pricing 169–85
weather derivatives 8, 272
  adjusting for warming and cooling trends 129
  basic 30–35
  evaluation 21
  and Heating Oil Partners LP (HOP) 273–5
  modelling and valuation 127–50
  payoff distribution 128–9, 130–31
  and weather insurance products 26
weather disturbances 151
weather forecasts 169, 273
  and daily models 177–8
  incorporation into pricing models 175–6
  and index-based pricing 176–7
  skilful timescales 151–3
  sources 171–2
weather hedging 6, 8, 19–21, 302
  Heating Oil Partners LP (HOP) 265–77
  reasons for not hedging 20–21
weather index dependencies 145
weather index distributions, validating 136–8
weather index insurance
  and collective mutual assistance groups 285
  and government disaster assistance 287–8
  sales to producers and agri-business 285–6
weather index modelling
  goodness-of-fit tests 137–8
  non-parametric 135
  parametric 135–6
weather indexes 8, 26–7, 29
  selection 27
  use in developing countries 285–8
weather insurance products, and weather derivatives 26
weather market 7
  entering 36–41
weather measurements 56–9
weather note securitisation 231–41
  characteristics of a successful note 235–7, 239
  credit of the issuer 237–8
weather predictability 7
weather probabilities, and derivative prices 12
weather protection 3–4, 6
weather risk, and realised losses 282–8
weather risk analysis 6
weather risk management
  for agriculture and agri-business, in developing countries 295–310
  end-users 300–1
  market for new 298–9
  and rural development 295–9
  seasonal weather forecasts in 160–63
Weather Risk Management Association 9, 290
  (2001) 228
weather risk market 3
  and climate 7–8
  and developing countries 281–2
  size 9
weather station history, and metadata 77–8
weather stations 49, 51, 74
  and precipitation risk 44–5
weather variables 59
weather variations
  cost 6
  and impact on revenue 9–12
weather-linked instruments, investing in 258
weather-sensitive businesses 17
wheat yield, and rainfall 45–6
Wilks, D.S. and R.L. Wilby (1999) 139
wind 202
  measurement 58–9
Wind Chill Factor 59
wind contracts 9
windstorm 87J, UK 61
WINRS notes 233, 235
winter temperatures 1890–2000, US 15
winter temperatures, Germany 13–14, 18
World Bank (2001) 295, 296, 299
World Meteorological Organization (WMO) 74, 299
World Meteorological Organization's World Weather Watch (WMO WWW) 118
WSI Energycast Trader (WSI) 153

Z

Zeng, L.
  (2000) 250, 251, 260
  (2001) 246, 257, 258
Zeroth Law of Thermodynamics 56
Zurbenko, L. *et al.* 85, 88